# APPLIED POLYMER RHEOLOGY

# APPLIED POLYMER RHEOLOGY

## Polymeric Fluids with Industrial Applications

Edited by

**Marianna Kontopoulou**
Queen's University
Kingston, Ontario, Canada

WILEY

A JOHN WILEY & SONS, INC., PUBLICATION

Published by John Wiley & Sons, Inc., Hoboken, New Jersey
Published simultaneously in Canada



***Library of Congress Cataloging-in-Publication Data:***

Applied polymer rheology : polymeric fluids with industrial applications / edited by Marianna Kontopoulou.
p. cm.
ISBN 978-0-470-41670-9 (hardback)
1. Polymers—Rheology. 2. Polymer melting. I. Kontopoulou, Marianna.
TP1150.A527 2012
547′.704541—dc23

2011026189

oBook ISBN: 978-1-118-14061-1
ePDF ISBN: 978-1-118-14058-1
ePub ISBN: 978-1-118-14060-4
eMobi ISBN: 978-1-118-14059-8

10 9 8 7 6 5 4 3 2 1

# CONTENTS

# PREFACE

The topic of rheology of polymeric fluids has been the subject of intense research and has attracted some of the brightest minds in fluid mechanics, polymer physics, and chemistry. There are two main venues through which this research has been disseminated: The first is through numerous state-of-the-art textbooks, written by the top scientists in the field. These cover extensively the science and fundamental principles of rheology; constitutive equations; principles of rheometry, measurement methods, and techniques. The other dissemination method is through articles published in peer reviewed journals. These cover a vast array of topics, from purely theoretical concepts to the applied rheology of specific polymeric fluids and melts. Many review papers on various topics of applied rheology are also available. Although very comprehensive, these publications are often not readily accessible to professionals, consultants and industrial practitioners, who because of time constraints are not necessarily in a position to follow the advanced rheological concepts presented in many of the scientific papers published in top peer-reviewed journals.

The purpose of this book is to present the state-of-the-art for specific polymeric systems of current industrial interest, as studied by well-recognized researchers who are experts in their fields. It targets professionals, industrial practitioners, as well as researchers and graduate students in the fields of polymer engineering, materials science, and chemical engineering who need to know the most recent developments pertaining to the rheology of various polymeric systems. This target audience would benefit from the existence of a book that explains the distinct rheological characteristics of various industrial systems and their impact on their processing characteristics and ultimate engineering properties. This textbook provides them with a comprehensive overview of the rheological characteristics of the polymeric fluids of interest to their application, so they can interpret their findings or plan their experiments. It is also intended as a useful tool for processors and equipment designers, who need to plan carefully the processing conditions of their materials based on their rheological characteristics.

When teaching graduate courses I have frequently encountered the challenge of finding suitable review papers in many industrially relevant areas that I could present to my students. I envision this book as a useful starting point in the research of graduate students in the fields of polymer science and engineering, materials science, and chemical engineering, who once exposed to a basic course on rheology will be ready to venture on the study of the polymeric fluids of interest in their graduate work. This book provides them with an overview of what has been done to date and, I hope, some insight on what needs to be done in the future.

Even though some basic concepts of rheology are introduced in Chapter 1, this textbook assumes a basic knowledge of rheometry and familiarity with the fundamental concepts of rheology. For in-depth coverage of these topics, the reader is referred to the many excellent books that cover the theoretical principles of rheology, fundamental concepts, constitutive equations and principles of rheometry.

This book comprises 11 chapters, organized loosely in subtopics. The first chapter establishes basic relations between the rheological properties and processability of polymer melts. Knowledge of the rheology of polymer melts can serve as a tool in predicting their behavior during processing. This chapter provides the readers with an appreciation of the importance of rheology to processing. It also includes some fundamental theory of rheology and constitutive equations. Processing aids, which are used extensively to improve the processability of polyolefins and other polymers are presented in Chapter 2. Chapter 3 covers the rheology of branched polyolefins, which have attracted great interest in recent years, because of their applicability in important industrial processes, such as thermoforming, extrusion coating, and foaming. The presence of branching affects profoundly the rheology and thus the processing characteristics of these polyolefins. This chapter is divided in two subsections, describing the rheology of branched polyethylene (PE) and polypropylene (PP). Filled polymers and composites are the topic of the next three chapters. These include the rheology of fiber suspensions, polymer nanocomposites, and wood fiber composites. Addition of solid particulates significantly affects the rheological properties and thus the processability of the melts. These effects are discussed extensively in Chapters 4–6. Chapter 4 introduces the theoretical framework on which the analysis of these materials is based. Chapters 7–10 include systems with microstructure, such as block copolymers, liquid crystalline polymers, electrorheological fluids, and reactive blends. Chapter 11 deals with a novel topic, that of paste extrusion of polytetrafluoroethylene paste.

I would like to thank all the contributors who provided their expertise and enthusiasm to this project and Wiley for making this work possible. I would like to dedicate this book to the memory of my father, Antonios Kontopoulos, who was a committed academic and educator.

# CONTRIBUTORS


Donald G. Baird, Department of Chemical Engineering, Virginia Polytechnic Institute and State University, Blacksburg, VA, USA

Claire Barrès, Université de Lyon, INSA de Lyon, Ingénierie des Matériaux Polymères, Villeurbanne Cedex, France

Philippe Cassagnau, Université de Lyon, Ingénierie des Matériaux Polymères, Villeurbanne, France

Tirtha Chatterjee, Materials Research Laboratory, University of California, Santa Barbara, CA

Hyoung Jin Choi, Department of Polymer Science and Engineering, Inha University, Incheon, Korea

Michel Cloitre, ESPCI ParisTech, Matière Molle et Chimie, Paris, France

Aaron P. R. Eberle, NIST Center for Neutron Research, National Institute of Standards and Technology, Gaithersburg, MD, USA and Department of Chemical Engineering, Virginia Polytechnic Institute and State University, Blacksburg, VA, USA

Fei Fei Fang, Department of Polymer Science and Engineering, Inha University, Incheon, Korea

Alexandros D. Gotsis, Technical University of Crete, Hania, Greece

Savvas G. Hatzikiriakos, Department of Chemical and Biological Engineering, The University of British Columbia, Vancouver, BC, Canada

Jiasong He, Beijing National Laboratory for Molecular Sciences (BNLMS), Key Laboratory of Engineering Plastics, Joint Laboratory of Polymer Science and Materials, Institute of Chemistry, Chinese Academy of Sciences, Beijing, China

Ramanan Krishnamoorti, Department of Chemical and Biomolecular Engineering, University of Houston, Houston, TX, USA

Tieqi Li, NOVA Chemicals Corporation, Calgary, AB, Canada

Grégory Martin, Hutchinson S.A., Centre de Recherche, Rue Gustave Nourry, Chalette-sur-Loing, France

Kalman B. Migler, Polymers Division, National Institute of Standards and Technology, Gaithersburg, MD, USA

Kevin Ortman, Department of Chemical Engineering, Virginia Polytechnic Institute and State University, Blacksburg, VA, USA

Nickolas Polychronopoulos, Polydynamics, Inc., Dundas, ON, Canada

John Vlachopoulos, Department of Chemical Engineering, McMaster University, Hamilton, ON, Canada

Dimitris Vlassopoulos, FORTH, Institute of Electronic Structure and Laser, Heraklion, Crete, Greece; University of Crete, Department of Materials Science and Technology, Heraklion, Crete, Greece

# CHAPTER 1

# BASIC CONCEPTS IN POLYMER MELT RHEOLOGY AND THEIR IMPORTANCE IN PROCESSING

JOHN VLACHOPOULOS[1] and NICKOLAS POLYCHRONOPOULOS[2]
[1] Department of Chemical Engineering, McMaster University, Hamilton, ON, Canada
[2] Polydynamics, Inc., Dundas, ON, Canada

## CONTENTS



*Applied Polymer Rheology: Polymeric Fluids with Industrial Applications*, First Edition.
Edited by Marianna Kontopoulou.

## 1.1 INTRODUCTION

The vast majority of the nearly 250 million tons of polymers produced annually are thermoplastics, which are melted by heating, shaped by flowing through dies or filling of molds, and subsequently solidified into final useful products. The melt processing of thermoplastics is, to a large extent, determined by flow behavior, which in turn depends on structure and the processing conditions of temperature and pressure. The relations of structure, processing, and properties of polymers are of paramount importance in all their applications as film, sheet, pipe, tubing, profiles, wire and cable coatings, containers, appliance housings, automotive, and aerospace parts and as numerous other products.

Molten polymers exhibit rather unusual flow behavior and some rather counterintuitive phenomena, as it is explained in several books [1–14] and numerous publications. In fact, the majority of publications in rheology, the science of deformation and flow of matter, deal with polymer melts and solutions. First, rheological measurements are necessary for process equipment design. Knowing the viscosity as a function of shear rate permits the calculation of pressures and production rates. Process troubleshooting and process optimization can be carried out by detailed rheological characterization, as it will become apparent in the subsequent sections and chapters of this book. Some rheological measurements are sensitive indicators of polymer structure, such as molecular weight, molecular weight distribution, and chain branching. In fact, the sensitivity and frequently simplicity of some rheological characterization methods are the main reasons for the success on rheology in polymer processing. Textbooks on polymer processing [15–25] are mainly devoted to describing how rheological measurements can be combined with fluid mechanical and heat-transfer principles for the prediction of how molten polymers flow through extruders, channels, dies, molds, and other types of equipment.

## 1.2 VISCOSITY OF POLYMER MELTS

Viscosity represents the resistance to shearing, i.e., flow of imaginary fluid slices like the motion of a deck of cards. Referring to Figure 1.1, we can define

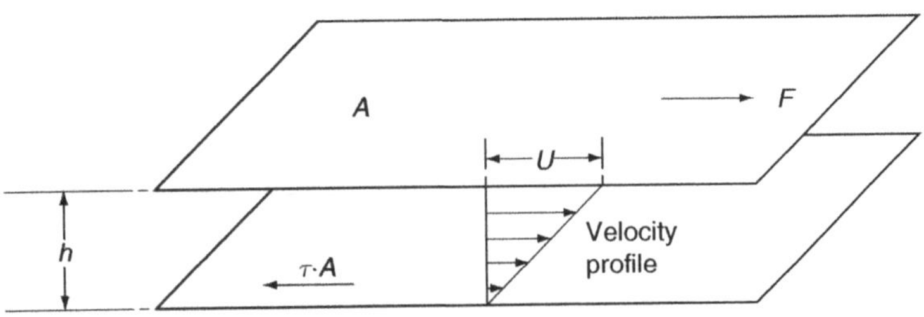


**FIGURE 1.1** Simple shear flow.

viscosity as the ratio of the imposed *shear stress* (force $F$, applied tangentially, divided by the area $A$), and the shear rate (velocity $U$, divided by the gap $h$)

$$\eta = \frac{\text{Shear Stress}}{\text{Shear Rate}} = \frac{F/A}{U/h} = \frac{\tau}{\dot{\gamma}} \tag{1.1}$$

This relation is known as Newton's Law of Viscosity. When the viscosity is independent of the shear rate, a fluid is called Newtonian. Molten polymers have viscosities dependent on shear rate, exhibit several other unusual flow properties, and are referred to as non-Newtonian.

One remarkable property of polymeric liquids is their *shear-thinning* behavior (also known as pseudo-plastic behavior). As the shear rate increases, the viscosity decreases, as shown in Figure 1.2. This reduction of viscosity is due to molecular alignments and disentanglements of the long polymer chains. The higher the shear rate, the easier it is for polymers to flow through dies and process equipment.

The most frequently used model to express the shear-thinning behavior of polymers is the power law:

$$\eta = m\dot{\gamma}^{n-1} \tag{1.2}$$

This expression is a straight line when plotted on double logarithmic coordinates. The value of the consistency index $m$ can be obtained from the intercept at $\dot{\gamma} = 1$. The exponent $n - 1$ is the slope, because

$$\log \eta = \log m + (n - 1) \log \dot{\gamma} \tag{1.3}$$

This model is not suitable for fitting low shear rate data because the viscosity predicted approaches infinity as the shear rate goes to zero, while polymeric liquids exhibit a characteristic Newtonian plateau at very low shear rates.

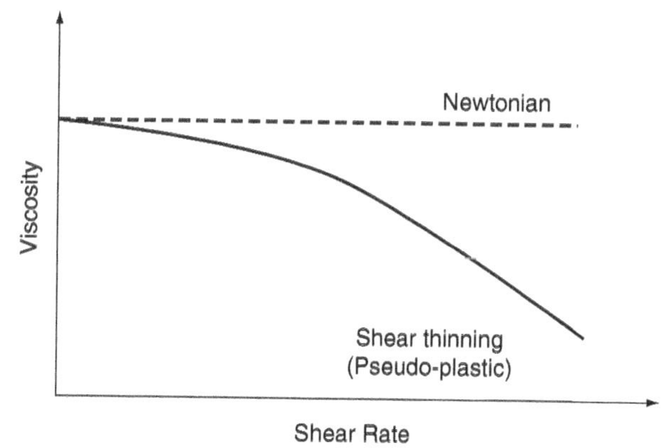


**FIGURE 1.2** Newtonian and shear-thinning viscosity behavior.

For $n = 1$, the power law model reduces to Newton's law (constant viscosity). As $n$ decreases, the polymer becomes more shear thinning. The power law exponent of commercial polymers varies between 0.8 (for some polycarbonate, PC, grades) and 0.2 (for some rubber compounds). For various polyethylene (PE) grades, the range is $0.3 < n < 0.6$, and depends on molecular weight and chain branching and does not change much with temperature. The consistency $m$ varies a lot with temperature and under usual processing conditions the consistency index for the most common molten polymers varies between 1000 $\mathrm{Pa \cdot s}^n$ (for some polyethylene terephthalate, PET, resins) to 100,000 $\mathrm{Pa \cdot s}^n$ for highly viscous rigid polyvinyl chloride (PVC). The value depends on chain mobility and molecular weight.

During single-screw extrusion, shear rates may reach 200 $\mathrm{s}^{-1}$ in the screw channel near the barrel wall, and much higher between the flight tips and the barrel. At the die lip exit the shear rate can reach or exceed 1000 $\mathrm{s}^{-1}$. During cavity filling in injection molding shear rates can reach 10,000 $\mathrm{s}^{-1}$ and in some wire-coating applications shear rates may exceed 100,000 $\mathrm{s}^{-1}$. However, in the process of rotational molding shear rates are much less than 1 $\mathrm{s}^{-1}$.

Melt index (MI), melt flow index (MFI), or melt flow rate (MFR) (for polypropylene) refers to the grams per 10 min pushed out of a die of prescribed dimensions according to an ASTM Standard [4] under the action of a specified load, as shown in Figure 1.3. For PE (ASTM D-1238) the load is 2.16 kg and the die dimensions are $D = 2.095$ mm and $L = 8$ mm. The experiment is carried out

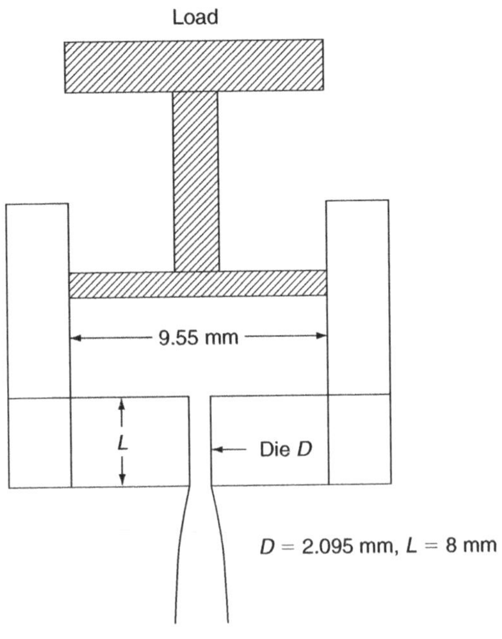


**FIGURE 1.3** Schematic of a melt indexer.

at 190°C. For PP, the same load and die dimensions are used, but the experiment is carried out at 230°C.

When measuring the melt index with a 2.16-kg load, the wall shear stress can be calculated to be $\tau_w = 1.94 \times 10^4$ Pa, and the wall shear rate approximately $\dot{\gamma} = (1838/\rho) \times MI$, where $\rho$ is the melt density in kg/m$^3$. Assuming $\rho = 766\,\text{kg/m}^3$ for a typical PE melt, we get $\dot{\gamma} = 2.4 \times MI$. A low melt index means a high molecular weight, highly viscous polymer. A high melt index means a low molecular weight, low viscosity polymer. When the melt index is <1, the material is said to have a fractional melt index. Such materials are used for film extrusion. For some film grades MI can be <0.1. Most extrusion PE grades seldom exceed MI = 12, however, for injection molding, MI is usually in the range of 5–50.

The MI (inversely) corresponds to just one point on a viscosity curve (usually at low shear rates). Frequently, the so-called high load melt index (HLMI) is measured, usually with either a 2.16-kg or a 10-kg load on the melt indexer. From two points it is possible to obtain a power law viscosity fit, but such approximations should always be used with extreme caution due to inherent inaccuracies in the measurement method.

For fully developed pressure driven axial flow ($z$ direction) of polymeric fluid in a tube of radius $R$, the velocity profile is quasi-parabolic, the shear stress varies linearly with the radius and the shear rate nonlinearly as shown in Figure 1.4.

For Newtonian fluids, the wall shear rate is given by

$$\dot{\gamma}_w = \frac{4Q}{\pi R^3} \tag{1.4}$$

This relation is referred to as apparent shear rate when used in non-Newtonian flows. A correction is necessary (Rabinowitsch correction) for shear thinning fluids. For the power law model, the true (Rabinowitsch corrected) shear rate becomes

$$\dot{\gamma}_w = \frac{3n+1}{4n}\,\frac{4Q}{\pi R^3} \tag{1.5}$$

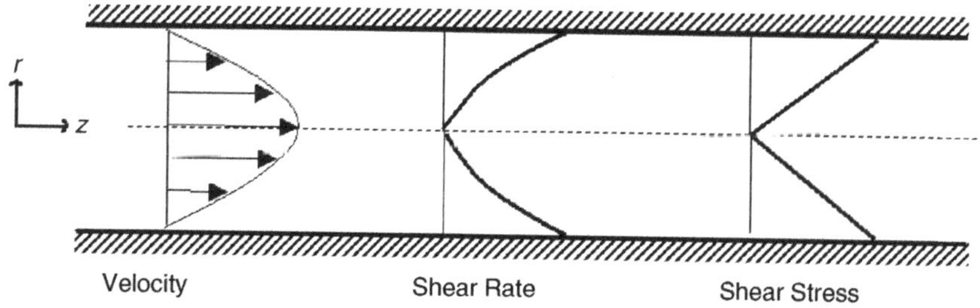


**FIGURE 1.4** Velocity, shear rate, and shear stress profiles for pressure-driven flow through a tube of radius $R$.

This means that for a material with a power law exponent $n = 0.4$ (very common for several commercial polymer grades), the relation between apparent and true shear rates is given by

$$\dot{\gamma}_{\text{true}} = 1.375 \times \dot{\gamma}_{\text{apparent}} \tag{1.6}$$

The above equations are used in capillary viscometry for measurement of viscosity from pressure drop $\Delta P$ versus flow rate $Q$ data.

The pressure drop $\Delta P$ is measured in the reservoir of the viscometer, and at the entrance to the capillary there is an excess pressure drop $\Delta P_e$. The easiest way to determine the excess pressure drop $\Delta P_e$ is to carry out the experiment using a twin-bore viscometer having two capillaries. One of them has very short length ($L = 0$), and the pressure drop recorded is essentially the excess pressure due to the entrance (orifice). This correction is necessary when capillaries are relatively short ($L/R < 50$) and is known as the Bagley correction. The Bagley correction is usually expressed as

$$n_{\text{B}} = \frac{\Delta P_e}{2\tau_w} \tag{1.7}$$

The Bagley correction ($n_{\text{B}}$) may reach perhaps 20 when polymeric materials are extruded near the critical stress for sharkskin. For a Newtonian fluid the value for $n_{\text{B}}$ is 0.587. For information on the Bagley correction and entrance and exit pressure drops the reader is referred to the monographs by Han [7].

Without the Rabinowitsch and Bagley corrections there can be significant errors in viscosity data obtained from capillary instruments.

In addition to the power law model, two other expressions are frequently used for better fitting of data over the entire range and to include the Newtonian plateau at low shear rates.

1. The Carreau–Yasuda model:

$$\eta = \eta_o\left(1 + (\lambda\dot{\gamma})^a\right)^{\frac{n-1}{a}} \tag{1.8}$$

   where $\eta_o$ is the viscosity at zero shear and $\lambda$, $a$, and $n$ are fitted parameters.

2. The Cross model:

$$\eta = \frac{\eta_o}{1 + (\lambda\dot{\gamma})^{1-n}} \tag{1.9}$$

   where $\eta_o$ is the zero shear viscosity and $\lambda$ and $n$ are fitted parameters. Note that in this model when $\lambda = 1/\dot{\gamma}, \eta = \eta_o/2$

Capillary viscometers are usually used for the shear rate range from $1\,\text{s}^{-1}$ to $3000\,\text{s}^{-1}$. The lower limit is determined by the ability of pressure gauges to measure low pressures, with accuracy, and the high limit is determined by instabilities, wall slip, and viscous heating phenomena. Rotational viscometers are usually used for the range $10^{-2}$ to $5\,s^{-1}$. At higher rotational speeds, secondary flows and instabilities may occur that invalidate the simple shear assumption. For more information about viscosity measurements, the reader is referred to Macosko [2].

The viscosity of polymer melts varies with temperature and for most polyolefins it obeys the Arrhenius relation is the form

$$\eta = \eta_{ref} \exp\left[\frac{E}{R}\left(\frac{1}{T} - \frac{1}{T_{ref}}\right)\right] \tag{1.10}$$

where $E$ is the activation energy, $R$ the gas constant and $T_{ref}$ is the reference temperature.

In polymer processing this relation is frequently simplified into a simple exponential, which applies over a shorter temperature range but it is good enough for most extrusion calculations and simulations

$$\eta = \eta_{ref} \cdot \exp(-b \cdot \Delta T) \tag{1.11}$$

The temperature sensitivity coefficient $b$ is usually between 0.01 and $0.1\,°C^{-1}$. For a linear polymer, high density polyethylene (HDPE) the value of $b$ is roughly 0.01, while for a branched low density polyethylene (LDPE) it may reach 0.03.

The viscosity increases with pressure in the form

$$\eta(p) = \eta(0) \cdot \exp(\alpha \cdot p) \tag{1.12}$$

The coefficient is probably of the order $2 \times 10^{-8}\,Pa^{-1}$. This means for a pressure increase of 10 MPa the viscosity will go up by 22%. Cogswell [3] expresses pressure dependence in terms of an equivalent temperature change. His results suggest that applying 10 MPa is equivalent to decreasing the temperature by about 5 °C. Usually, pressure dependence of viscosity is not taken into consideration in extrusion, but it is necessary in injection molding calculations and simulations. Pressure drops in die extrusion may reach at most 50 MPa, while in injection molding cavity filling the pressure may reach 200 MPa.

The effects of factors such as shear rate, molecular weight distribution, pressure, filler, temperature and additives on viscosity are summarized in Figure 1.5, following Cogswell [3]. Linear narrow molecular weight distribution polymers, such as metallocene catalyzed polyethylenes, are more viscous than their broad distribution counterparts. Fillers may increase viscosity (greatly). Pressure increases viscosity (negligible under usual extrusion conditions, but important in injection molding). Various additives, such as processing aids and lubricants, are available and are designed to decrease viscosity. The zero shear viscosity increases dramatically with the weight average molecular weight:

$$\eta_o = \text{const}\, M_w^{3.4} \tag{1.13}$$

For some metallocene catalyzed PEs with long chain branching, the exponent might be much higher (perhaps 6.0).

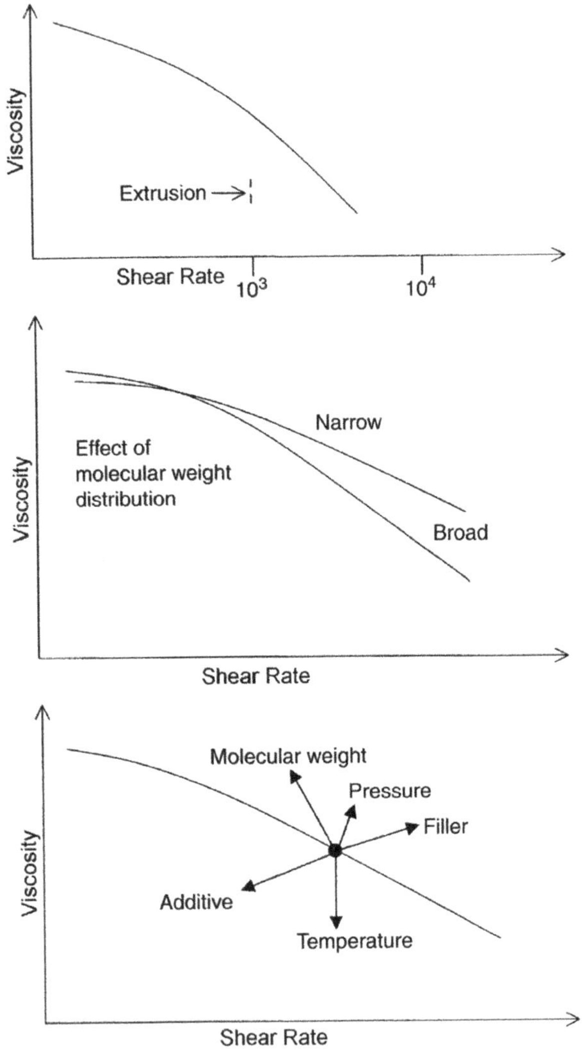


**FIGURE 1.5** The influence of various parameters on polymer viscosity.

In the above discussion of viscosity measurements, the assumption is made that the no-slip condition on the die wall is valid. This is, however, not always the case. In fact, at shear stress levels of about 0.1 MPa for PE, slip occurs. Wall slip is related to the sharkskin phenomenon [26, 27]. Wall slip is measured by the Mooney method in which the apparent shear rate ($4Q/\pi R^3$) is plotted against $1/R$ for several capillaries having different radii. In the absence of slip, the plot is horizontal. The slope of the line is equal to 4 × (slip velocity), as explained by Dealy and Wissbrun [5].

## 1.3 EXTENSIONAL VISCOSITY AND MELT STRENGTH

Extensional (or elongational) viscosity is the resistance of a fluid to extension [1, 2]. While stretching a low-viscosity fluid like water is difficult to imagine, polymer melts exhibit measurable resistance. In fact, about 100 years ago, Trouton measured the stretching and shearing resistance of stiff liquids, including pitch, and found that the extensional to shear viscosity ratio is equal to 3.

$$\frac{\eta_e}{\eta} = 3 \tag{1.14}$$

This relation, known as the Trouton ratio, is valid for all Newtonian fluids and has a rigorous theoretical basis that confirms Trouton's experiments.

Measuring elongational viscosity is considerably more difficult than measuring shear viscosity. One device used involves capillary extrusion and subsequent stretching with a pair of rollers. The maximum force required to break the extruded strand is referred to as *melt strength*. In practice, the terms *extensional viscosity* and *melt strength* are sometimes confused. Figure 1.6. shows extensional viscosity as a function of stretch rate ($\dot{\varepsilon}$), and compares it to the shear viscosity as a function of shear rate ($\dot{\gamma}$). Melt strength is an engineering measure of resistance to extension. Several extrusion processes involve extension, such as film blowing, melt spinning, thermoforming, and blow molding.

The excess pressure drop encountered in flow from a large reservoir to a smaller diameter capillary is due to elongational viscosity. In fact, Cogswell [3]

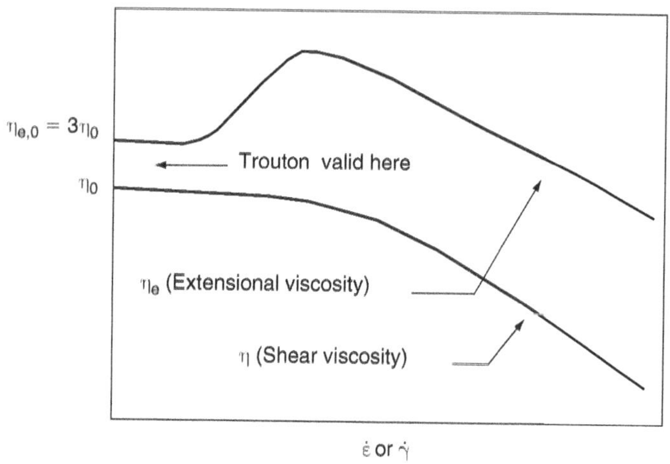


**FIGURE 1.6** Extensional and shear viscosity as a function of stretch and shear rate, respectively.

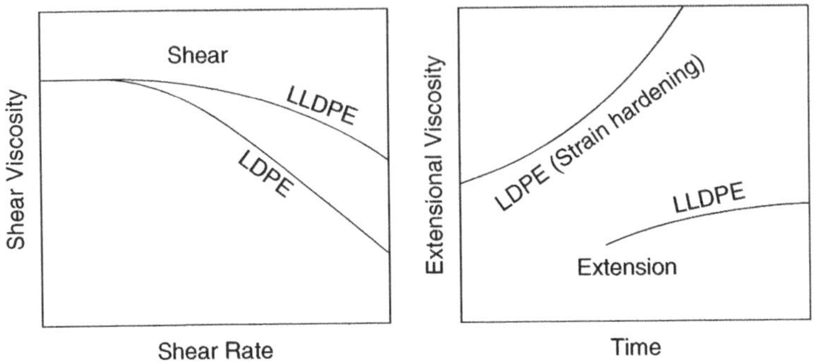


**FIGURE 1.7** Schematic representation of LDPE and LLDPE behavior in shear and extension.

has developed a method for measurement of elongational viscosity $\eta_e$ from excess pressure drop $\Delta P_e$ (i.e., the Bagley correction):

$$\eta_e = \frac{9(n+1)^2(\Delta P_e)^2}{32\eta\dot{\gamma}^2} \tag{1.15}$$

$$\dot{\varepsilon} = \frac{4\eta\dot{\gamma}^2}{3(n+1)\Delta P_e} \tag{1.16}$$

Shear and extensional viscosity measurements reveal that LLDPE (which is linear) is "stiffer" than LDPE (branched) in shear, but "softer" in extension. In extension, the linear LLDPE chains slide by without getting entangled. However, the long branches of the LDPE chains result in significantly larger resistance in extension. In the film-blowing process, LDPE bubbles exhibit more stability because of their high extensional viscosity. Typical LDPE and LLDPE behavior in shear and extension is shown in Figure 1.7. LDPE is often blended with LLDPE to improve the melt strength and consequently bubble stability in film blowing. Most polypropylene (PP) grades are known to exhibit very low melt strength. However, recent advances in polymer chemistry have led to the production of some high melt strength PP grades (with long chain branching).

Measurements of elongational viscosity have been plagued by experimental complexities, lack of repeatability, and considerable inaccuracies. The recently developed SER rheometer [12, 28] seems to have put these problems at rest and it is expected to play a significant role in the analysis and optimization of polymer processes in the near future.

## 1.4 NORMAL STRESS DIFFERENCES AND EXTRUDATE SWELL

Stress is defined as force divided by the area on which it acts. It has units of $N/m^2$ (Pascal, Pa) in SI. When a force is acting tangentially on a surface, the corresponding stress is referred to as *shear stress*. When a force is perpendicular (normal) to a surface, it is termed *normal stress*. Pressure is a normal stress. When a fluid flows through a conduit, it is acted by the normal (pressure) forces, and it exerts both normal and shear (stress) forces on the conduit walls. For flow through a planar die the shear stress is zero at the mid-plane and maximum at the wall, while the corresponding velocity profile is quasi-parabolic.

Weissenberg discovered in the 1940 s [1, 2] that polymer solutions and melts, when subjected to shearing, tend to develop normal stresses that are unequal in the $x$ (direction of flow), $y$, and $z$ (normal directions) planes, which are added to or subtracted from the local pressure. They are generated because a polymer's long molecular chains exhibit anisotropic or nonuniform properties when they flow. Any further explanation of the physical origin of normal stresses is likely to be controversial. When an (elastic) polymer solution or melt flows along a pressure gradient, it is less compressed in the direction of flow than in the other two normal directions.

The first normal stress difference $N_1$ is defined as the total normal stress in the direction of the flow ($\tau_{xx}$) minus the perpendicular ($\tau_{yy}$) stress.

$$N_1 = \sigma_{xx} - \sigma_{yy} = (-P + \tau_{xx}) - (-P + \tau_{yy}) = \tau_{xx} - \tau_{yy} \tag{1.17}$$

The second normal stress difference is

$$N_2 = \sigma_{yy} - \sigma_{zz} = (-P + \tau_{yy}) - (-P + \tau_{zz}) = \tau_{yy} - \tau_{zz} \tag{1.18}$$

We use normal stress differences rather than just normal stresses to remove the value of the pressure present. Experiments show that $N_1$ is positive for usual polymers (i.e., extensive, while the compressive pressure forces are negative). $N_2$ is negative and of the order of 20% of $N_1$ for most common polymers. $N_1$ is very sensitive to the high molecular weight tail of a polymer. Broad molecular weight distribution polymers exhibit high $N_1$ values.

The normal stress differences can be very large in high shear-rate extrusion through the lips of a die. Some authors suggest a variation for the normal stress difference at the wall in the form

$$N_{1w} = A\tau_w^b \tag{1.19}$$

The stress ratio

$$S_R = \frac{N_{1w}}{2\tau_w} \tag{1.20}$$

can reach a value of 10 or more at the onset of melt fracture.

The rod-climbing effect observed by Weissenberg when a cylinder rotates in a polymeric liquid is due to some sort of "strangulation" force exerted by the extended polymer chains (Fig. 1.8a), which results in an upward movement

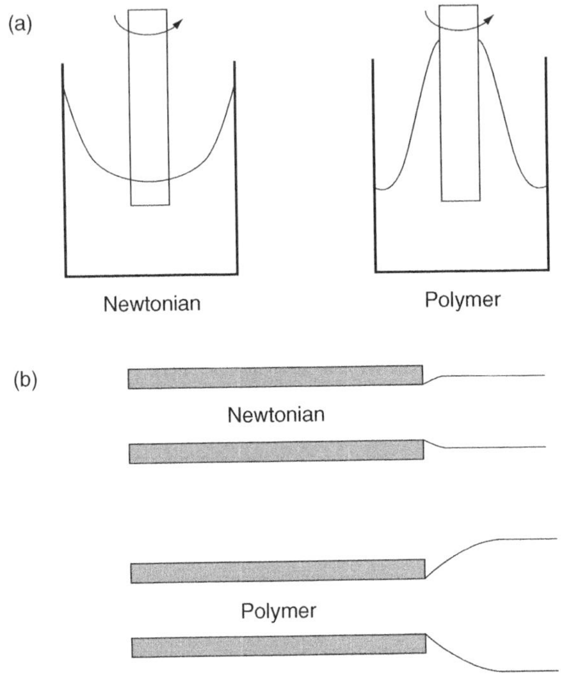


**FIGURE 1.8** (a) Rod climbing (Weissenberg) effect in polymeric fluids, (b) Extrudate swell.

normal to the direction of rotation (normal stress difference). The extrudate swell phenomenon [1, 29] (Fig. 1.8b) is due mainly to the contraction of the exiting polymer that is under extension in the die due to $N_1$. The uneven extension/compression in the various directions results in a number of unusual flow patterns and instabilities. The secondary flow patterns in square channels observed by Dooley and co-workers [30, 31] are due to the second normal stress difference. Bird et al. [1] state: "A fluid that's macromolecular is really quite weird, in particular the big normal stresses the fluid possesses give rise to effects quite spectacular."

In extrusion through dies the extrudate diameter ($d$) is larger than the die diameter ($D$). Extrudate swell ratios ($d/D$) reach values of 400% or more, under certain conditions. This phenomenon (also known as die swell) has been studied by several researchers. While the primary mechanism is release of normal stresses at the exit, other effects are also important. Extrudate swell is largest for zero length dies (i.e., orifices). It decreases for the same throughput with increasing die length due to fading memory as the residence time in the die increases. Even Newtonian fluids exhibit some swell exiting dies (13% for round extrudates, 19% for planar extrudates). This Newtonian swell is due to streamline rearrangement at the exit. The swell ratio can be influenced by thermal effects due to viscosity differences between the walls and die center. Maximum thermal swell can be obtained when a hot polymer flows through a die with colder walls. Swell ratio of about 5% on top of other mechanisms can be obtained from temperature differences. For linear polymers the swell ratio increases as the molecular weight distribution broadens.

Several attempts have been made to predict extrudate swell through equations relating the swell ratio $d/D$ (extrudate diameter/die diameter) to the first normal stress difference at the wall $N_{1w}$. Based on the theory of rubber elasticity, the following equation is obtained [29]

$$N_{1w} = 2\tau_w\left(3\left[\left(\frac{d}{D}\right)^4 + 2\left(\frac{d}{D}\right)^{-2} - 3\right]\right)^{\frac{1}{2}} \tag{1.21}$$

Based on stress release for a Maxwell fluid (described later in the chapter) exiting from a die, Tanner's equation can be derived [6, 29]

$$N_{1w} = 2\sqrt{2}\,\tau_w\left[\left(\frac{d}{D} - 0.13\right)^6 - 1\right]^{\frac{1}{2}} \tag{1.22}$$

Although this equation has a more rigorous derivation and theoretical basis, the rubber elasticity theory is believed to give better predictions. Unambiguous evaluations are virtually impossible to carry out because there are no reliable methods for measuring $N_{1w}$ at high shear rates and stresses. At low shear rates (up to perhaps $1\,s^{-1}$) the first normal stress difference can be determined by measuring the separation force that develops in a cone-and-plate instrument due to the Weissenberg effect.

## 1.5 STRESS RELAXATION AND DYNAMIC MEASUREMENTS

When flow stops, the stresses become immediately zero for small molecule Newtonian fluids like water or glycerin. For polymer melts, the stresses decay exponentially after flow stops. Stress relaxation can be measured in a parallel plate or a cone-and-plate rheometer by applying a given shear rate level (rotation speed/gap) and measuring the stress decay after the rotation is brought to an abrupt stop. Such tests, however, are not performed routinely, because of experimental limitations associated with abrupt stopping of strain and stress measurement decay over more than three orders of magnitude with one transducer.

Dynamic measurements involve the response of a material to an imposed sinusoidal stress or strain on a parallel plate or cone-and-plate instrument. A perfectly elastic material that behaves like a steel spring, by imposition of extension (strain), would develop stresses that would be in-phase with the strain, because

$$\text{Stress}(\tau) = \text{Modulus}(\text{G}) \times \text{Strain}(\gamma) \tag{1.23}$$

However, for a Newtonian fluid subjected to a sinusoidal strain, the stress and strain will not be in phase because of the time derivative (strain rate) involved

$$\tau = \eta\dot{\gamma} \tag{1.24}$$

$$\begin{aligned} \tau = \eta\frac{d\gamma}{dt} = \eta\frac{d}{dt}(\gamma_o \sin \omega t) &= \eta\omega\gamma_o \cos \omega t \\ &= \eta\omega\gamma_o \sin(\omega t + 90^\circ) \end{aligned} \tag{1.25}$$

where $\omega$ is frequency of oscillation. That is, a Newtonian fluid would exhibit 90° phase difference between stress and strain. Polymeric liquids that are partly viscous and partly elastic (viscoelastic) will be $0 \leq \varphi \leq 90°$ out of phase.

We can define

$$\begin{aligned} G'(\omega) &= \frac{\text{in-phase stress}}{\text{maximum strain}} \quad \begin{matrix}\text{storage}\\ \text{modulus}\\ \text{(elastic part)}\end{matrix} \\ G''(\omega) &= \frac{\text{out-of-phase stress}}{\text{maximum strain}} \quad \begin{matrix}\text{loss}\\ \text{modulus}\\ \text{(viscous part)}\end{matrix} \end{aligned} \tag{1.26}$$

where $\omega$ ranges usually from 0.01 to 500 rad/s. Larger $G'$ implies more elasticity. Further, we can define the dynamic viscosity

$$\begin{aligned} \eta' &= \frac{G''}{\omega} \\ \eta'' &= \frac{G'}{\omega} \end{aligned} \tag{1.27}$$

and the magnitude of the complex viscosity

$$|\eta^*| = (\eta'^2 + \eta''^2)^{1/2} \tag{1.28}$$

An empirical relationship called the Cox-Merz rule states that the shear rate dependence of the steady state viscosity $\eta$ is equal to the frequency dependence of the complex viscosity $\eta^*$, that is

$$\eta(\dot{\gamma}) = |\eta^*(\omega)| \tag{1.29}$$

The usefulness of this rule, which holds for most conventional polymers, is that while steady measurements of shear viscosity are virtually impossible for shear

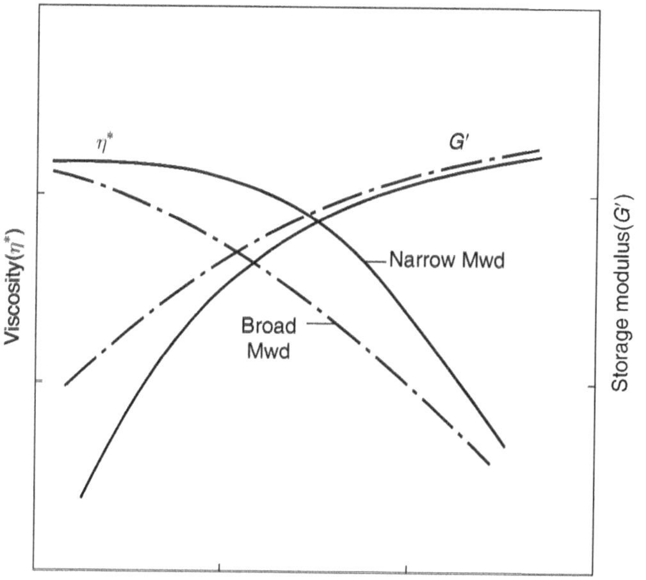


**FIGURE 1.9** Storage modulus $G'$ and dynamic viscosity $\eta^*$ behavior of broad and narrow molecular weight distribution polymers.

rates larger than 5/s with rotational instruments, the dynamic measurements can easily be carried out up to 500 rad/s (corresponds to shear rate of 500 $s^{-1}$) or even higher. Thus the full range of viscosity needed in extrusion can be covered.

Some typical results involving narrow and broad molecular weight distribution samples are shown in Figure 1.9. The relative behavior of $G'$ versus $\omega$ can be used to identify whether a sample is of narrow or broad molecular weight distribution [5]. In fact, from the crossover point where $G'=G''$, it is possible to get a surprisingly good estimate of the polydispersity $M_w/M_n$ for PP [32]. For such experiments to be meaningful the imposed strain amplitude must be low, so that the measured $G'$ and $G''$ values do not vary with the strain, but they are intrinsic properties of the polymer structure. This is the region of the so-called linear viscoelasticity.

Another interesting result is the relation between storage modulus and first normal stress difference at very small deformations ($\omega \rightarrow 0$, $\gamma \rightarrow 0$)

$$2G' = N_1 \tag{1.30}$$

For higher frequencies an expression developed by Laun [33, 5] is used:

$$N_1 = 2\frac{G'}{\omega^2}\left[1 + \left(\frac{G'}{G''}\right)^2\right]^{0.7} \tag{1.31}$$

It is possible to measure $N_1$ using a cone-and-plate rotational rheometer. The Weissenberg effect results in a separating force between the cone and the plate which can be measured to give $N_1$.

Instruments capable of measuring $N_1$ require high precision construction and very sensitive force gauges. However, measurement of $G'$, the storage modulus, can be carried out more easily.

## 1.6 CONSTITUTIVE EQUATIONS

Constitutive equations are relations between stresses and strains (deformations). In its simplest form, the Newtonian equation is a linear relation between the shear stress and the shear rate

$$\tau = \eta\dot{\gamma} = \eta\frac{dv_x}{dy} \tag{1.32}$$

This is valid for simple shear flow between two flat plates, as explained earlier in this chapter, and it is directly applicable to unidirectional flows. In polymer processing, however, most interesting flow problems require two- or three-dimensional analyses, of creeping (low Reynolds number, Re << 1) flows. For incompressible steady flow the mass conservation equation is

$$\nabla \cdot \overline{v} = 0 \tag{1.33}$$

or

$$\frac{\partial v_x}{\partial x} + \frac{\partial v_y}{\partial y} + \frac{\partial v_z}{\partial z} = 0 \tag{1.34}$$

and the momentum equation

$$0 = -\nabla P + \nabla \cdot \overline{\overline{\tau}} \tag{1.35}$$

where $P$ is the pressure (which is a scalar) and $\overline{\overline{\tau}}$ the stress, which is a (second order) tensor

$$\overline{\overline{\tau}} \Rightarrow \begin{pmatrix} \tau_{xx} & \tau_{xy} & \tau_{xz} \\ \tau_{yx} & \tau_{yy} & \tau_{yz} \\ \tau_{zx} & \tau_{zy} & \tau_{zz} \end{pmatrix} \tag{1.36}$$

To generalize the Newtonian equation in three dimensions we must propose a linear relation between stress components and strain rate components. The strain rate tensor is

$$\overline{\overline{D}} = \frac{1}{2}(\nabla\overline{v} + \nabla\overline{v}^T) = \frac{1}{2}\left(\frac{\partial v_i}{\partial x_j} + \frac{\partial v_j}{\partial x_i}\right) \tag{1.37}$$

$$\overline{\overline{D}} \Rightarrow \begin{pmatrix} D_{xx} & D_{xy} & D_{xz} \\ D_{yx} & D_{yy} & D_{yz} \\ D_{zx} & D_{zy} & D_{zz} \end{pmatrix} \tag{1.38}$$

where $D_{xx} = \frac{1}{2}\left(\frac{\partial v_x}{\partial x} + \frac{\partial v_x}{\partial x}\right) = \frac{\partial v_x}{\partial x}$, $D_{xy} = \frac{1}{2}\left(\frac{\partial v_x}{\partial y} + \frac{\partial v_y}{\partial x}\right)$, and similarly the other components can be written out explicitly in terms of the components in the $x$, $y$, and $z$ directions.

The Newtonian constitutive equation is generalized in the form

$$\bar{\bar{\tau}} = \eta(2\bar{\bar{D}}) \tag{1.39}$$

This means that $\tau_{xx} = \eta(2D_{xx}) = 2\eta\frac{\partial v_x}{\partial x}$, $\tau_{xy} = \eta(2D_{xy}) = \eta\left(\frac{\partial v_x}{\partial y} + \frac{\partial v_y}{\partial x}\right)$ etc.

The models expressing shear thinning behavior of polymer melts (power – law, Carreau–Yasuda and Cross) are generalized by replacing $\dot{\gamma}$ by the function $(II_D)$ which is the second invariant of the strain rate tensor $2\bar{\bar{D}}$. It is called *invariant*, because this quantity remains unchanged under rotation of the coordinate axes. It is given in rectangular, cylindrical, and spherical coordinates in Table 1.1. Thus we have the generalized power law equation written as

$$\eta = m(II_D)^{\frac{n-1}{2}} \tag{1.40}$$

**TABLE 1.1 The Second Invariant Strain Rate Tensor in Rectangular, Cylindrical, and Spherical Coordinates.**

Rectangular

$$II_D = 2\left[\left(\frac{\partial v_x}{\partial x}\right)^2 + \left(\frac{\partial v_y}{\partial y}\right)^2 + \left(\frac{\partial v_z}{\partial z}\right)^2\right] + \left[\frac{\partial v_y}{\partial x} + \frac{\partial v_x}{\partial y}\right]^2 + \left[\frac{\partial v_z}{\partial y} + \frac{\partial v_y}{\partial z}\right]^2 + \left[\frac{\partial v_x}{\partial z} + \frac{\partial v_z}{\partial x}\right]^2$$

Cylindrical

$$II_D = 2\left[\left(\frac{\partial v_r}{\partial r}\right)^2 + \left(\frac{1}{r}\frac{\partial v_\theta}{\partial \theta} + \frac{v_r}{r}\right)^2 + \left(\frac{\partial v_z}{\partial z}\right)^2\right] + \left[r\frac{\partial}{\partial r}\left(\frac{v_\theta}{r}\right) + \frac{1}{r}\frac{\partial v_r}{\partial \theta}\right]^2 + \left[\frac{1}{r}\frac{\partial v_z}{\partial \theta} + \frac{\partial v_\theta}{\partial z}\right]^2 + \left[\frac{\partial v_r}{\partial z} + \frac{\partial v_z}{\partial r}\right]^2$$

Spherical

$$II_D = 2\left[\left(\frac{\partial v_r}{\partial r}\right)^2 + \left(\frac{1}{r}\frac{\partial v_\theta}{\partial \theta} + \frac{v_r}{r}\right)^2 + \left(\frac{1}{r\sin\theta}\frac{\partial v_\phi}{\partial \phi} + \frac{v_r}{r} + \frac{v_\theta \cot\theta}{r}\right)^2\right] + \left[r\frac{\partial}{\partial r}\left(\frac{v_\theta}{r}\right) + \frac{1}{r}\frac{\partial v_r}{\partial \theta}\right]^2 + \left[\frac{\sin\theta}{r}\frac{\partial}{\partial \theta}\left(\frac{v_\phi}{\sin\theta}\right) + \frac{1}{r\sin\theta}\frac{\partial v_\theta}{\partial \phi}\right]^2 + \left[\frac{1}{r\sin\theta}\frac{\partial v_r}{\partial \phi} + r\frac{\partial}{\partial r}\left(\frac{v_\phi}{r}\right)\right]^2$$

Using the expression of the second invariant in rectangular coordinates, it can easily be shown that for simple shear flow ($x$-velocity only, varying in $y$-direction only) we have

$$\eta = m\left(\frac{\partial v_x}{\partial x}\right)^n \tag{1.41}$$

The above fluid model, which expresses simply the shear thinning behavior, is referred to as the generalized Newtonian fluid (GNF) in the rheological literature. This model cannot explain any of the viscoelastic flow phenomena, such as stress relaxation, normal stresses, or extrudate swell.

The simplest way to develop viscoelastic constitutive equations is to combine a model for an elastic solid

$$\tau = G\gamma_{\text{solid}} \tag{1.42}$$

where $\gamma_{\text{solid}}$ is the strain, with that for a Newtonian fluid

$$\tau = \eta\dot{\gamma}_{\text{fluid}} \tag{1.43}$$

By differentiating the elastic solid equation and adding the two strain rates, we get

$$\frac{\dot{\tau}}{G} + \frac{\tau}{\eta} = \dot{\gamma} \tag{1.44}$$

or

$$\tau + \lambda\dot{\tau} = \eta\dot{\gamma} \tag{1.45}$$

where $\lambda = \eta/G$ has dimensions of time (relaxation constant).

This is known as the Maxwell model and shows that the viscoelastic nature of polymers can be described by viscosity and a relaxation time. This model can be generalized in two dimensions as follows

$$\tau_{xx} + \lambda\left[\frac{\partial\tau_{xx}}{\partial t} + v_x\frac{\partial\tau_{xx}}{\partial x} + v_y\frac{\partial\tau_{xx}}{\partial y} - 2\frac{\partial v_x}{\partial x}\tau_{xx} - 2\frac{\partial v_x}{\partial y}\tau_{yx}\right] = 2\eta\frac{\partial v_x}{\partial x} \tag{1.46}$$

$$\tau_{yy} + \lambda\left[\frac{\partial\tau_{yy}}{\partial t} + v_x\frac{\partial\tau_{yy}}{\partial x} + v_y\frac{\partial\tau_{yy}}{\partial y} - 2\frac{\partial v_y}{\partial x}\tau_{xy} - 2\frac{\partial v_y}{\partial y}\tau_{yy}\right] = 2\eta\frac{\partial v_y}{\partial y} \tag{1.47}$$

$$\tau_{xy} + \lambda\left[\frac{\partial\tau_{xy}}{\partial t} + v_x\frac{\partial\tau_{xy}}{\partial x} + v_y\frac{\partial\tau_{xy}}{\partial y} - \frac{\partial v_x}{\partial y}\tau_{yy} - \frac{\partial v_y}{\partial x}\tau_{xx}\right] = \eta\left[\frac{\partial v_x}{\partial y} + \frac{\partial v_y}{\partial x}\right] \tag{1.48}$$

$$\tau_{xy} = \tau_{yx}(\text{the stress tensor is symmetric}) \tag{1.49}$$

Actually, the above is one of the possible generalizations that satisfy mathematical invariance. In general, the Maxwell model is written as

$$\bar{\bar{\tau}} + \lambda\frac{\Delta\bar{\bar{\tau}}}{\Delta t} = 2\eta\bar{\bar{D}} \tag{1.50}$$

where the derivative $\Delta/\Delta t$ is referred to as the upper convective derivative as given in the above equations in two dimensions. There is also the lower convective derivative as well as other forms (e.g., co-rotational) which are the subject of specialized handbooks [1, 6, 7, 12] and numerous articles.

There are several possible extensions and generalizations of simple models. For example the convected Maxwell model can be written in the form

$$\bar{\bar{\tau}} + \lambda(II_D)\frac{\Delta \bar{\bar{\tau}}}{\Delta t} = 2\eta(II_D)\,\bar{\bar{D}} \tag{1.51}$$

where $\lambda(II_D)$ and $\eta(II_D)$ represent the relaxation time and the viscosity, respectively; both of them are functions of the second invariant of the strain rate tensor. In this form, the model is known as White–Metzner. When a generalization is proposed the key criterion is to satisfy the principle of material indifference, which states that the predicted response of a material must be the same for all observers irrespective of their coordinate system of reference. As a consequence of this, in the development of constitutive equations, a coordinate system that moves, rotates, and deforms with the material should be used. This requirement results in mathematically complex constitutive equations. Despite their complexity and the requirement of fitting of numerous parameters most viscoelastic constitutive equations fail to predict most of the unusual rheological phenomena exhibited by polymeric liquids. The most successful constitutive equation is the so-called *K-BKZ* integral model inspired by the theory of rubber elasticity [7] and involves more than two dozen experimentally fitted parameters (e.g., Ref. 34). Current trends involve the development of models based on macromolecular motions. De Gennes proposed the snake-like motion of polymer chains called reptation [2, 12] and deduced from scaling relations that the zero shear viscosity must be $\eta_o \approx M^{3.0}$, while experiments give $\eta_o \approx M^{3.4}$ ($M$ is the molecular weight). Based on the reptation concept Doi and Edwards [2, 12] developed a constitutive equation that leaves much to be desired before it can be used for predicting viscoelastic flow phenomena. Several attempts were made to fix the Doi–Edwards theory [12]. The most talked about viscoelastic model recently, is the pom-pom polymer model, developed by McLeish and Larson [12, 35]. The motivation for its development was that the *K-BKZ* equation fails to predict the observed degree of strain hardening in planar extension when certain functions are adjusted to fit the observed degree of strain softening in shear. The failure to describe the rheology of long-branched polymers suggests that some new molecular insight is needed into the nonlinear relaxation processes that occur in such melts under flow. The pom-pom model uses an H-polymer structure, in which molecules contain just two branch points of chosen functionality and a backbone that links the two pom-poms. The pom-pom model exhibits rheological behavior remarkably similar to that of branched commercial melts like LDPE. It shows strain hardening in extension and strain softening in shear. It can describe both planar and uniaxial extension. The constitutive equation is an integro differential. For successful application at least 32 parameters must be obtained by fitting experimental rheological data. Of course, fitting 32 or more parameters in a complicated constitutive equation is a mathematical challenge.

Modeling polymer viscoelastic behavior has always been a controversial subject. While viscoelastic constitutive equations have contributed toward understanding of various deformation mechanisms and flow, they unfortunately have *not* provided us with quantitative predicting power for polymer process and equipment design. Very often the predictions depend on the model used for the computations and are not corroborated with experimental observations. Some viscoelastic problems can be solved with the appropriate constitutive equations, but this is still an area of academic research with very limited practical applications at the moment.

For design of polymer process equipment such as extruders, dies, and molds, computer simulations are carried out in two or three dimensions, using the GNF model. The Carreau–Yasuda and Cross models have a clear advantage over the power law because they can capture the viscosity behavior from the Newtonian plateau at low shear rates to high shear regions with substantial shear thinning.

## 1.7 PROBLEM SOLVING WITH THE HELP OF RHEOLOGY

### 1.7.1 Using the Melt Flow Index (MFI)

The MFI, a simple flow test, is used for quality control purposes and, as explained earlier in this chapter, gives a rough estimate of viscosity and indirectly the approximate average molecular weight. Despite the misgivings expressed by academic rheologists, more than 100 million tons of HDPE, LDPE, LLDPE, PP, PS, and other thermoplastics are sold annually, on the basis of solid density and the MFI, under standard conditions (2.16 kg load and 190° for polyethylene). The conversion of polymer pellets into film, sheet, pipe, profiles, automotive parts, containers, appliance housings, and numerous other products is carried out mainly in small and medium enterprises, which number over 100,000 around the world. Such companies do not possess the instruments or the know-how to carry out accurate rheological measurements and to fully get the benefit of such characterization. Frequently, original equipment manufacturers are asked to design and build processing equipment (extruders, dies and molds) without any rheological information other than the MFI. Consequently some short of zero-shear viscosity or consistency index in the power law model has to be estimated, and a power law exponent has to be chosen for any calculations or simulations. Processors extrude, injection mold, or otherwise process huge quantities of plastics without any additional controls on the incoming raw materials (pellets or powders).

Measurement of a high load melt index (usually with 21.6 or 10 kg) can provide an approximate second point on a viscosity curve, and the power law model parameters can be easily obtained [15], but they will not be very accurate due to experimental limitations of the melt indexer. The ratio HLMFI:MFI gives an indication of the degree of shear thinning, and it can be used to differentiate between two grades of the same polymer type or to assess processability. For example, the higher the HLMFI:MFI ratio the more the shear thinning and the lower the pressures and viscous heat generated.

The MFI is used for selecting a polymer grade for a particular application, depending on whether a larger or a smaller average molecular weight is more suitable. This in turn can be related to end use properties such as part failure. For example, polymers with higher molecular weight are known to have better environmental stress cracking resistance (ESCR). Consequently, polymers with lower MFI should be processed for parts required to have better ESCR. Similarly, polymers with lower MFI will produce stronger films for packaging and stronger fibers for ropes than their higher MFI counterparts.

Generally, the MFI test (preferably with two loads) can be used to give an indication on the average molecular weight, the viscosity and shear thinning, and the end-use properties, before more sophisticated and more expensive characterizations commence.

### 1.7.2 Using Viscosity Data

Accurate measurements of viscosity are absolutely necessary for computer simulations and rational equipment design. The data can easily be fitted to the Carreau–Yasuda or Cross viscosity models and subsequently be used in numerical analysis of molten polymer flows through channels and process equipment, by solving the appropriate differential equations for conservation of mass, momentum and energy. The Carreau–Yasuda and Cross viscosity curves are smooth from very low to very high shear rates, and the various numerical analysis schemes (finite differences, finite elements, boundary elements) do not offer any special problems associated with the viscosity functions. The power law model, which permits closed-form solutions for some unidirectional and lubrication approximation flow problems, may give some numerical difficulties (stability, convergence) due to the fact that the viscosity tends to infinity as the shear rate approaches zero. Obviously, there is no advantage in choosing the simple power law model whenever finite differences or finite elements are involved.

The viscosity curve itself can be used for polymer grade identification purposes. The zero-shear viscosity is related to the average molecular weight as discussed earlier in this chapter. To obtain an accurate value of the zero-shear viscosity the measurements have to be carried out at shear rates of $10^{-3}\,s^{-1}$ or less, which is very difficult in steady mode, and consequently the dynamic measurements are preferable. The degree of shear thinning can be used for differentiation between linear polymers of narrow and broad molecular weight distribution (more shear thinning). Also, more shear thinning implies a higher degree of long chain branching. Strong temperature dependence of viscosity indicates a branched polyethylene, while a weak dependence indicates a linear polyethylene (the temperature sensitivity factor in the exponential model is about $b = 0.01$ for HDPE and $b = 0.02 - 0.03$ for LDPE).

From the shape of the viscosity curve more information can be inferred relating to other polymer characteristics and properties. Higher shear thinning implies broader molecular weight distribution, which in turn implies more elasticity, and this is turn implies, larger first normal stress difference, larger extrudate swell, and longer stress relaxation. The polymer elasticity is also related to several end-use properties, such as warpage (more) and environmental stress cracking resistance (higher).

### 1.7.3 Using the Storage Modulus $G'$

Storage modulus measurement is the most reliable method for determining the elasticity of a polymer melt, which in turn is related to several other processing and end-use properties. Higher $G'$ at low frequencies (higher elasticity) implies a longer high molecular weight tail in the molecular weight distribution curve. During processing, the more elastic polymers tend to develop larger normal stresses which result in larger extrudate swell. Also, upon cooling of the finished product more stresses, are frozen in, which may be released on reheating, and this results in more warpage. The environmental stress cracking resistance is usually higher for the more elastic polymers. Differences between polymers that cannot be detected using the melt flow index and the viscosity curve may show up very clearly in comparing $G'$ data, especially at very low frequencies.

### 1.7.4 Using Extensional Viscosity

Several processes involve extensional flow, such film blowing, blow molding, thermoforming, and fiber spinning. The polymers to be used in such processing operations should, therefore, be characterized by their extensional properties. LDPE (branched) exhibits high extensional viscosity and strong strain hardening and, consequently, produces stable bubbles in film blowing. LLDPE exhibits lower extensional viscosity than LDPE without any significant strain hardening and produces bubbles that may be prone to instabilities. For such materials the cooling and bubble handling systems downstream of the die lips have to be expertly designed to reduce or eliminate the instabilities. Conventional PP exhibits very low resistance in extension and cannot be processed in conventional film blowing lines. A blown film line for such a material requires extrusion downwards, cooling, reheating and then formation of a bubble. However, newer PP grades with long chain branching can be processed in conventional blown film lines, because the long chain branching provides the required level of extensional viscosity.

In the process of thermoforming the level of extensional viscosity is very important. Very low extensional viscosity materials are not thermoformable. However, when the extensional viscosity is very high, it may be difficult for the sheet to form in the exact shape of corners or other intricate mold geometry.

In foaming of thermoplastics usually low extensional viscosity usually results in bubble collapse and poor bubble structure. Very high extensional viscosity may result in difficulties in the bubble formation process.

### 1.7.5 Troubleshooting Co-Extrusion

There are two rheological phenomena that manifest themselves during the flow of immiscible polymer melts through dies: layer nonuniformity and interfacial instability [21, 24, 36, 37].

Layer nonuniformity in coextrusion flows is caused mainly by the less viscous polymer going to the high shear region (i.e., the wall) thereby producing encapsulation. Partial encapsulation can occur in common types and sizes of dies, and complete encapsulation is possible for extremely long dies. Differences in polymer wall adhesion and viscoelastic characteristics can also be contributing factors. Weak secondary flows caused by viscoelastic effects (from the second normal stress difference) have been demonstrated to produce layer nonuniformities even when co-extruding differently colored polymer streams of the same polymer [31]. This defect can be reduced by choosing materials with the smallest possible differences in viscosity and viscoelasticity ($G'$, $G''$, extrudate swell), or by adjusting the stream temperatures to bring the polymer viscosities closer to one another.

Layer nonuniformity can also arise in feedblock co-extrusion, in which melt streams are merged into a single stream in a feedblock before entering the flat die. Uneven flow leakage from the flat die manifold to the downstream die sections can lead to encapsulation of the more viscous polymer by the less viscous, or even the reverse! Feedblock profiling is used to counteract the natural tendency for encapsulation due to viscosity differences [38]. This involves contouring the feedblock flow passages for regions of high or low volumetric throughput, as shown in Figure 1.10. Feedblock profiling combined with eliminating uneven flow leakage from the feeding section of a flat die (or the use of this leakage to counteract the natural tendency for encapsulation) can be used to produce layer-to-layer uniformity in the extrudate. The problem is much more complex when co-extruding many layers, as profiling for one layer will disrupt the other layers. The influence of a feedblock design change is virtually impossible to predict accurately, at present. Powerful computer machinery would be needed to determine the interfaces and interactions with the walls. This is not an exercise for routine equipment design purposes.

Interfacial instability in co-extrusion refers to two common defects consisting of highly irregular or sometimes regular waviness that appears in co-extruded structures at the polymer–polymer interface. The effect is to significantly reduce the optical quality of coextruded film. It is an internal defect, which distinguishes it from sharkskin, which is a surface defect.

The most frequently encountered interfacial instability is zigzag (also known as die-land) instability [36, 37], which appears as chevrons pointing in the flow direction. It is initiated in the die land (which is the parallel wall die section) and is characterized by a critical interfacial shear stress, in the range of 40–80 kPa (roughly a quarter to a half of the critical wall shear stress level for sharkskin). This problem can arise even if adjacent layers are of the same material. The mechanism responsible has not been conclusively identified. Apparently there is amplification of certain disturbance wavelengths under high stress conditions.

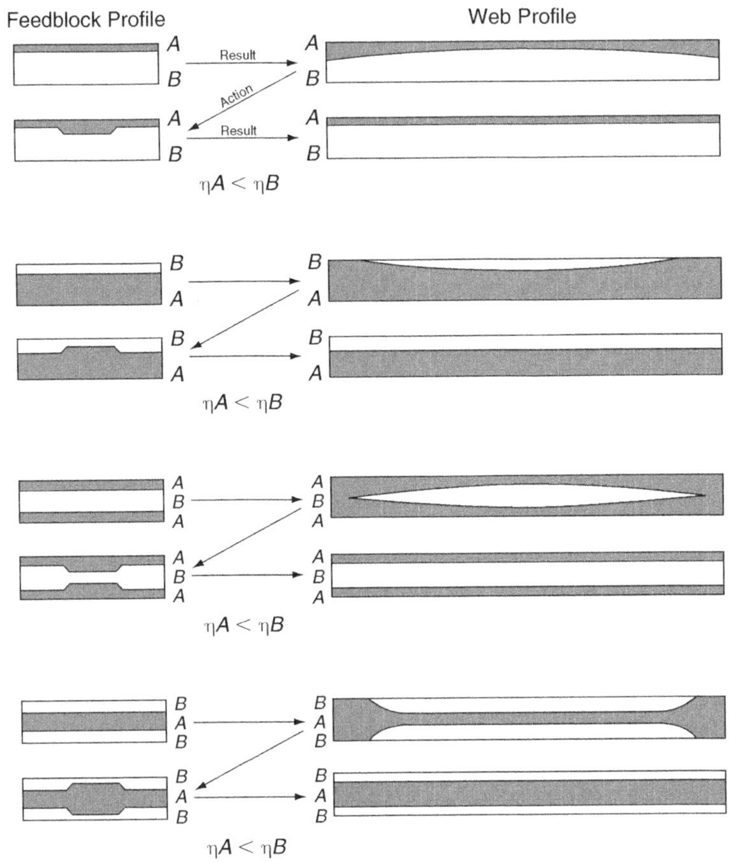


**FIGURE 1.10** Feedblock profiling and the resultant effects.

Viscoelasticity is probably a contributing factor, i.e., the interfacial normal stress difference is important. Unfortunately, this is impossible to measure and difficult to calculate accurately. The most reliable means of diagnosing zigzag instability at present is to calculate interfacial shear stress using simulation software. Zigzag instability problems are remedied by reducing interfacial shear stresses.

### 1.7.6 Using Computer Simulation and Rheological Data

Computer simulation can be very effective in process troubleshooting. First requirement is the existence of reliable viscosity data over a wide range of shear rates, e.g., for extrusion the viscosity must be known from less than $1\,s^{-1}$ to more than $1000\,s^{-1}$, while in injection molding the shear rates may reach or exceed $10{,}000\,s^{-1}$. The second step is for the data to be fitted to a suitable viscosity model such as power law, Carreau–Yasuda, or Cross, with temperature dependence. The third step is to carry out the simulation and perhaps several what-ifs, so possible causes of observed problems can be identified. The simulation data are then compared to whatever observations and measurements are available from the factory floor.

Of course, it all depends on how good the computer software is. If we assume that the software is capable of providing reasonably error-free results, then local pressure, shear rates, shear stresses, temperatures, and streamline patterns are very useful in the troubleshooting process.

Process engineers working on a production line may be biased about a process problem or product defect. Usually, when there are visible extrudate defects they are blamed to sharkskin–melt fracture. However, other problems such as material degradation due to excessive residence time at very high temperatures or die flow lines might be responsible. If the computed die lip shear stress is low ($<0.1$ MPa), then sharkskin–melt fracture can be excluded and other causes must be sought. If there is a region of nearly stagnant flow region (low wall shear rates) this might be the source of polymer degradation. In transparent film production, if there are wall shear rates $<5\ s^{-1}$ anywhere, they are likely to result in degradation related defects, which may appear as black spots, haziness, die lines, or streaks.

Rheological data are of crucial importance, especially when comparing different production batches with differences in the incoming raw material. Again, computer simulations can provide the local pressures, shear rates, shear stresses, and flow patterns, and potential root causes of the problems and defects can be identified.

In special cases, viscoelastic constitutive equations, which can be fitted to reliable rheological data, can be used for simulations. Even if reliable flow simulation is not feasible with a viscoelastic constitutive equation, we can use flow patterns based on GNF and then use the viscoelastic equation to determine post-processing effects, such as shrinkage and warpage. It is also possible for the user to make reasonable approximations and limit a viscoelastic flow analysis to only a region of interest, which might be a die lip exit in extrusion, a sudden restriction in injection molding, or sheet thinning near a corner in thermoforming. Despite the limited predictive ability of the viscoelastic constitutive equations, certain trends might be identified.

## 1.8 CONCLUDING REMARKS

The rheological characterization of molten polymers is of paramount importance in polymer processing for several reasons. First, viscosity measurements are necessary for equipment design. It is important that the viscosity be measured over the full shear range rate expected in the process. Second, rheological measurements can be used to identify or to differentiate various polymer grades. For such purposes it might be necessary to measure, in addition to the steady shear viscosity, the extensional viscosity and the storage and loss moduli.

Occasionally, measurements of normal stress differences might provide significant insight into the materials or the processes.

Rheological measurements are usually more sensitive and easier than other methods of characterization, and they are excellent tools for process troubleshooting and optimization purposes. The viscoelastic models provide significant insight into polymer behavior, but they are not used in routine equipment design, due to mathematical complexities and limited predictive power for industrial operating conditions.

## REFERENCES


1. R.B. Bird, R.C. Armstrong, and O. Hassager, *Dynamics of Polymeric Liquids*, vol. 1, Wiley, New York, 1987.
2. C.W. Macosko, *Rheology: Principles, Measurements and Applications*, VCH Publishers, New York, 1994.
3. F.N. Cogswell, *Polymer Melt Rheology*, Woodhead Publishing, Cambridge, UK, 1996.
4. A.V. Chenoy and D.R. Saini, *Thermoplastic Melt Rheology and Processing*, Marcel Dekker, New York, 1996.
5. J.M. Dealy and K.F. Wissbrun, *Melt Rheology and Its Role in Plastics Processing*, Chapman & Hall, London, 1996.
6. R.I. Tanner, *Engineering Rheology*, Oxford Engineering Science, Oxford, UK, 2000.
7. C.D. Han, *Rheology and Processing of Polymeric Material*, vols. 1–2, Oxford University Press, Oxford, UK, 2007.
8. A.Y. Malkin and A.I. Isayev, *Rheology*, ChemTech Publishing, Toronto, 2006.
9. W.W. Graessley, *Polymeric Liquids and Networks: Dynamics and Rheology*, Garland Science, London, 2008.
10. J.L. White, *Principles of Polymer Engineering Rheology*, Wiley, New York, 1990.
11. M.T. Shaw and W.J. MacKnight, *Introduction to Polymer Viscoelasticity*, Wiley, New York, 2005.
12. J.M. Dealy and R.G. Larson, *Structure and Rheology of Molten Polymers*, Hanser, Munich, 2006.
13. R.K. Gupta, *Polymer and Composite Rheology*, Marcel Dekker, New York, 2000.
14. J.L. Leblanc, *Filled Polymers*, CRC Press, Boca Raton, 2010.
15. J. Vlachopoulos and J.R. Wagner, eds., *The SPE Guide on Extrusion Technology and Troubleshooting*, Society of Plastics Engineers, Brookfield, CT, 2001.
16. S. Middleman, *Fundamentals of Polymer Processing*, McGraw-Hill, New York, 1977.
17. Z. Tadmor and C.G. Gogos, *Principles of Polymer Processing*, Wiley, New York, 2006.
18. R.G. Griskey, *Polymer Process Engineering*, Chapman & Hall, New York, 1995.
19. M.M. Denn, *Polymer Melt Processing*, Cambridge University Press, Cambridge, UK, 2008.

20. T. Kanai and G.A. Campbell, eds., *Film Processing*, Hanser, Munich, 1999.
21. J.R. Wagner, ed., *Multilayer Flexible Packaging*, Elsevier, Amsterdam, 2010.
22. D.G. Baird and D.J. Collias, *Polymer Processing Principles and Design*, Wiley, New York, 1998.
23. C.I. Chung, *Extrusion of Polymers*, Hanser, Munich, 2000.
24. T.J. Butler, ed., *Film Extrusion Manual*, TAPPI Press, Atlanta, GA, 2005.
25. M. Kamal, A. Isayef, and S. Liu, *Injection Molding Fundamentals and Applications*, Hanser, Munich, 2009.
26. M.M. Denn, *Ann. Rev. Fluid Mech.*, 33, 265 (2001).
27. S.G. Hatzikiriakos and K. Migler, eds., *Polymer Processing Instabilities: Control and Understanding*, Marcel Dekker, New York, 2004.
28. M. Sentmanat, B.N. Wang, and G.H. McKinley, *J. Rheol.*, 49, 585 (2005).
29. J. Vlachopoulos, *Rev. Def. Beh. Mat.*, 3, 219 (1981).
30. B. Debbaut, T. Avalosse, J. Dooley, and K. Hughes, *J. Non-Newt. Fluid Mech.*, 69, 255 (1997).
31. J. Dooley, "*Viscoelastic Flow Effects in Multilayer Polymer Coextrusion*," doctoral thesis, University of Eindhoven, Netherlands, 2002.
32. S.W. Shang, *Adv. Polym. Tech.*, 12, 389 (1993).
33. H.M. Laun, *J. Rheol.*, 30, 459 (1986).
34. E. Mitsoulis, *J. Non-Newt. Fluid M.*, 97, 13 (2001).
35. T.C.B. McLeish and R.G Larson, *J. Rheol.*, 42, 81 (1998).
36. R. Shroff and H. Mavridis, *Plast. Techn.*, 37, #2, 5 (1998).
37. J. Perdikoulias and C. Tzoganakis, *Plast. Eng.*, 52, #4, 41 (1996).
38. P. Cloeren, in *Proceedings of S.P.E. Extrusion Division RETEC*, Toronto, Canada, 1993.

# CHAPTER 2

# POLYMER PROCESSING ADDITIVES FOR MELT FRACTURE CONTOL

SAVVAS G. HATZIKIRIAKOS[1] and KALMAN B. MIGLER[2]

[1] Department of Chemical and Biological Engineering, The University of British Columbia, Vancouver, BC, Canada

[2] Polymers Division, National Institute of Standards and Technology, Gaithersburg, MD, USA

## CONTENTS



*Applied Polymer Rheology: Polymeric Fluids with Industrial Applications*, First Edition. Edited by Marianna Kontopoulou.

## 2.1 INTRODUCTION

Commercial polymer processing operations, such as fiber spinning, film blowing, blow molding, profile extrusion, and various coating flows, suffer from the onset of flow instabilities, which manifest themselves as distortions on the surface of the extruded product [1–4]. These phenomena, collectively known as *melt fracture*, pose a limitation to the rate production that threatens the economic feasibility of the processes [1–8].

In extrusion processes, the extrudate distortions are first observed when the volumetric flow rate exceeds a critical value. The severity of the distortions (amplitude and periodicity and/or irregularity) increases with increasing flow rate [1, 9–12]. Thus the production rates of commercially acceptable products are limited to the ones below these critical values. To overcome these difficulties and render the processes economically feasible, polymer processing aids (PPAs) are frequently used. PPAs can eliminate the flow instabilities known as sharkskin melt fracture and stick-slip or postpone them to higher flow rates. The end result is an increase of the productivity as well as an energy cost reduction, while high product quality is maintained.

The purpose of this chapter is to examine processing aids that have been investigated over the past several decades and to discuss their performance in melt fracture elimination, so as to elucidate the role of these materials in instability elimination and polymer processability improvement. The factors that determine the additive performance, such as concentration, dispersion quality, and interactions between additives, additive and polymer, and additive and die surface, are also examined. The various mechanisms by which different types of processing aids help eliminate the instabilities are reviewed as well.

The first additives discussed in this chapter are fluoropolymer based and their use is primarily for polyolefin extrusion processes [13–15]. Additional examples of other conventional processing aids such as stearates, silicon based polymers, hyperbranched polyethylenes, and various polymer blend combinations are also discussed. Newer boron nitride (BN)-based PPAs are examined in detail as well [16]. However, the most commercially relevant example is the use of fluoropolymer based PPAs to eliminate sharkskin melt fracture in linear low-density polyethylene (LLDPE) industry; therefore, this chapter is focused more on these PPAs.

## 2.2 FLUOROPOLYMER PPA TECHNOLOGY

The fluoropolymer PPA technology was discovered and developed in the early 1960s by DuPont Canada. It was introduced to the newly developed LLDPE to eliminate sharkskin melt fracture. The discovery was made accidentally when the scientists of DuPont tested an LLDPE on a blown film line extruder that had just previously been used for fluoropolymer extrusion. While sharkskin melt fracture was expected to be seen, the film came out clear and free of any distortions. Then it was realized that the extruder was not purged properly, and the small amount of fluoropolymer still present was acting as a processing aid. It was only after a few hours that sharkskin came back again into the film appearance.

A widely used fluoropolymer additive for LLDPE is a copolymer of vinylidene fluoride ($VF_2$) and hexafluoropropylene and is often also referred to as fluoroelastomer, although it is not cross-linked. The general chemical structure of this is as follows:

$$-(CF_2-\underset{\displaystyle CF3}{\underset{|}{CF}})_n-(CF_2-CH_2)_m-$$

In $VF_2$ the hydrogens become acidic due to the electronegativity of the fluorine. The resulting dipole moment of this monomer gives it an attraction with the oxides in the steel. By itself the $VF_2$ is crystalline; the incorporation of hexafluoropropylene breaks up the crystallinity and makes the material an uncross-linked elastomer. The hexafluoropropylene also has poor interaction with the nonpolar polyethylene, reducing the adhesion between these two polymers. The molecular mass and viscosity are comparable to the polyethylenes that they typically are used with, and they are normally molten at processing temperatures. New generations of processing aids use combinations of this fluoropolymer with polyethylene glycol or other low molecular mass compounds.

In typical industrial usage, fluoropolymers are added to the polymer in small quantities (<0.1% of mass fraction of PPA in polymer); they are highly immiscible with polyolefins, and the resulting material is a dilute two-phase blend. Some brand names of fluoropolymers used as PPAs are DYNAMAR®, VITON®, KYNAR®, TEFLON®, and DFL® (dry film type[1]).

Often a two-step procedure is used to make the blend, in which a master batch is first prepared and pelletized, and then mixed with the process

1. Certain equipment, instruments, or materials are identified in this paper to adequately specify experimental details. Such identification does not imply recommendation by the National Institute of Standards and Technology nor does it imply the materials are necessarily the best available for the purpose.

polymer; this is an effective route for preparing homogeneous dispersions [13, 14]. In the case of VITON, diluting the fluoropolymer in acetone before mixing was found to be beneficial [15]. In addition to optimizing the additive level, there are other factors that critically affect the additive performance, including interactions with other additives, dispersion quality, droplet size, and the rheology (some of these are discussed later). Many industrial patents have been issued for the incorporation of various processing aids in extrusion processes [16–18], which justifies the ongoing intense academic and industrial research in this area.

Beyond elimination of some extrusion instabilities, other benefits from the use of fluoropolymers include gel reduction (reduction of cross-linked, oxidized, and unmelted/unmixed gels that form in the extruder), reduced die buildup or die drool, decrease of the induction time for color change, and to some extent elimination of draw resonance in film casting, fiber spinning, and film blowing operations [19].

## 2.3 MECHANISMS OF INSTABILITIES, SLIPPAGE AND THE ROLE OF PPAS

### 2.3.1 Sharkskin and Other Melt Fracture Phenomena

Sharkskin and other melt fracture phenomena have been the subject of study of many scientific papers over the past 60 years [1–10], and they have been reviewed many times [2, 3], most lately by Migler [20]. We mention here only a few salient features of them, particularly those relevant to the fluoropolymers. Figure 2.1 shows the flow curve of a high density polyethylene (HDPE) relating wall shear stress to apparent wall shear rate in the presence and absence of a fluoropolymer additive obtained by means of a constant speed pressure driven capillary rheometer [21]. The apparent shear rate, $\dot{\gamma}_A$, is defined as:

$$\dot{\gamma}_A \equiv \frac{32Q}{\pi D^3} \tag{2.1}$$

where $Q$ is the volumetric flow rate, and $D$ is the diameter of the capillary. The wall shear stress, $\sigma_w$, is defined as:

$$\sigma_w = \frac{\Delta P - \Delta P_{end}}{4L/D} \tag{2.2}$$

where $L$ is the length of the capillary, $\Delta P$ is the total pressure drop driving the flow, and $\Delta P_{end}$ is the end correction for the pressure drop. Four distinct flow regimes can be identified in Figure 2.1. Initially, the flow is stable, and the shear stress increases following a power-law dependence with apparent shear rate and appearing linear on a log–log plot. In this regime the extrudate has a smooth

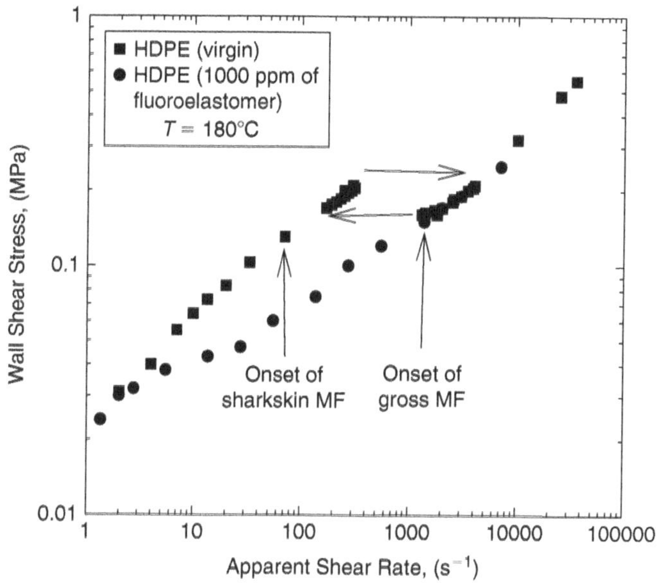


**FIGURE 2.1** Flow curve of HDPE extruded in a pressure-driven capillary rheometer in its virgin form and in the presence of a fluoroelastomer.

and glossy appearance. When the shear stress reaches a first critical value, small-amplitude, periodic distortions appear on the extrudate surface. This instability is accompanied in many cases by a change in the slope of the flow curve [1, 2, 10]. It is observed up to the end point of the left branch of the flow curve (often referred to as the low flow rate branch) and is generally referred to as *sharkskin* or *surface melt fracture.* The amplitude and wavelength of the periodic distortions is an increasing function of shear rate. It has been shown to initiate at the air–polymer–wall interface as the polymer exits the die [20]. The boundary condition between the wall and the polymer melt is crucial to the nature of flow instability occurring during the extrusion, but it has been somewhat controversial. It is not clear whether stick, partial slip, or stick-slip conditions exist. As discussed by Migler [20], while wall slippage may occur, it is not a necessary condition for sharkskin, as the latter has been observed in the absence of slip. Some scenarios for the boundary condition are shown in Figure 2.2. When directly introduced to the die wall, fluoropolymer additives clearly cause (or enhance) wall slippage since their interaction with the mainstream polymer is weaker than that with the wall (Fig. 2.2).

A second instability occurs when the shear stress exceeds a second critical value (shown in Figure 2.1 by the highest point of the low flow rate branch) and only within a certain range of apparent flow rate values. It corresponds to the region separating the two branches of the flow curve (shown by the two parallel

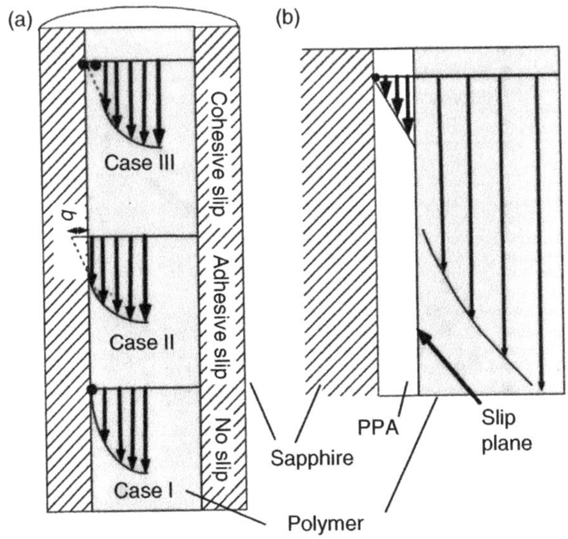


**FIGURE 2.2** (a) Possible boundary conditions at the polymer–wall interface, using a sapphire cover. Case I, no-slip; case II, polymer-wall slippage (adhesive); case III, internal slip (cohesive). (b) In the case of a fluoropolymer preferentially wetting the wall, slippage may occur at the polymer–polymer interface. Reprinted with permission from Ref. 29.

arrows), and is usually referred to as *stick-slip instability* or *spurt phenomenon* or *oscillating melt fracture*. In this regime, the flow ceases to be stable when a controlled rate capillary rheometer is used. Instead, pressure, and thus shear stress, oscillates between two extreme values and alternate distorted and smooth zones appear on the extrudate surface [1–3, 11, 21]. This phenomenon has been reviewed recently by Georgiou [22].

Finally, at even higher apparent shear rate values, the mass flow rate becomes again stable, but the extrudate exhibits gross, irregular, and chaotic distortions. This type of instability, is referred to as *gross melt fracture*, and corresponds to the right branch of the flow curve. An excellent review of the phenomenon has been reported by Dealy and Kim [23].

### 2.3.2 Slippage

To be effective, the fluoropolymer additives dispersed into the primary polymer must do two things. First, they must coat the die wall, in particular at the exit. Second, they must induce slippage between themselves and the primary polymer. As shown in Figure 2.1, in the absence of the PPA (fluoropolymer in the present case) the sharkskin, stick-slip and gross melt fracture regimes are all

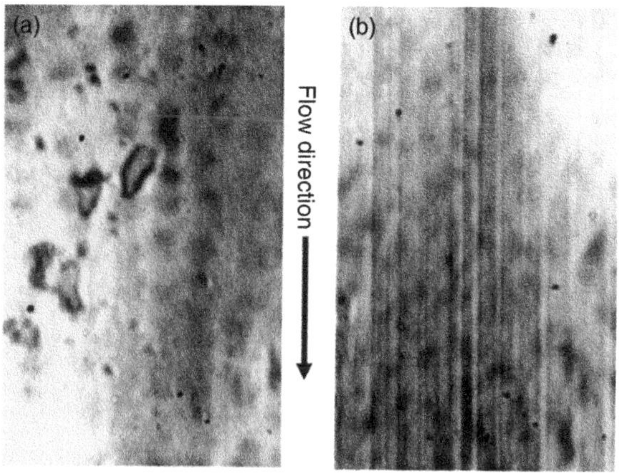


**FIGURE 2.3** Direct visualization of the PPA coating of the die wall. While during the extrusion of LLDPE the polymer–wall interface does not exhibit any special features except for a few gel-like particles moving slowly along it (a), the a characteristic streak-like nature of the interface develops during the extrusion of the 0.1% PPA/LLDPE blend (b). Reprinted with permission from Ref. 29.

present. In the presence of the PPA, both the sharkskin and the stick-slip instability disappear, while the gross melt fracture regime still occurs. Similar results for a variety of polyolefins have been published [1, 5, 10, 12, 14, 17, 20, 21, 24–26]. The reduced shear stress at a given throughput is a sign of slippage. However, in one study it was found that a DFL fluoropolymer PPA promoted adhesion, while having no effect on the flow curve [27]. BN also has no effect on the flow curve, although it may eliminate sharkskin melt fracture [16, 18].

An additional benefit of the fluoropolymers can be deduced from Figure 2.1. As a result of the slippage in the presence of the fluoropolymer, at a given apparent shear rate, the shear stress is reduced. Consequently, the power/torque requirement for a given production rate is lowered, and the production efficiency is increased [13, 27] while the product maintains its high quality.

Recently, Migler et al. [28–31] employed stroboscopic optical microscopy to image the coating process of the PPA onto the capillary wall. They found that the PPAs migrate to the surface of the die, creating a thin coating that acted as a lubricant. In the case of extrusion of a 0.1% PPA–LLDPE blend, a new phenomenon, the formation of streaks on the die surface, that gradually propagate downstream, was observed (Fig. 2.3). As the onset of streaks was concomitant with a pronounced decrease in shear stress and elimination of sharkskin, it was attributed to the formation of the fluoropolymer layer. By monitoring the flow profiles during the extrusion of LLDPE with and without the PPA (Fig. 2.4), the authors demonstrated that the slip occurred at the interface between the two polymers, indicating that the polymers were fully disentangled.

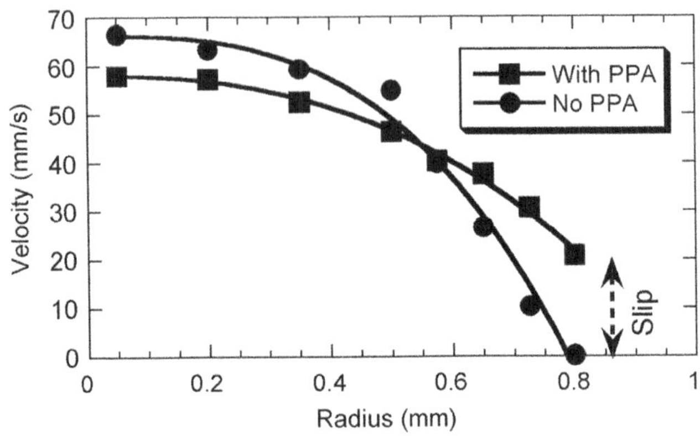


**FIGURE 2.4** Flow velocimetry measurements of LLDPE with and without PPA. Measurements were conducted at 210 °C and a throughput of 3.9 g/min, using a sapphire die with $L/D = 15.9$ and $D = 1.6$ mm. The uncertainty in the radial direction is due to the finite depth of field (50 μm); the relative standard deviation of the velocities is 5%, Reprinted with permission from Ref. 29.

### 2.3.3 Precoating Dies with PPAs

To separate the study of the slippage from the process of coating formation, the walls are often precoated with the fluoropolymer using solution casting methods and subsequent evaporation and annealing. Hatzikiriakos and Dealy [32] applied two different fluoropolymer PPAs to coat the surface of the plates in a sliding plate rheometer, using polyethylene and polypropylene (PP) [33]. They reported that the coatings resulted in a significant decrease of the critical shear stress for the onset of wall slip for both polyolefins and an increase of the slip velocity. The critical shear stress for the onset of slip, which is 0.1 MPa for the clean plate, was as low as 0.027 MPa.

In addition, Anastasiades and Hatzikiriakos [34] used a sessile drop method to measure the work of adhesion between HDPE–LLDPE in clean as well as fluoropolymer-coated stainless steel. They found that the fluoroelastomers (FEs) reduce the work of adhesion between polyethylene and the surface. These findings support the hypothesis that slip is a result of adhesive failure and that the slip velocity and critical shear stress at the onset of slip depend on the work of adhesion. Slip models taking into account these phenomena have been developed by Hatzikiriakos [35], Hill et al. [36], and Stewart [37].

### 2.3.4 Coating Process of Fluoropolymers

When an additive is present (either dispersed in the polymer or as a coating), proper die conditioning is needed before any equilibrium measurements. Kazatchkov et al. [33] found that when PP was extruded through Teflon®-coated capillary dies, the shear stress was reduced gradually. A typical example

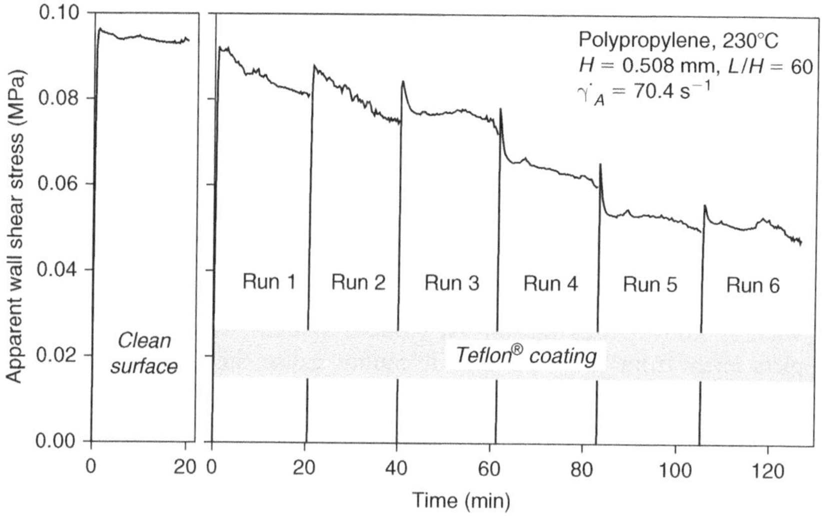


**FIGURE 2.5** The effect of a Teflon® coating on the shear stress during continuous extrusion of polypropylene (transient capillary experiments) using a slit die with a length-to-height ratio of $L/H = 60$.

is depicted in Figure 2.5. Each run corresponds to the extrusion of a full capillary barrel. This indicates that the fluoropolymer coating was initially imperfect (perhaps porous as is typical for solution casting methods) and after a long induction time, it was smoothened out, providing better lubrication and larger pressure reduction.

For the PPAs to be effective, they must first coat the die wall; in fact, the major metric used to evaluate the efficacy of a given PPA is the time it takes to completely eliminate sharkskin. There are numerous factors that influence the coating; however, the *rate* at which the PPA coats the die wall as well as the characteristics of the coating are less explored [15, 29, 30, 38]. For example, it was shown that increase in the concentration of the PPA, improving its dispersion quality and achieving higher flow rates reduces coating times [26, 38–40]. Recently it was shown by Oriani and Chapman [41] and Oriani [42] that larger droplets of the PPA eliminate sharkskin faster. The kinetics of the coating process as a function of droplet size was directly measured by Bigio et al. [43]. The overall coating mechanism is only now being addressed in active research, as described below.

One factor that must influence the effectiveness of the coating is the relative affinity of the polymers for the wall. It is reasonable to conclude that if the fluoropolymer has a greater affinity for the wall compared to processed polymer then it will naturally tend to coat it. As described earlier, a commonly used fluoropolymer contains a $VF_2$ comonomer, which does have a strong affinity to the oxides present in the steel. For example, fluoropolymers will displace nonpolar hydrocarbons with low work of adhesion but will not displace polar polymers such as nylon, polyesters or poly(methyl-methacrylate) from metal surfaces, because of the very high value for the work of adhesion of these polymers to metals.

In addition to the relative affinity of the coating to the die material, there is an important component from the flow kinetics that drives droplets (the minority phase) to the surface during extrusion. In a two-phase material, there are two mechanisms that may cause the droplets to migrate in a capillary flow. The first one is known as cross-stream migration and has been studied experimentally [44–46] as well as via numerical simulations [47–51]. However, based on what is known thus far for Newtonian systems, this mechanism drives droplets *away* from the wall, thus it cannot cause the coating. The second mechanism is one whereby the lower viscosity component coats the wall. However, this mechanism is unlikely to cause the coating for several reasons: First, the concentrations used here are well below those where this low viscosity coating mechanism operates. Second, Shih [44, 52] was able to create a coating in which ethylene–propylene–1,4-hexadiene terpolymer (EPDM) is the minor phase in a fluoropolymer as well as the reverse, where the fluoropolymer is the minor phase in (EPDM) [44, 52] and found that either phase can coat the wall when in the minority, even though one of them is of higher viscosity. Third, in the case of fluoropolymer additives, lowering their viscosity generally results in poorer performance. Thus we must consider other mechanisms involved in the coating formation.

Kanu and Shaw reported an effect of entrance angles, whereby a sharp entrance angle enhanced the droplet migration to the wall [15], pointing to the importance of the entrance region. Furthermore, it is known that the fluoropolymer need be present only at the very exit to eliminate sharkskin [20]. The hypothesis that the coating may first form at the very *entrance* of the die was further advanced in the work of Kharchenko et al. [53], using the following protocol: Pure LLDPE was extruded through a capillary rheometer, followed by a 0.1% blend of fluoropolymer in LLDPE. Figure 2.6 shows the sharkskin fraction along with the reduced pressure values, versus the extrusion time for each load of the blend. After 2 min the flow was stopped, and the die was reversed at a time when the pressure just started to decrease, implying that some coating had occurred, but there was still 100% sharkskin in the extrudate. Therefore, assuming that the fluoropolymer first coated the die at the entrance region, then when the flow was turned back on, the coating would be present at the exit, exactly where sharkskin is initiated [20]. Indeed, as the extrusion was recommenced after the reversal, the sharkskin disappeared *instantaneously* and *completely*, while the reduced pressure remained unchanged. This shows that in the time just before the flow stopped, the fluoropolymer had fully coated the die entrance, but the die exit contained no fluoropolymer. In contrast, when the experiment was conducted normally (without a reversal), there was a *gradual* and *slower* elimination of sharkskin, suggesting that the coating that started forming at the die entrance would require some *finite* time to migrate toward the die exit.

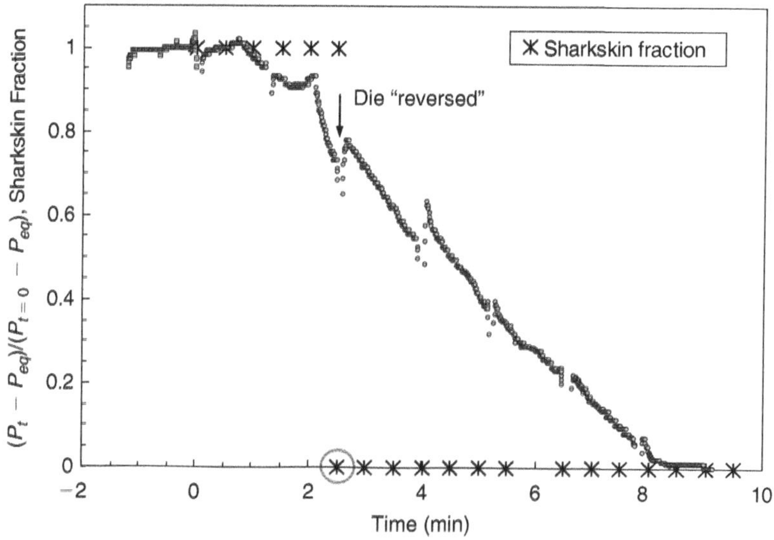


**FIGURE 2.6** Elucidation of the coating mechanism via a die reversal experiment. Measurements are conducted using a 0.1% PPA–PE blend at $\dot{\gamma} = 225\ s^{-1}$ and $T = 180\ °C$. $P_t$ is the pressure developed during extrusion of the PPA–LLDPE blend; $P_{eq}$ is the pressure at steady state when the PPA coating is fully developed, and $P_{t=0}$ is the pressure of pure PE. The uncertainty in the $y$-axis is estimated to be 5% based on repeated measurements. Reprinted with permission from Ref. 53.

The idea of the entrance-to-exit PPA migration was further reinforced by extruding LLDPE (certainly, under conditions of strong sharkskin) through the cylindrical sapphire die initially precoated with the fluoropolymer layer at its entrance [53]. In that experiment, the first sharkskin-free streak appeared *after* the pressure dropped to its minimum (≈ 80% of the initial value). This suggests that the PPA material propagated along the die wall with a certain migration speed of its front, depending on the flow rate and the initial thickness of the PPA coating applied to the die entrance. Therefore, a plausible mechanism of PPA coating can be envisaged as follows. During the extrusion of PPA–LLDPE blends, PPA droplets first adsorb at the die entrance (which explains the reduced pressure), and then they migrate as streaks toward the die exit under the influence of the shear field.

### 2.3.5 Coating Thickness and a Simple Coating Model

The issue of coating thickness has recently been explored with an imaging technique based on the frustrated total internal reflection [53]. The reflectivity data in that study conformed to streak-like nature of the PPA coating observed in earlier works through direct stroboscopic microscopy visualization [29]. It was found that the steady state PPA coating thickness developed during the

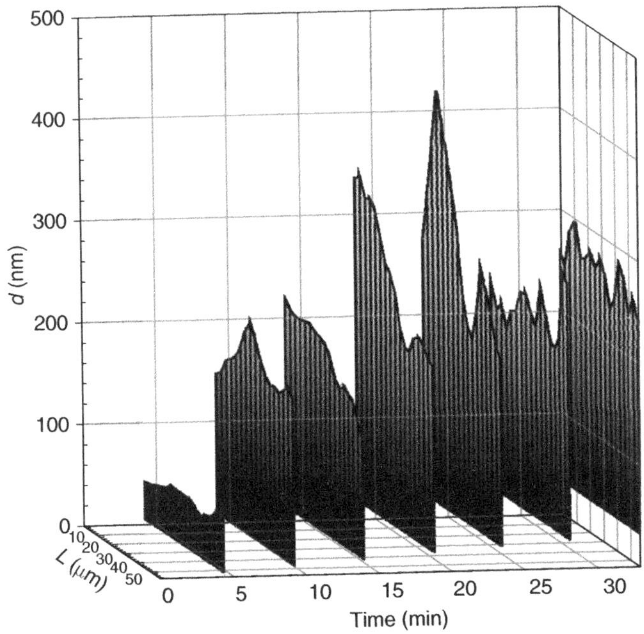


**FIGURE 2.7** Progression of PP coating thickness as a function of time during the extrusion of a 0.1% PPA–PE blend at $\dot{\gamma} = 215\,\mathrm{s}^{-1}$ and $T = 180°\mathrm{C}$ $d$ is the coating thickness, and $L$ is the die circumferential length. The uncertainty in the $y$-axis is estimated to be 20% based on repeated measurements. Reprinted with permission from Ref. 53.

extrusion of PPA–LLDPE at the die exit is on the order of only 150–450 nm, depending on the PPA concentration (Fig. 2.7). It should be noted that very thin coatings of about the size of a few PPA molecules (or about 25 nm) are sufficient in elimination of the melt fracture.

A simple coating model developed in [43, 53] is based on the balance between the mass flow rate of the fluoropolymer that coats the die ($Q_{\mathrm{in}}$) and the mass flow rate of the fluoropolymer leaving the die due to convection downstream ($Q_{\mathrm{out}}$). It assumes that a droplet of radius $S$ will coat the wall if its streamline is within a distance $S$ from the wall, otherwise it will not coat it. This is shown in Figure 2.8, where upon contact with the wall, the PPA droplet sticks and spreads downstream, thereby coating it. The mass flow rate of the coating material leaving the die is based on a linear velocity profile within the fluoropolymer layer. This allows us to write down the equations for the mass flow rates of the fluoropolymer in and out of the die:

$$Q_{\mathrm{in}} = \int_0^{2\pi}\int_0^{S} C[v_{\mathrm{PE}}(r')]R\,dr'\,\mathrm{d}\phi = 2RSC\pi\left(\frac{\dot{\gamma}_{PE}S}{2} + V_S\right) \qquad (2.3)$$

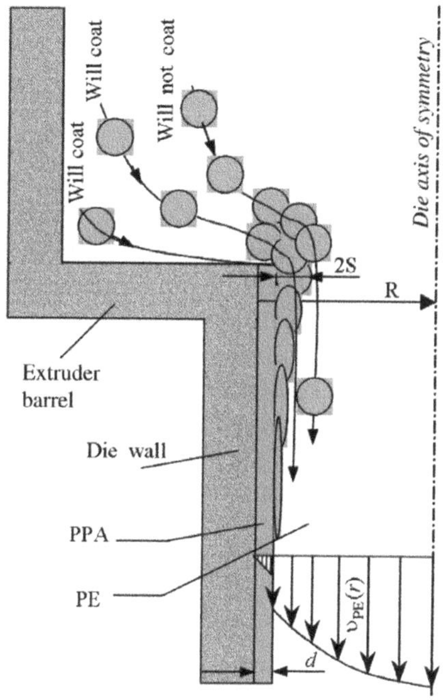


**FIGURE 2.8** Proposed coating mechanism for PPA–PE flow in a circular die. Reprinted with permission from Ref. 53.

$$Q_{out} = \int_0^{2\pi}\int_0^{d} \rho \upsilon_{PPA}(r')dr' R d\phi = R\rho\dot{\gamma}_{PPA}d^2\pi \tag{2.4}$$

where $r'$ is the distance from the wall, $R$ is the die radius, $d$ is the die diameter, $C$ is the bulk concentration of the PPA in the PPA–PE blend, $\rho$ is the density of the PPA, and $\upsilon_{PE}(r')$ is the velocity of the polyethylene near the wall given by $\upsilon_{PE}(r') = \dot{\gamma}_{PE}(R, t)r' + V_S(\dot{\gamma}, t)$. The slippage at the PE–PPA interface is taken into account through the slippage velocity $V_S(\dot{\gamma}, t)$. $V_S$ is a function of $t$ because at the beginning of the process, the PE is in direct contact with the wall and we expect little to no slippage, whereas in steady state when the wall is fully coated, we anticipate strong slip. In steady state the right-hand sides of (2.3) and (2.4) can be equated:

$$2CS\left(\frac{\dot{\gamma}_{PE}S}{2} + V_S\right) = \rho\dot{\gamma}_{PPA}d^2 \tag{2.5}$$

At the beginning of the process, the ratio $\dot{\gamma}_{PE}\, S/2$ is an important factor for the coating kinetics. However, as the system reaches steady state and the die walls are fully coated, $V_s$ is much larger than $\dot{\gamma}_{PE}\ S/2$ in (2.5) and $d$ can be expressed as:

$$d = \left(\frac{2CV_S S}{\rho\dot{\gamma}_{PPA}}\right)^{0.5} \tag{2.6}$$

The slippage velocity $V_S$ scales nearly linearly with apparent shear rate [27]: $V_S = (0.1\,\text{mm})\dot{\gamma}$. Predicted stead-state coatings using the model are found to compare well to the experimentally measured ones [43, 53]. An important observation drawn from (2.4) is that the steady state coating thickness at the exit appears to be independent of die diameter and scales with the droplet size, which is consistent with the recent work by Oriani and Chapman, who found that larger droplets coat the walls significantly faster than smaller ones [41, 42]. As most of the droplets do not coat the wall and pass directly to the finished product, the upper limit on droplet size is presumably dictated by the largest size acceptable to the final product.

Note that (2.6) shows that the thickness (and coating time) is improved with increasing concentration. However, an excessive amount of lubricant should be avoided, first due to the cost of the PPA and second because it might lead to excessive lubrication of the extruder barrel, causing the polymer to slip in the extruder screw, which can have undesired effects [13].

### 2.3.6 Use of Fluoropolymer PPA in Combination with Interfacial Agents

Co-additives are compounds used to enhance the effectiveness of PPAs. Commercially available poly(ethylene)glycol (PEG) [54–64] and biodegradable aliphatic polyesters, such as polycaprolactone (PCL) when used as co-additives greatly enhance the performance of conventional PPAs [41] through a synergistic mechanism. They do not improve polyolefin processing when used *alone*, although they are often used to amplify mechanical properties of starch-based composites [65]. They are believed to work by encapsulating the fluoropolymer droplets and are thus called interfacial agents (IA).

For a given concentration of the PPA, co-additives allow a reduction in the time for complete elimination of the melt fracture and can also be used in lesser amounts if the PPA rheology is modified. If the PEG:fluoropolymer ratio becomes too high, a loss in the output capacity of an extruder [41, 42, 66] occurs because there is an excess of PEG that does not coat the PPA droplets but instead contributes to the slippage in the feed zone of the extruder.

### 2.3.7 Interactions with Other Additives

PPAs may interact positively or negatively with other additives used in polymer processing [67]. These include antioxidants, anti-block agents, pigments, and hindered amine light stabilizers (HALS). Most of the studies in the literature refer to interactions of fluoropolymer-based processing aids with other additives; these are reviewed in this section.

Particles contained in anti-block and pigment additives can have an especially deleterious effect on the fluoropolymer PPA efficiency. This is essentially due to the fact that fluoropolymer molecules adsorb easily on the surface of these particles and thus hindering their migration to the wall. In addition, such fillers can abrade the PPA from the die surface, preventing the formation of the coating on the die surface; as a result higher extrusion pressures are obtained. In such cases a higher amount of PPA is needed to maintain the PPA efficiency [19, 60]. Priester et al. [14] have also found that anti-block agents shift the conditioning curve, such that the induction time, or time needed for obtaining steady-state is prolonged. A higher additive concentration is required to obtain the same conditioning curve as in the case of no additive. Similarly to anti-blocks, $TiO_2$ pigments have been reported to interfere with fluoropolymer PPAs through an abrasion and adsorption mechanism [19, 64].

HALS are also known to interact with PPAs [19, 68]. Although the mechanism is not known, from the practical point of view caution must be exercised when using HALS at high temperatures, as the interaction increases. It has been reported that some HALS have a deleterious effect on the performance of PPAs at high temperature in terms of melt fracture elimination [19, 67, 68]. Combined master-batches augment the interference with the performance of PPA and therefore in such cases separate master-batches are recommended.

Acid neutralizers such as various types of stearates can also compete with PPAs in terms of their affinity to the die wall. Stearates can behave as processing aids themselves, as will be discussed below [69].

Apart from the negative effects that additives might have on the efficiency of fluoropolymer PPAs, positive effects have also been reported. For example, lack of anti-oxidants (AOs) in the polymer reduces the ability of the PPA to perform [67]. Essentially the degradation products of the polymer (due to lack of AOs) prevent the PPA coating formation. As discussed above in detail, the use of interfacial agents such as PEG also improves the PPA performance through a synergistic mechanism.

## 2.4 BORON NITRIDE–BASED PROCESSING AIDS

Boron nitride is a foam nucleating agent, and when added to the polymer melt can act as a very effective processing aid, as discovered by DuPont researchers who were working on the extrusion of foamed fluoroelastomers. Specifically Randa et al accidentally discovered that the use of BN may eliminate gross melt fracture phenomena in the extrusion of fluoroelastomers and polyolefins [16, 18]. More details on this discovery can be found elsewhere [70].

The present section discusses the use of BN and its combinations with fluoropolymers as a processing aid for the extrusion of fluoropolymers and

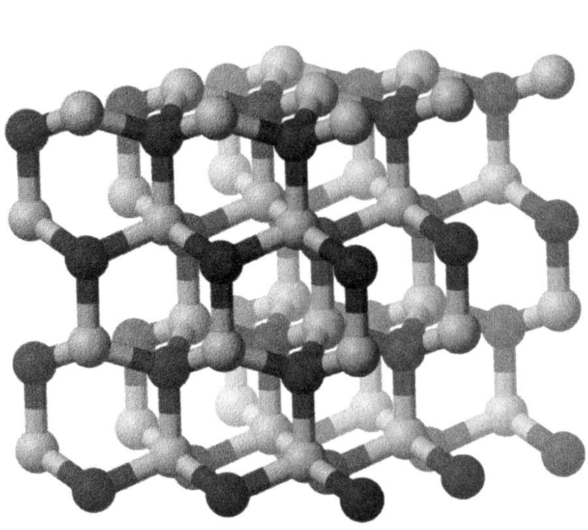

**FIGURE 2.9** Typical structure of boron nitride. Reprinted with permission from Ref. 18.

polyolefins [16, 18, 70]. BN is also used as a solid lubricant in many applications. Its structure resembles that of graphite (Fig. 2.9).

### 2.4.1 Effects of BN on Polyethylene and Fluoropolymer Processing

Even though the use of small amounts of BN (typically up to 1 wt %) into polyethylenes has practically no effect on their flow curves, it has been found to eliminate sharkskin and stick-slip melt fracture in the capillary extrusion of LLDPE [18, 71]. Once in the gross melt fracture regime, the presence of BN may eliminate gross distortions, although it fails in eliminating surface distortions at such high rates. In other words, gross distorted extrudates of virgin resin appear as sharkskinned in the presence of BN at rates well within the gross melt fracture region. This can also be seen from the experiments of Lee [72–74] who performed capillary experiments using tungsten carbide and BN hot-pressed capillary dies (Fig. 2.10). The gross distortions (irregular and large distortions) obtained in the extrusion of PE through the tungsten carbide die at $2007\,s^{-1}$ (Fig. 2.10d) in fact become surface distortions (relatively small and regular) when BN is added into the base resin [74]. It is noted that surface and gross melt fracture are two separate phenomena. The former originates at the exit [2, 20], whereas the latter at the entrance to the die [23]. Therefore, at high rates one may easily see the gross distortions and ignore the small scale ones that possibly imply sharkskin. Clearly BN can eliminate the gross and irregular distortions but not the surface and regular ones at high shear rates. The latter originate at the exit and significant lubrication is needed to eliminate them.

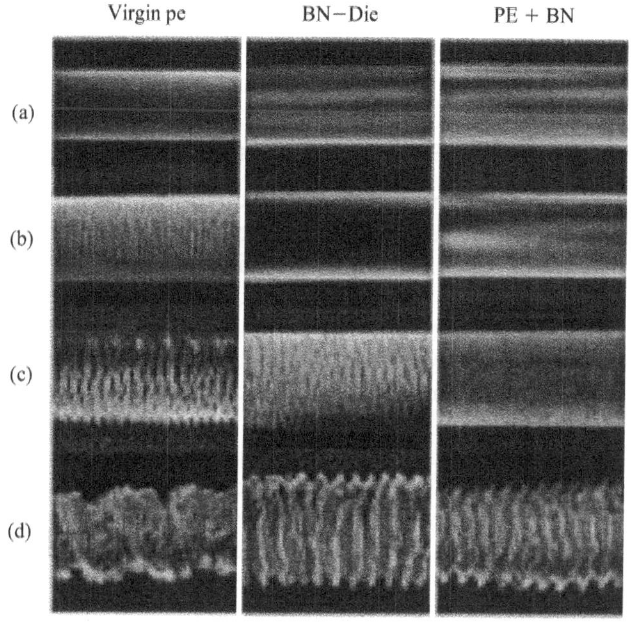


**FIGURE 2.10** The effect of the addition of BN on the melt fracture behavior of polyethylene at various shear rates in capillary experiments: (a) $115.4\,s^{-1}$, (b) $224.1\,s^{-1}$, (c) $851.3\,s^{-1}$, and (d) $2007.4\,s^{-1}$. Reprinted with permission from Ref. 73.

The effect of the BN addition on the processability of PEs has also been tested in continuous wire coating extrusion [16, 18, 75], blow molding [76, 77], film blowing [78], and fiber spinning [79]. In all these processes, the experimental results have concluded that the use of BN is beneficial. In other words, the use of BN not only can eliminate surface melt fracture but can also postpone the gross melt fracture at high shear rates. Further experiments have shown that the maximal shear rate yielding a smooth extrudate is sensitive to the BN content [18, 71, 75, 80]. It seems that there exists an optimum amount of BN that can attain the full benefit from its use. High concentrations of BN facilitate the presence of agglomerated particles, which has a detrimental effect on the BN performance as a processing aid.

BN was found beneficial to the processing of fluoropolymers. It has been reported that the addition of BN into tetrafluoroethylene–hexafluoropropylene (FEP) resins may eliminate gross melt fracture and postpone it to very high shear rates [16]. This was examined into more detail by Rosenbaoum [18, 71, 81]. Several types of FEP and copolymers of tetrafluoroethylene–hexafluoropropylene and perfluoropropylvinylether fluoropolymers were studied. The effects of BN addition were found beneficial in all cases [71].

Induction time effects using BN in polyethylene and fluoropolymer matrices have also been reported [71, 82–84]. In other words, the elimination of instabilities is not instantaneous on startup of the extrusion process. In many cases, several minutes are needed for the effect to attain its full strength, i.e., gradual improvement of the extrudate appearance.

### 2.4.2 Effects of Particle Size and Dispersion

The effects of particle size and the dispersion quality of BN into the processability were also studied in detail [85]. It had been reported that the optimum particle size is in the range of 5–20 μm. In addition, the absence of agglomerated particles is a requirement for optimum performance of BN as a processing aid. Possibly the presence of agglomerated particles ranging from 100 to 400 μm cannot provide the necessary lubrication at the die exit (change from a local stick-slip to a continuous smooth boundary condition) in order to eliminate sharkskin melt fracture.

The quality of the dispersion is also a requirement for the optimum performance of BN as a processing aid [71]. Dry blending of BN into the polymer matrix resulted into a poor performance of BN as processing aid [71]. Masterbatches containing 5 or 10% by mass should be first prepared, either by using a twin-screw extruder or a single-screw extruder if the polymer is available in powder form or in flakes. BN may be dispersed easily into polyethylene and fluoropolymer matrices [71–85].

### 2.4.3 Effects of Surface Energy

Surface energy of BN powders plays a significant role in eliminating melt fracture phenomena in the extrusion of polymers [79, 85–88]. In particular, Yip [85] and Seth [88] performed experiments with two types of BN having the same particle size, degree of agglomeration, and degree of dispersion in a LLDPE. It was found that BN with a small amount of boron oxides could not effectively eliminate melt fracture phenomena and failed at small shear rates. This was attributed to the effect of surface energy.

Seth [88] and Seth et al. [86, 87] examined a number of BN powders as processing aids by measuring their surface energy. A capillary rise technique based on Washburn's equation was used [89]. Both the dispersive and nondispersive (polar) components of surface energy were determined. The values of the surface energy of BN powders have been found to correlate well with the critical shear rate for the onset of melt fracture, indicating the importance of surface energy in the procedure of selecting an effective processing aid. In particular, the critical shear rate for the onset of melt fracture was found to increase with the ratio of dispersive component to nondispersive (polar) component of the surface energy (Fig. 2.11). This trend was confirmed by Rathod et al. [90] by using a sessile drop method technique to assess the surface energy of several BN powders. He found that the presence of oxides increases the nondispersive component (NDC) of the surface energy of polymers while leaving the dispersive (DC) to be about

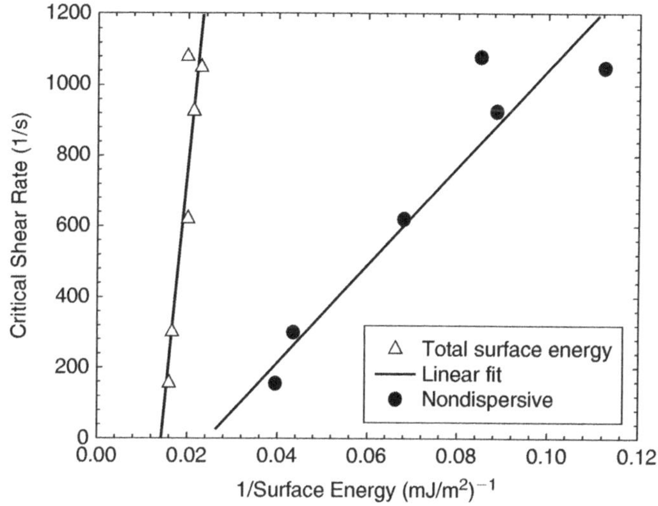


**FIGURE 2.11** Relation between the dispersive and nondispersive components of surface energy of BN powders to critical shear rate for onset of gross melt fracture. Reprinted with permission from Ref. 86.

the same ($36.5 \pm 0.6\,mJ/m^2$). Then they identified that the critical shear rate for the onset of melt fracture correlates well with the nondispersive component of the surface energy. A nondispersive component of $8–12\,mJ/m^2$ is ideal for optimizing the performance of BN as processing aid in polyolefins. Higher values ($23–27\,mJ/m^2$) were found to have a detrimental effect in the performance of BN as a processing aid. Thus it can be concluded that the higher the NDC of the surface energy, the lower the critical shear rate for the onset of melt fracture.

### 2.4.4 Combinations of BN with Fluoroelastomers

As discussed before, fluoropolymers can best work in the sharkskin and stick-slip melt fracture regions eliminating melt fracture and reducing the pressure drop. On the other hand, BN works in the gross melt fracture region, where fluoropolymers are not effective. Their combinations demonstrate the synergistic effects resulting in a more effective processing aids. Rozenbaoum [71] first combined a Teflon® processing aid (Teflon APA) with BN in the processing of a metallocene polyethylene (Exact™ 3128). He reported that the presence of both BN and fluoropolymer together postponed the onset of melt fracture to apparent shear rates much higher than those attained when the two components were used individually. Similar results have been reported by others upon combining BN with Teflon processing aids or with fluoroelastomers such as Viton® and Dynamar® [18, 71, 76, 79, 91–94].

Pruss et al. [78] have examined the use of BN-based processing aids (possibly combination of BN with fluoroelastomers since a throughput increase is reported) in the film-blowing process of a metallocene LLDPE. Extrusions with the use of BN have resulted in an over twofold increase in volumetric flow rate compared to the rate obtained by using a conventional fluoropolymer. In addition, they have reported improvements in film quality, such as better caliper control, mitigation of film streaking, control of coefficient of friction, enhanced gloss, reduced haze, and enhanced heat sealability when BN is used as a processing aid. On the other hand, mechanical properties such as yield strength and Graves tear remain the same.

### 2.4.5 Mechanisms of BN Action

The mechanism by which the addition of BN eliminates gross melt fracture in polyolefins was studied by Kazatchkov et al. [95] by using visualization experiments (laser-speckle technique). They have used a transparent die made out of fused quartz to visualize the change in the flow patterns at the entrance to a capillary die (site of initiation of gross melt fracture). A summary of the findings are as follows: At low shear rates the polymer enters the capillary at a higher entry angle, and the vortex in the corner of the reservoir is larger. As the shear rate increases, the streamlines bend more near the entrance to the capillary. At the highest shear rates, gross melt fracture is observed, and the streamlines seem to be discontinuous. The flow in the entry region appears to be separated into several layers, and each layer moves with its own velocity. At regular time intervals, different in each layer, the motion stops for a brief period (local stick-slip within the bulk of the melt). BN eliminates this discontinuous movement of the layers. The presence of BN over these layers within the bulk material where stick-slip was originally observed, provides proper lubrication between these singular layers (change of stick-slip or intermittent fluid layer motion to a continuous slippage between those fluid layers), thus eliminating stick-slip phenomena within the bulk of the polymer. The motion is rather continuous in the presence of BN.

## 2.5 OTHER PROCESSING AIDS

### 2.5.1 Stearates

Stearates, such as calcium and zinc stearates are present in several commercial resins of both linear and long-chain branched polyethylenes [69]. There is strong evidence that stearates promote slip and aid in the reduction of extrudate instabilities in the case of metallocene low-density polyethylene copolymers containing a small degree of long chain branching [69]. In typical transient capillary extrusion experiments of linear low-density polyethylenes (with no stearates), the load rises initially with time until it reaches a steady

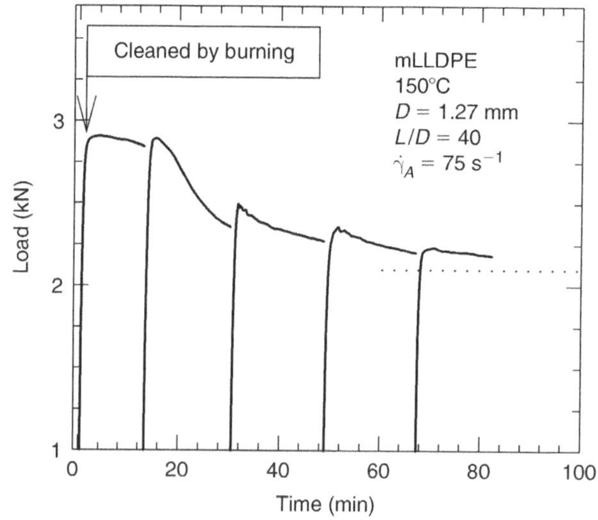


**FIGURE 2.12** The effect of stearates on the extrusion load in the continuous capillary extrusion of a metallocene LLDPE. Reprinted with permission from Ref. 69.

state. However, in the case of metallocene polyolefins that contain stearates, the die-conditioning phenomenon described in Section 2.3.4 is observed. An example is given in Figure 2.12 [69]. It can be seen that the load (thus pressure and shear stress) needed to extrude the polymer through a capillary die initially rises and then starts dropping. Each load corresponds to the extrusion of a full barrel. After several consecutive capillary extrusions, i.e., by reloading the reservoir of the rheometer, the load attained its steady state value, indicating that complete conditioning of the die surface has been achieved. Comparison of different resins has shown that the conditioning is longer as the amount of long chain branching is increased [69].

### 2.5.2 Silicon-Based Additives

Silicon-containing polymers and polydimethyl siloxane (PDMS) oils have been used as polymer processing aids [19]. Since these fluids have a low surface tension, they tend to reside on the polymeric surfaces and thus lower their frictional properties. However, their surface activity has limited the use of these materials in articles that are printed, and/or painted. The high molecular weight of PDMS that is a solid at room temperature has been reported to be efficient in the extrusion processes of LLDPE and PP [96, 97]. It has been essentially reported that use of such additives can significantly reduce surface roughness in extruded polymers as tapes. Stearates have also been tested as PPAs in the extrusion of poly(methyl methacrylate) resins [98] with beneficial results to a certain extent.

Kulikov and Hornung [99] and Kulikov et al. [100] have used several low and high molecular weight silanols cured with boric acid as die coatings or as additives to base LLDPE to investigate the impact on sharkskin delay. They have reported delay of flow instabilities such as sharkskin melt fracture to rates of extrusion 25–35 times higher than without PPA and about 40% less extrusion pressure. Filling the silanols with particulates having plate-like particles (kaolin, mica, BN) helped delay the flow instabilities further to even higher throughputs.

### 2.5.3 Hyperbranched Polymers

Hong et al. [101, 102] have shown that addition of small amounts of hyperbranched hexadecanote-terminated polyesters (HBPs), synthesized based on Boltorn dendritic additives, significantly improved fiber extrusion and film-blowing processing of LLDPE. Sharkskin elimination and substantial reduction in extruder power were achieved on addition of only 0.5% of HBP. At the same time, the loss in the mechanical properties (tensile strength and elongation at break) was apparently minimal. By using X-ray photoelectron spectroscopy and transmission electron microscopy, the authors showed that HBP had a tendency to migrate to the wall surfaces, where it preferentially accumulated forming a phase-separated layer. The mechanism of fracture elimination by HBP was advocated to arise from either the reduced viscosity or the interfacial slip. More recently, Wang et al. [103] demonstrated that HBP forms a phase-separated system with LLDPE. The droplets of HBP (minor component) can migrate to the die surface and form a lubricating layer promoting extrudate slippage, reducing pressure drop and shear stresses and thus improving extrudate distortion.

### 2.5.4 Other Polymer Blends

Hydrocarbons blended with the process polymer, can also act as PPAs. This technique might involve not only a small quantity of PPA dispersed into a process polymer but also polymer blending to ratios of the same order of magnitude. Although the notion of processing aid has a meaning at small concentrations (typically $<1\%$ PPA mass fraction in resin), blending at higher concentrations might also improve processing. Examples from the literature are discussed below.

Blending of PP with poly(ethylene vinyl acetate) (EVA) for flow instability reduction of PP was examined using capillary rheometry [104]. It was shown that in these blends the critical shear rate for the onset of fracture increased with EVA concentration. Mechanical property measurements showed that while pure EVA had completely elastomeric behavior, blends containing 50% PP and higher exhibited thermoplastic behavior with yield stress that increased with PP content. It was thus concluded that EVA–PP blends containing 15–20% EVA provided a good balance between the mechanical properties and melt fracture behavior.

Fujiyama and Kawasaki [105] studied the capillary flow properties of blends composed with various ratios of PP and HDPE with low (L), medium (M), and high (H) melt flow indexes (MFIs). They found that the flow curves of a number of different blends were located between the flow curves of the two virgin components. In the blends for which PP had a lower or equal MFI than the PE, the slope of the curves increased with decreasing PP content. The critical shear rates and the critical shear stresses for the onset of melt fracture of the blends were close to the ones determined by logarithmic additivity of the values that correspond to individual components.

The blending of two incompatible polymers, a fluoropolymer and EPDM, in various proportions, to improve the performance of one or the other polymer was experimentally examined by Shih [44]). The addition of a small amounts of fluoropolymer (as low as 0.4%) in EPDM, significantly reduced the shear stress and improved the appearance of the extrudate. EPDM improved the processing of the fluoropolymer in a similar fashion. This effect was more pronounced as the fluoropolymer concentration increased up to concentrations of 5%; no improvement was observed with further increase of fluoropolymer concentration. It was experimentally determined that, during the extrusion of blends, the polymer used in small concentration created a buildup layer on the die wall, indicating that a phase separation occurred in which the polymer used as additive accumulated on the wall.

Rosenbaum et al. [106] studied the use of PE as a processing aid in the extrusion of a fluoropolymer FEP (tetra-fluoro-ethylene-polypropylene) resin. Capillary extrusion experiments of LLDPE dispersed in small quantity (0.1 wt %) in FEP 4100 indicated that LLDPE is an efficient PPA for eliminating melt fracture and stick-slip instability in the extrusion of fluoropolymers (Fig. 2.13). As a result of LLDPE addition, the apparent flow curve is shifted to lower shear stresses, allowing for smooth extrudates up to $1000\,s^{-1}$. It was suggested that LLDPE droplets, which have a lower viscosity than FEP, migrate to the wall. Furthermore, the low friction coefficient at the FEP-PE interface results in the slip of the extruded FEP over the thin PE-layer. Consequently, this results in shear stress reduction and in instability suppression.

The gradual wall-coating process discussed before for fluoropolymers and stearates (plotted in Figures 2.5, 2.6, and 2.12) was also reported in this case (polyethylene in the extrusion of fluoropolymers) (Fig. 2.13). The time required to obtain steady-state response was found to depend on the process and geometry parameters. It decreased with increasing apparent shear rate, and with decreasing diameter or length to diameter ratio (Fig. 2.14). The latter suggests that this time is proportional to the time for complete surface coating. A decrease in the diameter or the *L:D* ratio reduces the area to be coated, and an increase in the apparent shear rate increases the rate of the coating process. In both cases, the time required for complete coating and steady-state operation is reduced.

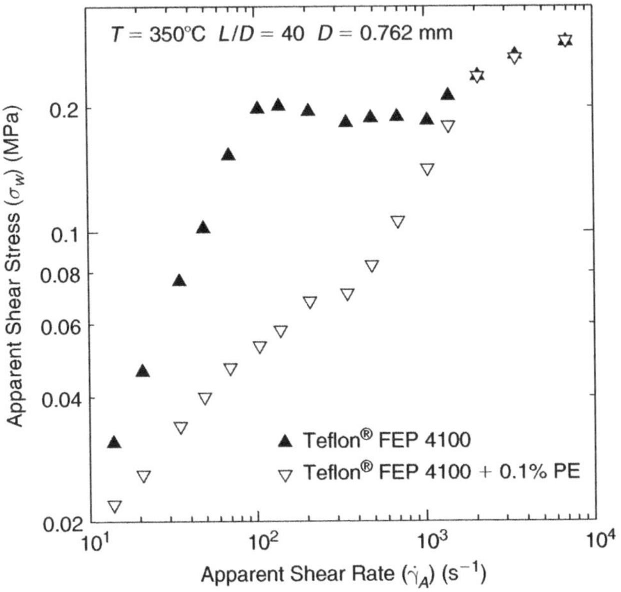


**FIGURE 2.13** The effect of a small amount of PE (Union Carbide, GRSN 7047) on the flow curve of an FEP resin at 350°C. Reprinted with permission from Ref. 106.

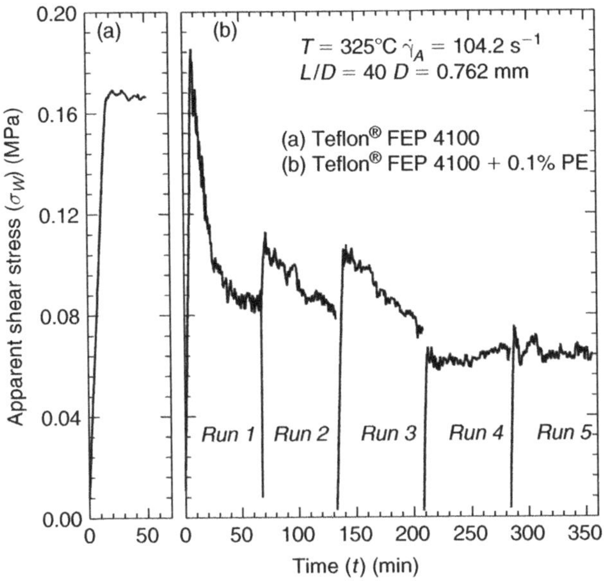


**FIGURE 2.14** Transient capillary extrusion: the effect of the addition of 0.1% of polyethylene on the transient response in the capillary extrusion of FEP 4100 at 325 °C, $\dot{\gamma}_A = 104.2\,\mathrm{s}^{-1}$, $L/D = 40$, and $D = 0.762$ mm. Reprinted with permission from Ref. 106.

## 2.6 CONCLUSIONS

Product quality and processing windows during extrusion of molten polymers are limited by the various flow instabilities that are observed at high production rates. These instabilities are attributed to the complex rheological and constitutive character of polymers in general, and have been scrutinized by researchers both experimentally and theoretically. While there are still unresolved issues in comprehending these phenomena, the industry demands for process optimization dictate the employment of processing aids for product quality improvement and energy requirement reduction.

Fluoropolymers have been traditionally used by the industry for elimination of the sharkskin and the stick-slip instabilities. Used as coatings, or more effectively in dispersion with the processing polymer, they typically increase the slip velocity of the extrudate, therefore reducing the energy requirements for a particular flow rate. Furthermore, they shift the critical shear rates for the onset of instabilities up to the point of the onset of gross melt fracture, therefore permitting higher production rates. Preferential coating of the walls of the die entrance by the fluoropolymer droplets, with consequent propagation of the fluoropolymer streaks to the die exit and lubrication at the polymer wall interface was determined to be the dominant mechanism of instability suppression. Fluoropolymer performance is highly dependent on the dispersion quality; this is achieved using larger droplets in the size range of 1 to 5 μm, thoroughly mixed with the polymer by means of a master batch. Optimization of the additive level, choice of the appropriate additive with the desired interactions with the polymer and/or the wall and other additives, and use of clean surfaces are also important.

Boron nitride–based processing aids have been found to be effective in the extrusion of both fluoropolymers and polyolefins. For BN to be an effective processing aid the following requirements apply: (1) to have a fine particle size in the range of 5 to 20 μm, (2) to have no agglomerated particles, (3) to be finely dispersed into the resin under processing by appropriately preparing a concentrate first before final dissolution, (4) to be used at its optimum concentration, and (5) to have proper surface chemistry with minimum nondispersive surface energy component if BN is to be used alone (essentially no boron oxides on the surface). More recently, it has been shown that other plate-like particles such as nanoclays (montmorillonite clays), molybdenum disulfide, kaolin, and mica particles might exhibit similar properties as processing aids [100, 112].

Stearates, which are typically present in several commercial polymer resins, are believed to act as PPAs in the case of long-chain branched metallocene polyolefins and other polymers. Processing of these metallocene resins, which have no other additives but stearates, is very similar to the processing of fluoropolymers containing resins. Die conditioning associated with a steady-state slip enhancement and instability elimination within a range of shear rates is observed.

Polymer blending is another traditional method for improving processing. Similar to fluoropolymers, blending of hydrocarbons, or addition of a hydrocarbon in a fluoropolymer, can lead to significant reduction of the power requirement and elimination of the sharkskin and the stick-slip instability. Hydrocarbon blending results in processing behavior and mechanical properties that are intermediate between the polymers used. The mechanism by which a hydrocarbon enhances the processability of fluoropolymers was shown to be the same as in the case of polymer processing with fluoropolymers as processing aids.

Complete comprehension of these extrusion instabilities—as well as of the role of the additives in eliminating the instabilities—is important for process optimization. Elaborate techniques employed recently for flow characterization in extrusion processes, such as ATR/FTIR [113], fluorescence label chains for velocity measurements [114], NMR velocimetry [115], laser Doppler velocimetry [116], detailed flow visualization through quartz–sapphire–glass capillary [29, 31, 95, 117] and frustrated total internal reflectance [53] can prove valuable in studying the mechanisms of flow instability elimination when processing aids are present.

## 2.7 ACKNOWLEDGMENTS

Savvas G Hatzikiriakos would like to dedicate this chapter to the memory of Dr. Charles W. Stewart, a researcher who made several important contributions to the subject of processing aids.

## REFERENCES

1. A.V. Ramamurthy, *J. Rheol.* 30, 337–357 (1986).
2. M.M. Denn, *Ann. Rev. Fluid Mech.*, 33, 265–297 (2001).
3. R.G. Larson, *Rheol. Acta.*, 31, 213–263 (1992).
4. C.D. Han, *Rheology in Polymer Processing*, Academic, New York, 1976.
5. G.V. Vinogradov and A.Y. Malkin, *Rheology of Polymers*, Khimia, Moscow, 1980.
6. D. De Kee and K.F. Wissbrun. *Physics Today*, 51, 24–29 (1998).
7. L.L. Blyler and A.C. Hart. *Polym. Eng. Sci.*, 10, 193–203 (1970).
8. J. Galt and B. Maxwell, *Plast. Eng.*, 42, 115–132, (1964).
9. B.T. Atwood and W.R. Schowalter. *Rheol. Acta*, 28, 134–146 (1989).
10. S.G. Hatzikiriakos and J.M. Dealy, *J. Rheol.*, 36, 703–741, 1992.
11. S.G. Hatzikiriakos and J.M. Dealy, *J. Rheol.*, 36, 845–884, 1992.
12. J.M. Lupton and R.W. Regester, *Polym. Eng. Sci.*, 5, 235–241, 1965.
13. A. Rudin, J.E. Blacklock, S. Nam, and A.T. Worm, *Proc. SPE ANTEC*, 32, 1154–1158 (1986).
14. D.E. Priester, K.M. Stika, G.R. Chapman, et al. *Proc. SPE ANTEC*, 39, 2528–2532 (1993).

15. R.C. Kanu and M.T. Shaw, *Polym. Eng. Sci.*, 22, 507–511 (1982).
16. M.D. Buckmaster, D.L. Henry, and S.K. Randa, U.S. Pat. 5,688,457 (1997).
17. S.J. Kurtz, T.R. Blakeslee, and L.S. Scarola, U.S. Pat. 4,282,177 (1981).
18. E.E. Rosenbaum, S.K. Randa, S.G. Hatzikiriakos, and C.W. Stewart, U.S. Patent 00481792001.
19. S.E. Amos, G.M. Giacoletto, J.H. Horns, et al., in *Plastics Additives Handbook*, Hanser Gardner Publications, 5th ed., New York, 2000.
20. K.B. Migler in *Polymer Processing Instabilities: Control and Understanding*, Marcel Dekker, New York, 2005.
21. S.G. Hatzikiriakos, a Wall Slip and Melt Fracture of HDPES PhD. Thesis, Department of Chemical Engineering, McGill University, Montreal, 1991.
22. G. Georgiou, in *Polymer Processing Instabilities: Control and Understanding*, Marcel Dekker, 2005.
23. J.M. Dealy and S. Kim, in *Polymer Processing Instabilities: Control and Understanding*, Marcel Dekker, New York, 2005.
24. S.G. Hatzikiriakos, P. Hong, W. Ho, and C.W. Stewart, *J. Appl. Polym. Sci.*, 55, 595–603 (1995).
25. K.C. Xing and H.P. Schreiber. *Polym. Eng. Sci.*, 36, 387–393 (1996).
26. R.J. Athey, R.C. Thamm, R.D. Souffle, and G.R. Chapman, *Proc. SPE ANTEC*, 32, 1149–1152 (1986).
27. S.G. Hatzikiriakos, C.W. Stewart, and J.M. Dealy, *Intern. Polym. Process.*, 8, 30–35, (1993).
28. K.B. Migler, C.L. Gettinger, V.P. Thalacker, and R. Conway, *Proc. SPE ANTEC*, 57, 3128–3131 (1999).
29. K.B. Migler, C. Lavallée, M.P. Dillonet, et al., *J. Rheol.*, 45, 565–581 (2001).
30. K.B. Migler, C. Lavallée, M.P. Dillonet, et al., *Proc. SPE ANTEC*, 59, 1132–1136 (2001).
31. K.B. Migler, Y. Son, F. Qiao, and K. Flynn, *J. Rheol.*, 46, 383–400 (2002).
32. S.G. Hatzikiriakos and J.M. Dealy, *Intern. Polym. Process*, 8, 36–44 (1993).
33. I.B. Kazatchkov, S.G. Hatzikiriakos, and C.W. Stewart, *Polym. Eng. Sci.*, 35, 1864–1871 (1995).
34. S.H. Anastasiadis and S.G. Hatzikiriakos, *J. Rheol.*, 42, 795–812 (1998).
35. S.G. Hatzikiriakos, *Intern. Polym. Process*, 8, 135–142 (1993).
36. D.A. Hill, T. Hasegawa, and M.M. Denn, *J. Rheol.*, 34, 891–918 (1990).
37. C.W. Stewart, *J. Rheol.*, 37, 499–513 (1993).
38. E.C. Achilleos, G. Georgiou, and S.G. Hatzikiriakos, *J. Vinyl. Add. Tech.*, 8, 7–24, (2002).
39. S.E. Amos, G.M. Giacoletto, J.H. Hornset et al., *Polymer Processing Aids (PPA)*. 4th ed., New York, 2001.
40. C. Lavallée and S.S. Woods, *Proc. SPE ANTEC*, 58, 2857–2861 (2000).
41. S.R. Oriani and G.R. Chapman, *Proc. SPE ANTEC*, 60, 6011–3018 (2002).
42. S.R. Oriani, *J. Plastic Film Sheeting*, 21, 179–198 (2005).
43. D. Bigio, M.G. Meillon, S.B. Kharchenko, et al., *J. Non-Newt. Fluid Mech.*, 131, 22–31 (2005).

44. C.K. Shih, *Polym. Eng. Sci.*, 16, 742–746 (1976).
45. D.R. Oliver, *Nature*, 194, 1269–1271 (1962).
46. B. Shizgal, H.L. Goldsmith, and S.G. Mason, *Can. J. Chem. Eng.*, 43, 97–101 (1965).
47. P.C.H. Chan and L.G. Leal, *J. Fluid Mech.*, 92, 131–170 (1979).
48. H. Zhou and C. Pozrikidis, *Phys. Fluids*, 6, 80–94 (1994).
49. C. Coulliette and C. Pozrikidis, *J. Fluid Mech.*, 60, 1–28, (1998).
50. X.F. Li and C. Pozrikidis, *I.J. Multiphase Flow*, 26, 1247–1279 (2000).
51. S. Mortazavi and G. Tryggvason, *J. Fluid Mech*, 411, 325–350 (2000).
52. C.K. Shih, *Capillary Extrusion and Mold Flow Characteristics of an Incompatible Blend of Two Elastomers*. Science and Technology of Polymer Processing. Cambridge: MIT Press, 1979.
53. S.B. Kharchenko, P.M. McGuiggan, and K.B. Migler, *J. Rheol.*, 47, 1523–1545 (2003).
54. P.S. Leung, E.D. Goddard, and F.H. Ancker, U.S. Pat. 4,857,593 (1989).
55. D. Duchesne and B.V. Johnson, U.S. Pat. 4,855,360 (1989).
56. D. Duchesne and B.V. Johnson, U.S. Pat. 5,015,693 (1989).
57. T.J. Blong, M.P. Greuel, and C. Lavallée, U.S. Pat. 5,830,947 (1998).
58. J.W. Taylor, S.K. Goyal, D.J. Aubee, and N.K.K. Bonet, U.S. Pat. 5,459,187 (1989)
59. T.J. Blong, K. Focquet, and C. Lavallée, *Proc. SPE ANTEC*, 6011–3018 (1997).
60. D. Duchesne, J.E. Blacklock, B.V. Johnson, and T.J. Blong, *Proc. SPE ANTEC*, 35, 1343–1347 (1989).
61. P.S. Leung, E.D. Goddard, and F.H. Ancker, U.S. Pat. 4,925,890 (1990).
62. K. Focquet, G. Dewitte, and S.E. Amos, Polymer processing additive having improved stability. U.S. Pat. 6,294,604 (2001).
63. S.R. Oriani and G.R. Chapman, Fundamentals of melt fracture elimination using fluoropolymer process aids. 2002.
64. G.R. Chapman and S.R. Oriani, Paper Presented at the SPE Annual Technical Conference, Nashville, 2003.
65. P. Matzinos, V. Tserki, C. Gianikouriset et al., *Eur. Polym. J.* 38, 1713–1720 (2002).
66. K.R. Slusarz and S.E. Amos, *SPE ANTEC Tech. Papers*, 60, 2115–2119 (2002).
67. T.J. Blong, K. Focquet, and C. Lavallée, *Proc. SPE ANTEC*, 43, 3011–3018 (1997).
68. S.S. Woods, R.E. King, and C. Lavallée, *TAPPI Laminations and Coatings Conference Proceedings*, 1115–1124 (2000).
69. S.G. Hatzikiriakos, I.B. Kazatchkov, and D. Vlassopoulos. *J. Rheol.*, 41, 1299–1316 (1997).
70. S.G. Hatzikiriakos, in *Polymer Processing Instabilities: Control and Understanding*, Marcel Dekker, New York, 2005.
71. E.E. Rozenbaoum, "Rheology and Processability of FEP Resins for Wire Coating," Ph.D. dissertation, University of British Columbia, Vancouver, BC, 1998.
72. S.M Lee, J.G Kim and J.W. Lee. *Proc. SPE ANTEC*, 46, 2862–2866, (2000).
73. S.M. Lee and J.W. Lee. *Proc. SPE ANTEC*, 45, 1223–1227, (2001).
74. S.M. Lee, G. Nam, and J.W. Lee. *Adv. Polym. Tech.*, 22, 343–354 (2003).

75. E.E. Rozenbaum, S.K. Randa, S.G. Hatzikiriakos, et al., *Proc. SPE ANTEC*, 44, 952–956 (1998).
76. F. Yip, R. DiRaddo, and S.G. Hatzikiriakos, *J. Vinyl Add. Tech.*, 6, 196–204 (2000).
77. M. Seth, F. Yip, and S.G. Hatzikiriakos, *Proc. SPE ANTEC*, 47, 2649–2653 (2001).
78. E.A. Pruss, S.K. Randa, S.S. Lyle, and T.M. Clere, *Proc. SPE ANTEC*, 46, 2864–2868 (2002).
79. R. Vogel, S.G. Hatzikiriakos, H. Brunig, et al., *Intern. Polym. Proc.*, 113, 67–74, (2003).
80. K. Yip, E.E. Rozenbaum, S.K. Randa, et al., *Proc. SPE ANTEC*, 45, 1223–1227 (1999).
81. E.E. Rozenbaum, S.G. Hatzikiriakos, and C.W. Stewart, *Proc. SPE ANTEC*, 41, 1111–1115 (1995).
82. F. Yip, S.G. Hatzikiriakos, and T.M. Clere. *J. Vinyl Add. Tech.* 6, 113–118 (2000).
83. F. Yip, S.G. Hatzikiriakos, and T.M. Clere. *Proc. Polyolefins*, 46, 563–570 (2000).
84. F. Yip, S.G. Hatzikiriakos, and T.M. Clere, *Proc. SPE ANTEC*, 46, 2852–2857 (2000).
85. F. Yip, "The Effect of Boron Nitride on the Rheology and Processability of Polymers." MSc thesis, University of British Columbia, Vancouver, BC, 1999.
86. M. Seth, S.G. Hatzikiriakos, and T. Clere, *Polym. Eng. Sci.*, 42, 743–752 (2002).
87. M. Seth, S.G. Hatzikiriakos, and T. Clere, *Proc. SPE ANTEC*, 47, 2634–2638 (2001).
88. M. Seth, "The Role of Surface Energy of Boron Nitride on the Gross Melt Fracture Elimination of Polymers, MSc thesis, University of British Columbia, Vancouver, BC, 2001.
89. J.S. Laskowski, *Coal Flotation and Fine Coal Utilization*, 1st ed., Elsevier, New York, (2001).
90. N. Rathod and S.G. Hatzikiriakos, *Proc. SPE ANTEC*, CD-ROM, paper# 17, 49, (2003).
91. S. Randa, C.W. Stewart, E.E. Rozenbaoum, and S.G. Hatzikiriakos. U.S. Pat. 0048179 (2001).
92. S. Randa, C.W. Stewart, E.E. Rozenbaoum, and S.G. Hatzikiriakos, *Extrusion Aid Combination*, (WO01/46313A1) PCT, New York, 2001.
93. M. Seth and S.G. Hatzikiriakos. *J. Vinyl Add. Tech.* 7, 90–97 (2001).
94. F. Yip, E. Rozenbaoum, and S.G. Hatzikiriakos. *J. Plastic Film and Sheeting*, 16, 16–32 (2000).
95. I.B Kazatckov, F. Yip, and S.G. Hatzikiriakos, *Rheol. Acta*, 39, 583–594 (2000).
96. K.E. Lupton, P.G. Pape, and V.B. John, *RETEC Tech. Papers*, 93–106, 1988.
97. D.E. Hauenstein, D.J. Cimbalik, and P.G. Pape. *Proc. SPE ANTEC*, 21, 3002–3010 (1977).
98. C. Stamboulides and S.G. Hatzikiriakos, *Intern. Polym. Proc.*, 21, 155–163 (2006).
99. O. Kulikov and K. Hornung, *J. Non-Newt. Fluid Mech.*, 43, 169–188 (2004).
100. O. Kulikov, K. Hornung, and M. Wagner, *Rheol Acta*, 46, 741–754 (2007).
101. Y. Hong, S.J. Coombs, J.J. Cooper-Whiteet, et al., *Polymer* 41, 7705–7713 (2000).
102. Y. Hong, J.J. Cooper-White, M.E. Mackayet, et al., *J. Rheol.*, 43, 781–793 (1999).

103. J. Wang, M. Kontopoulou, Z. Ye, et al., *J. Rheol.*, 52, 243–260 (2008).
104. N. Montoya, J.D. Sierra, M. del Pilar Noriega, and T.A. Osswald, *Proc. SPE ANTEC*, 57, 1270–1275 (1999).
105. M. Fujiyama and Y. Kawasaki. *J. Appl. Polym. Sci.*, 42, 467–480 (1991).
106. E.E. Rosenbaum, S.G. Hatzikiriakos, and C.W. Stewart. *Intern. Polym. Process*, 10, 204–212 (1995).
107. G.M. Wise, M.M. Denn, A.T. Bellet, et al., *J. Rheol.*, 44, 549–567 (2000).
108. F. Legrand, J.M. Piau, and H. Hervet, *J. Rheol.*, 42, 1389–1402 (1998).
109. M.M. Britton, R.W. Mair, R.K. Lambert, and P.T. Callaghan, *J. Rheol.*, 43, 897–909, (1999).
110. H. Munstedt, M. Schmidt, and E. Wassner, *J. Rheol.*, 44, 413–428 (2000).
111. Y.W. Inn, R.J. Fischer, and M.T. Shaw, *Rheol. Acta*, 37, 573–582 (1998).
112. S.G. Hatzikiriakos, N. Rathod, and E.B. Muliawan, *Polym. Eng. Sci.*, 45 1098–1107 (2005).
113. G.M. Wise, M.M. Denn, A.T. Bellet, et al., *J. Rheol.*, 44, 549–567 (2000).
114. F. Legrand, J.M. Piau, and H. Hervet, *J. Rheol.*, 42, 1389–1402 (1998).
115. M.M. Britton, R.W. Mair, R.K. Lambert, and P.T. Callaghan, *J. Rheol.*, 43, 897–909 (1999).
116. H. Munstedt, M. Schmidt, and E. Wassner, *J. Rheol.*, 44, 413–428 (2000).
117. Y.W. Inn, R.J. Fischer, and M.T. Shaw, *Rheol. Acta.*, 37, 573–582 (1998).

# CHAPTER 3

# BRANCHED POLYOLEFINS

ALEXANDROS D. GOTSIS
Technical University of Crete, Hania, Greece

## CONTENTS



*Applied Polymer Rheology: Polymeric Fluids with Industrial Applications*, First Edition.
Edited by Marianna Kontopoulou.

## 3.1 INTRODUCTION

### 3.1.1 Polyolefins: Examples of Branched Polymers

Polyolefins are products of the polymerization of olefins (alkenes). Polyethylenes and polypropylene are common polyolefins, and they are the most widely used group of thermoplastic polymers today.

Isotactic polypropylene (iPP) is the commodity polymer with the largest single share in today's global market of polymeric materials. iPP has many advantageous properties when compared to the other polyolefins, PS and PVC: It has higher melting point and lower density, it shows excellent chemical resistance and high tensile modulus, and it costs less to produce and does not present the difficulties in recycling that are associated with PVC.

Polyethylenes have retained a large share of the market. A reason for this commercial resilience is their ability to be produced in a wide range of molecular weights and molecular weight distributions but also with a wide variability in their molecular structures, i.e., their degree and type of branching. The use of different olefins like butene, hexene, and octene as comonomers results in many different types of linear low density polyethylenes (LLDPE) with a broad range of properties. Metallocene catalysis in industrial polymerization processes has led to a generation of polyethylenes with specific and well-defined molecular structures.

Based on their monomeric units and their chain structures, the commercial polylefins produced today can be divided in the following subgroups [1]:

- Ethylene-based materials (PEs) produced in reactors under low-pressure conditions with transition metal catalysts of various types and showing a predominantly linear chain structure. This subgroup includes high-density PE (HDPE), medium-density PE (MDPE), LLDPE and other varieties, which are distinguished through the regulation of density and, subsequently, mechanical properties, through the incorporation of higher $\alpha$-olefines as comonomers. The linear nature of their polymeric chain can be disturbed in two ways: by longer comonomers like butene, hexene, or octene acting as short side chains and by catalysts that form active oligomers in situ and incorporate them further as long-chain branches (LCB).
- PEs produced in radical polymerization under high pressures with oxygen peroxides as chain initiators and showing predominantly branched chain structure. Because of their reduced crystallinity and density, these materials are termed low-density PEs (LDPE).
- Propylene-based polymers produced with transition-metal catalysts (PP) and its copolymers showing a linear chain structure with stereospecific arrangement of the propylene units. Recently sparse long-chain branches have been attached to the main chain of PP leading to some LCB–PPs (high melt strength PPs).
- Polyethylenes produced by metallocene catalysis (single site, constrained geometry) and having a mostly linear stereospecific chain structure with or without well-controlled sparse LCB (mPEs).
- Olefinic elastomers, based on transition metal or single site catalysts, with or without the incorporation of dienes, which make these materials cross-linkable.

The polyolefins show a large diversity in mechanical properties in both the solid and the melt states. This is due to a wide variety in their molecular architecture as described by their molecular weight $M_w$; their molecular weight distribution (MWD); and the number, distribution, and length of branches on their molecular chain backbone. The present chapter reviews the rheology of the polyolefins and mostly of LDPE, LLDPE, and PP. These materials differ qualitatively from each other mainly in their branched structure.

Polyethylene is one of the few widely available polymers that come with many different molecular chain architectures. LDPE is a typical polymer with a highly branched structure. There are many long-chain branches on the main chain and they may have branches themselves. The metallocene polyethylenes (mPE) have mostly linear chains, they may have short branches and a controlled number of LCB. HDPE consists of linear chains and has hardly any LCB. LLDPE has many short branches. PP is mainly linear or it has a few long-chain branches. These differences are reflected on their crystallinity and density in the solid, their mechanical properties and their rheology. The latter affects the ways that these materials can be processed and their applications. The approach of this chapter is based on the differences of the molecular structure of these types of polyolefins. LDPE is considered as a characteristicly densely long-chain branched polymer. Branched PP is usually a sparsely LCB polymer with degrees of branching of <1 branch per backbone chain. An LLDPE is a typical short-chain branched polymer. mPEs bridge the gap between LDPE and LLDPE or HDPE.

Long-chain branches are defined as branches of mostly the same chemical species as the long linear macromolecular chain on which they are rooted. These branches are long enough to be able to form at least 2–3 entanglements each, or they have lengths of at least 2.5 times the molecular weight at the onset of entanglements, $M_C$ [2]. This is in contrast to the short-chain branches (SCB), which have a length in the order of a few repeating units. Because of their length, the long branches act locally as individual molecules, are flexible, and produce entanglements, but are restrained in their motions by their roots on the backbone chain. The hydrodynamic volume of the branched chain, at least in theta conditions, is smaller than a linear chain with the same molecular weight. Thus long-branched chains could show a great enhancement of melt elasticity with a small, or even negative, increase in shear viscosity in comparison with linear chains of the same molecular mass. Short chains cannot provide extra entanglements. On the contrary, many short chains on a flexible chain could even reduce the main-chain flexibility protecting the individual backbone chain from its environment and thus reduce the possibility of entanglements and the melt elasticity [3].

### 3.1.2 Rheology, Melt Strength, Melt Strain Hardening, and Molecular Structure

The melt strength (MS) of a polymer is a measure of the tenacity of its melt. It is defined as the maximum (draw-down) force (in $cN$) by which a molten thread can be drawn under standard conditions before it breaks, e.g., during a Rheotens® measurement [4]. High values of MS are desired in processes where the material is stretched in its molten state, such as in film blowing, thermoforming and foaming. High melt strength is usually accompanied by strain hardening of the elongational viscosity [5]. Strain hardening is the acceleration of the elongational viscosity growth as a function of strain that takes place beyond some characteristic strain, e.g., in uniaxial elongational flow. It has been observed in several polymer melts.

Increasing the average molecular mass of a polymer (i.e., upon decreasing MFI) results in higher shear viscosity, as well as higher MS. The melt strength increases also when the MWD becomes broader. Nevertheless, upon addition of long chain branches on the polymer backbone, the MS increases much more dramatically than the shear viscosity [4, 6, 7]. For example, the MS of LDPE samples was found to be twice [3] or five times [8] higher than the melt strength of LLDPE and HDPE samples that had the same melt flow index (MFI). This effect is attributed to the strain hardening behavior of the elongational viscosity of the melt and is manifested more strongly in tree-type rather than comb-type LCB structures. Short branches do not seem to have the same effect. No differences in melt strength were seen by using different comonomers (1-butene, 1-hexene and 1-octene) in LLDPE [8]. It seems that all these side groups are too short to influence the elongational flow properties of the polymer melt and thus the melt strength.

The melt strength of the polyolefins increases strongly with decreasing MFI [7] as well as by widening the molecular weight distribution. But the strongest enhancement of the MS has been found on adding LCB: Branched polypropylenes obtained using electron beam irradiation may have 10 times higher melt strength than a linear PP with the same MFI [8]. The melt strength of branched PP increases with the increase of the weight average number of long chain branches per molecule, $B_n$ [8, 9].

The rheology of a polymer melt is strongly affected by its molecular architecture: chain branching and MWD. A broad or a bimodal MWD, e.g., can cause strain hardening in the melt elongational viscosity because it results in an increased number of entanglements and better network connectivity [10]. This has been shown by Takahashi et al. for blends of low and ultra high molecular weight PMMAs, and by for blends of polybutadienes [11, 12]. While blends with a narrow MWD come close to the linear viscoelastic limit for the elongational viscosity (i.e., $\eta_E = 3\eta_0$), the bimodal distributions show strain hardening. Similarly, PS samples with a bimodal MWD exhibit clear strain hardening, while samples of PS with a simple broad MWD show almost none [12].

In a thorough review on the effect of LCB on the linear viscoelasticity of polyolefins Vega et al. [13] indicate that the introduction of LCB induces higher levels of elasticity than what the broadening of the MWD of a melt of linear chains of a similar molecular weight does. On the other hand, SCB does not in general cause large elasticity increases. Yet, Vega et al. [14] reported that SCB resulted in higher zero shear rate viscosities, $\eta_0$, higher relaxation times, higher values of the elastic (storage) modulus, and higher activation energies of flow as compared to the linear polymers. Analysis, by size exclusion chromatography (SEC) coupled with intrinsic viscosity measurements revealed, however, that the SCB polymers used in that study actually also possessed some LCB. The enhanced elasticity, therefore, should primarily be attributed to these long branches.

Increasing LCB content in PE samples of comparable molecular weight and molecular weight distribution induces more pronounced shear thinning, higher zero shear viscosity, and much longer relaxation times (broader relaxation spectrum) [15–17]. The loss tangent (tan $\delta$) decreases, while the extrudate swell increases. The increase of the zero shear viscosity of samples made by modifying the same linear PP by the progressive addition of increasing amounts of long chain branching can be related to the degree of LCB, $B_n$ [18]. The activation energy of flow, $E_a$, also seems to increase with increasing branching content for some polyolefins [14, 19].

Strain hardening in extensional flows has been shown to be caused by long chain branches in several polymers and in particular in LDPE [e.g., 3, 10, 20, 21]. For example, the elongational viscosity, $\eta_E^+$, of long-chain branched LDPE is strongly strain hardening, while linear HDPE with similar shear viscosity shows much less strain hardening. The extent of strain hardening increases with increasing branching content [16, 18].

The number of entanglements per branch is a decisive parameter influencing both shear and elongational behavior [22]. Small amounts of long-chain branching increase both the zero shear viscosity and the strain hardening of the elongational viscosity, $\eta_E^+$ [23]. High values of $B_n$ can reduce $\eta_0$, while they still enhance the strain hardening of $\eta_E^+$ and the melt elasticity [24]. On the other hand, chain branches shorter than the entanglement length do not cause strain hardening [3].

Summarizing, LCB increases the possibility for entanglements in the polymeric melt and thus its elasticity. A similar effect, however, could result also from a broadening of the MWD. This is often observed in HDPE, which is a typical linear-chain polymer. Rheology has proved to be a reliable method for the verification of the existence of long branches on the polymeric chain [9, 25], and it is the easiest of the available characterization methods to implement. Gathering information from a number of different techniques is necessary for unambiguous answers to the molecular origin of the rheological changes.

## 3.2 LINEAR POLYETHYLENE (HDPE)

High density polyethylene shows the typical rheology of a (mostly) linear chain polymer. The major parameters that characterise its structure are the mass-average molecular weight, $M_w$ and the polydispersity index, $I_M = M_w/M_n$. For conventional HDPEs with a wide range of molecular weights and polydispersity indices it has been found [26, 27] that

$$\eta_0 = 3.4 \times 10^{-15} M_w^{3.6} \quad \text{(Pa.s)}. \tag{3.1}$$

The onset of shear thinning was found to decrease with the polydispersity index:

$$\dot{\gamma}_0 \sim I_M^{-3.0}. \tag{3.2}$$

The Cox-Merz rule, which relates the shear viscosity with the dynamic shear viscosity, seems to apply. Assuming that essentially only those molecular weights with reptation times shorter than the inverse of the shear rate will contribute effectively to the viscosity of that shear rate, then it can be shown that the shapes of $\eta^*(\omega)$ and $\eta(\dot{\gamma})$ are similar when $\omega \rightarrow \dot{\gamma}\sqrt{3}$ [28]:

$$\eta^*(\omega) = \eta(\dot{\gamma})|_{\omega = \dot{\gamma}/\sqrt{3}} \cdot \tag{3.3}$$

Muñoz-Escalona et al. [26], however, examined several metallocene-catalyzed monomodal and bimodal HDPEs and compared them with their conventionally produced homologes (similar molecular weights and polydispersities) and found that the metallocene PEs in general display higher $\eta_0$ values with a stronger dependence on molecular weight,

$$\eta_0 \sim M_w^{3.9} \tag{3.4}$$

and the polydispersity

$$\dot{\gamma}_0 \sim I_M^{-3.3}. \tag{3.5}$$

The power law index of the viscosity dependence on shear rate, $n$, increases with molecular weight for HDPE, but higher values for $n$ are observed for the metallocene-catalyzed bimodal HDPEs. Also the latter were more elastic (higher values of $G'$). These differences were presumably due to the presence of very small amounts of LCB that existed in the metallocene catalyzed samples.

## 3.3 SHORT-CHAIN BRANCHED POLYETHYLENE (LLDPE)

The linear low density polyethylene is made by the copolymerization of ethylene and a co-monomer (such as, e.g., octene). The long linear chain of LLDPE is composed of ethylene units with short side chains coming from the comonomer. Typically LLDPEs have no LCB and relatively narrow molecular weight distribution. Single-site catalysts can produce even narrower MWD. The length of the short side chain influences the mechanical properties: increasing from 2 to 6 carbon atoms may increase the tensile strength by a factor 2. The rheology of LLDPE is very similar to that of a linear chain polymer.

### 3.3.1 Shear Flow

Wood-Adams et al. [25] reported that the short-chain branching found in LLDPEs has no significant effect on the linear viscoelastic shear flow behavior of the melt. Linear polyolefins with the same main chain length have approximately the same hydrodynamic volume, regardless of the exact length of the short-chain branches.

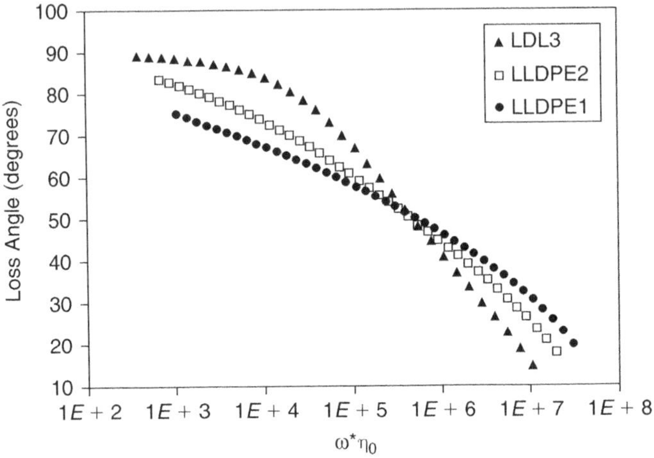


**FIGURE 3.1** Effect of polydispersity on the loss angle of three polyethylenes: $I_M(LDL3) < I_M(LLDPE2) < I_M(LLDPE1)$. Reprinted with permission from Ref. 25.

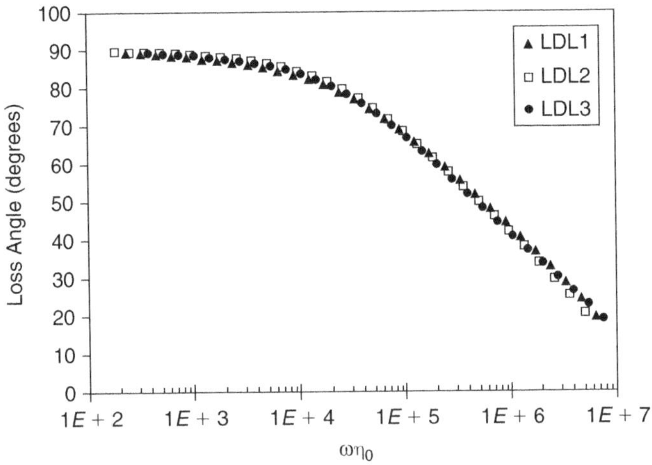


**FIGURE 3.2** Loss angle vs. frequency for three LLDPEs with increasing degree of SCB. Reprinted with permission from Ref. 25.

The effect of the polydispersity on the loss angle is shown in Figure 3.1. Increasing the polydispersity reduces the angle at low frequencies. The effect is inverted at higher frequencies, i.e., the curve becomes flatter as $I_M$ increases.

The activation energy for flow initially increases with the number of short branches and goes to saturation above a certain level (around 15% w/w of butene branches in the case reported by Wood-Adams and Costeux [29]). The viscoelasticity, as indicated by the loss angle does not seem to be influenced by the amount of short branches (Fig. 3.2) [25].

### 3.3.2 Extensional Flow

Because LLDPE has only short branches, it should show a linear behavior in elongational flow, i.e., its elongational viscosity at steady state should follow the Trouton law ($\eta_E \approx 3\eta_0$). No strain hardening should be expected then, if the measurement is conducted carefully.

However, Münstedt et al. [30] found that for specific LLDPEs in uniaxial elongational flow there was an indication of significant strain-hardening at large strains. This happened only at small strain rates and not at higher ones. This behavior was seen both in elongational stressing and creep experiments but not in shear flow. Two molecular processes could be distinguished in shear creep recovery experiments, indicating that the unexpected strain hardening in elongational flow was also the result of these two processes.

The two processes that cause this nonlinear response of the LLDPE are associated with the contributions of two (mass) components in the melt [31]: (1) A small fraction of a distinct linear component with higher molecular mass, and (2) the short-chain branched molecules. The linear component with the high molecular mass is immiscible with the short-chain branched macromolecules, giving a separate phase within the matrix of short chain branched molecules. At short times (deformations) the elongational flow is dominated by the properties of the (lower molecular mass) matrix, leading to short relaxation/retardation times and linear behavior. At longer times the deformation of the high $M_w$ phase comes into play resulting in a viscosity increase and strain hardening. Liquid-liquid phase separation in LLDPE should occur mainly in the case of very high branching degrees [32] but it has also been reported in LLDPE by e.g., [33, 34]. The high molecular mass component can also explain the large steady-state recoverable compliances found in shear and elongation for the LLDPE at low stresses.

## 3.4 LONG-CHAIN BRANCHED POLYETHYLENE (LDPE) AND METALLOCENE POLYETHYLENES (mPE)

LDPE was used in the standard IUPAC samples in the 1970s and 1990s for the characterization of the rheology of polymer melts, and a lot of rheological data exist today. It was chosen at the IUPAC study as an example of a *high-elasticity* melt, something that we understand now as being due to its highly (long-chain) branched structure. A good review of the rheology of LDPE has been given by Santamaria [35].

The use of metallocene catalysis has introduced a large number of new types of polyolefins with much better controlled molecular structure. The "constrained geometry"catalysts can produce samples with narrow molecular weight and introduce well controlled levels of long chain branching. There are, therefore, good samples now for systematic studies of the effects of LCB on the rheology of the polyolefin melts [25, 29, 36]. In the present section we will examine the LCB polyethylenes and compare the sparsely branched mPEs with LDPE.

### 3.4.1 Shear flow

***3.4.1.1 Shear Stresses*** The shear viscosity of the melt of LCB polyethylenes follows the trend of thermoplastic polymers with high values at low shear rates ($\eta_0$) and decreasing values as the shear rate increases (shear thinning). The well-known relation $\eta_0 = KM_w^{n_1}$ for the zero shear viscosity is followed by the LDPE melts, at molecular weights beyond the critical molecular weight but $n_1$ has usually a value higher than 3.4. In this case, $\eta_0$ of LCB polymers depends on both the molecular weight and the degree of LCB [37].

From the systematic studies on mPEs it has been found that long-chain branching affects the complex viscosity vs. frequency curve in four ways: (1) At low levels of branching the zero shear viscosity increases with $B_n$ for the same backbone molecular weight; it reaches a maximum and then decreases at higher values of $B_n$, (2) Shear thinning becomes stronger, (3) The transition zone between the zero shear viscosity and the power law zone is broadened, (4) Two points of inflection appear within the transition zone.

In general the presence of dense long-chain branches seem to result in a decrease in viscosity. On the other hand, an enhancement of zero shear viscosity is seen for high molecular weight polyethylene with moderate to low degrees of LCB. The discrepancy is resolved by examining two opposing effects of the long branches on the rheology of the melt. At low degrees of LCB the viscosity is enhanced because the entanglement possibility is increased. At higher degrees of LCB, the molecular size (hydrodynamic volume) is reduced because of the more compact form of the branched molecule. This effect becomes then more important and eventually it reduces the melt viscosity. At higher shear rates the shear thinning of the highly branched polyethylenes is stronger due to the effect of disentanglement and smaller coil dimensions.

The enhancement in the pseudoplasticity is due to the contribution of long branches to entanglements, which decreases as the shear rate increases. In addition to the steeper slope, LCB results in a delay of the onset of shear thinning, $\dot{\gamma}_0$. The shear compliance $J_0^e$ is very sensitive to the molecular weight distribution and on LCB. Graessley [38] has related the shear compliance with the zero shear viscosity and $\dot{\gamma}_0$ for branched polymers:

$$\eta_0 J_e^0 \dot{\gamma} = 0.6 \pm 0.2. \tag{3.6}$$

The points of inflection in the viscosity curve (Fig. 3.3) are indications of the existence of different relaxation modes and can be seen more clearly if the derivative of the viscosity $\eta^*(\omega)$ is used. Obviously, To use these points to characterize LCB, high-quality, noiseless data are needed for $\eta^*(\omega)$.

The distribution of relaxation times is broader in LDPE as compared with that of linear chain polymers. A broader molecular weight distribution and/or long-chain branching of the polymer backbone broadens the distribution of relaxation times associated with the entanglement slippage in the polymeric

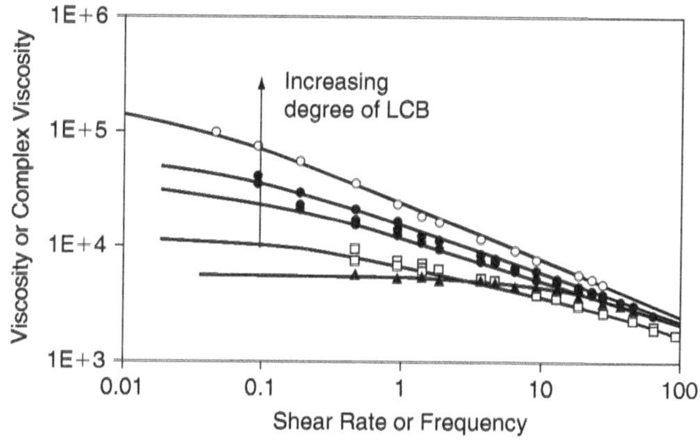


**FIGURE 3.3** The change of the form of the viscosity curve on increasing the degree of LCB in branched mPEs. Reprinted with permision from Ref. 25.

liquids. Even though the effect of MWD cannot be ruled out, the effect of LCB seems to be stronger.

The presence of LCB adds a relaxation mode at low frequency, which is absent in the linear material [36]. LCB, therefore, reduces the slope or adds a (hint of a) plateau to the curve of $G'(\omega)$ at low frequencies [23]. A more sensitive indicator of the changes in elasticity and the additional relaxation modes that LCB may induce is the loss angle ($\tan\delta = G''/G'$). Obviously, the lower the value of this angle, the more solid-like (i.e., elastic) the melt at the corresponding frequency. The curve of $\delta(\omega)$ of a linear polymer in a semi-log plot has a convex curvature: The loss angle remains at values close to 90° at low frequencies, as the melt is a viscous liquid at infinite times, and starts decreasing monotonically at higher frequencies, as the material responds as an elastic solid at infinite $\omega$. The presence of LCB introduces two extra inflection points in the curve, and the middle part of the curve becomes concave, sometimes leading to an intermediate plateau. The gradual change of convex to concave and the relative height of this plateau can be related qualitatively to the increase of LCB on the polymer chains (Fig. 3.4). Compare this to Figure 3.1 to see the differences between the effect of polydispersity and of LCB on the melt rheology.

Examining Figure 3.4, García-Franco et al. [39] noticed that the nondependence of the loss angle on frequency at the plateu is similar to what takes place at the gel point of polymer networks, as first described by Winter and Chambon [40]. At the gel point the dynamic moduli follow a scaling law of

$$G''(\omega), G'(\omega) \propto \omega^n, \qquad (3.7)$$

where $n$ is the the critical network relaxation exponent. The loss modulus at that point is

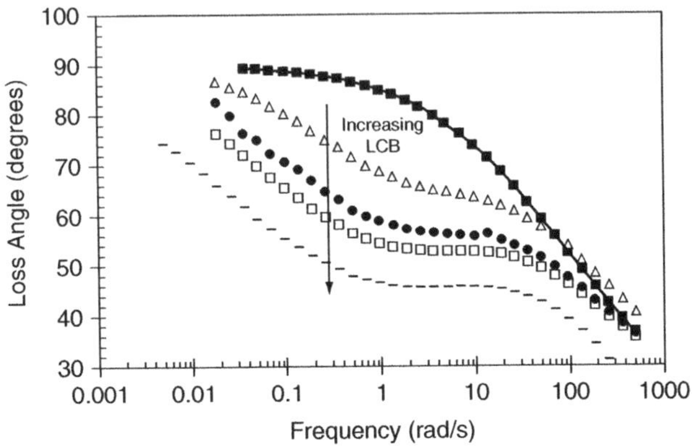


**FIGURE 3.4** The change of the form of the loss angle vs. frequency curve on increasing the degree of LCB in the polymer. Reprinted with permission from Ref. 25.

$$\tan \delta_c = \tan \frac{n\pi}{2} \tag{3.8}$$

The reader should keep in mind that these samples behave as physical gels only in a limited frequency (time) range. In contrast to true critical gels, the relaxation behavior observed here is bounded by two characteristic relaxation times, the two limits of the plateau, corresponding to crossover to different relaxation mechanisms. This probably has to do with the nonpermanent nature of the gel formed, which is essentially based on the entanglements involving the branches.

The relaxation modulus of such a gel is given by [40] as

$$G(t) = St^{-n}, \tag{3.9}$$

where $S$ is the gel stiffness. For the LCB polymers of Figure 3.4, the relaxation exponent of the temporary gel decreases with increasing branching, while the gel stiffness increases [39].

***3.4.1.2 Primary Normal Stress Difference*** The primary normal stress difference, $N_1$, involves the extra normal stresses that are measured for viscoelastic materials in simple shear flows [41]. They are the result of the reduction of the chain entropy as the chain is stretched in the flow field. Thus $N_1$ depends on the elasticity of the melt and is zero in Newtonian fluids.

The steady state normal stresses are usually expressed in the following form:

$$N_1 = \tau_{xx} - \tau_{yy} = -\Psi_1(\dot{\gamma})\dot{\gamma}_{yx}^2. \tag{3.10}$$

The normal stress coefficient $\Psi_1$ for a LDPE is given in Figure 3.5 as a function of $\dot{\gamma}$.

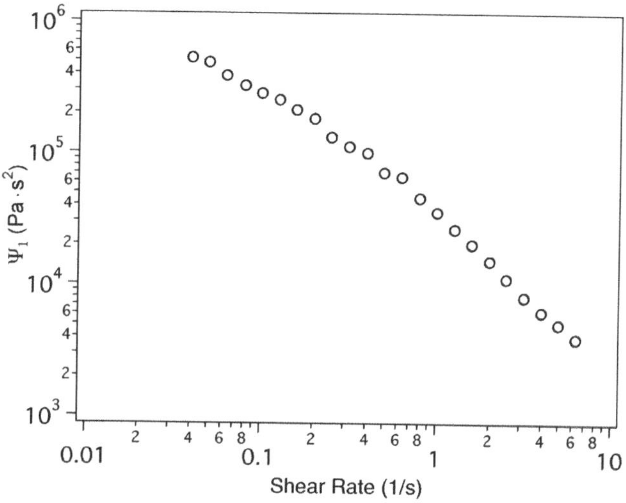


**FIGURE 3.5** The primary normal stress difference coefficient of a LDPE vs. the shear rate. Data from Ref. 42.

### 3.4.2 Thermorheology

Materials that follow the time-temperature superposition principle are said to be thermorheologically simple. Their shift factor, $a_T$, makes possible the superposition of dynamic moduli data obtained at several temperatures. The temperature dependence of the time scale shift factor of these materials far from $T_g$ follows an Arrhenius relation.

$$a_T = \exp\left\{\frac{E_a}{R}\left(\frac{1}{T} - \frac{1}{T_0}\right)\right\} \tag{3.11}$$

The activation energy, $E_a$, can be used as a measure of the temperature sensitivity of rheological material functions such as $\eta(T, \dot{\gamma})$ and $G'$.

For long-chain branched polymers, the effect of temperature often cannot be described by a single time shift factor. LCB polymers are considered thermorheologically complex materials because they are essentially blends of molecules with different numbers of branches. Then the concept of a single activation energy for flow is not meaningful and one has to use a continuous activation energy spectrum, $E_a(\lambda)$ [29], or the activation energy evaluated from the temperature shifting of $\eta_0$ (time independent).

On the other hand, Stadler et al. [43] indicated that simple time-temperature superposition may be possible for conventional (high-pressure polymerization) LDPE. The absence of a thermorheological complexity for LDPE can be explained by the assumption that every molecule in typical LDPE is long-chain branched and that its branching structure is not much different from that of the

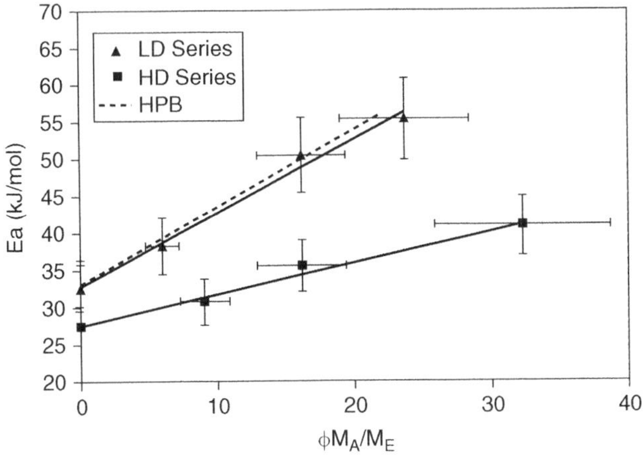


**FIGURE 3.6** The activation energy for flow based on the temperature shift of $\eta_0$ of several series of polyethylenes, as a function of the degree of LCB. Reprinted with permission from Ref. 29.

rest. The complexity, therefore, that is found in metallocene PEs is the result of the presence of fractions with different degrees of LCB in an essentially SCB polymeric matrix.

The activation energy for flow at a certain relaxation time increases with the degree of long-chain branching (Fig. 3.6). This is due to the additional energy that the long branches need to disentangle in shear flow. Moreover, there is a synergistic effect when both long and short branches are present. It is likely that it is the combination of long and short branches that gives high-pressure LDPE its high activation energy (ca. 58 kJ/mol) [29].

Graessley [44] suggests that the differences in the temperature dependence of the viscoelastic coefficients between linear and branched polymers are due to two mechanisms. One is based on the temperature dependence of the entanglement spacing. The other is based on differences in the possible paths for conformational relaxation: Linear chains rearrange by reptation, whereas branched chains rearrange by passing through relatively compact conformational states that may have different energies. Both mechanisms may be important in the case of highly branched LDPE.

Vega et al. [14] suggested a LCB index based on the difference of the activation energy for flow between a linear ($E_{a,l}$) and a LCB ($E_{a,b}$) polymer:

$$I_{LCB} = \frac{E_{a,b} - E_{a,l}}{E_{a,l}}. \tag{3.12}$$

For the case of polyethylene, $E_{a,l}$ would correspond to the activation energy of a HDPE (with $E_a \approx 24$ kJ).

### 3.4.3 Extensional Flow

Long-chain branches added on the linear chains increase the temporary network connectivity in the melt and reduce the rate of disentanglement. Thus the branched polymers show enhanced strain hardening of their viscosities in uniaxial elongational flow. In general, the more branches added on the chains and the longer these branches are, the steeper the stress growth at strains above 1 s.u. The viscosity growth curve of a highly branched melt is, therefore, strongly strain hardening [45] (Fig. 3.7).

The elongational viscosities of linear melts are essentially non-strain hardening. The extensional viscosity of LDPE, on the other hand, initially shows a simple monotonic growth as a function of the strain with a decreasing first derivative that compares well with the theory of linear viscoelasticity. Beyond a deformation of around 1 s.u., however, the curves show an increase in their slope as the viscosity growth accelerates; this is the region of strain hardening. At even higher strains the branched melts break more or less by elastic fracture. However, the force measured at those high strains is very low, and perhaps below the sensitivity of most rheometers. There are reports [46] that a steady state eventually comes and the viscosity takes a constant value. This value is much higher than the Trouton value of $3\eta_0$.

The strain hardening is stronger at higher strain rates. An increase in the degree of long-chain branching leads to a more pronounced strain hardening behavior. Some theories that predict the elongational flow behavior of LDPE with some accuracy are described in the following sections.

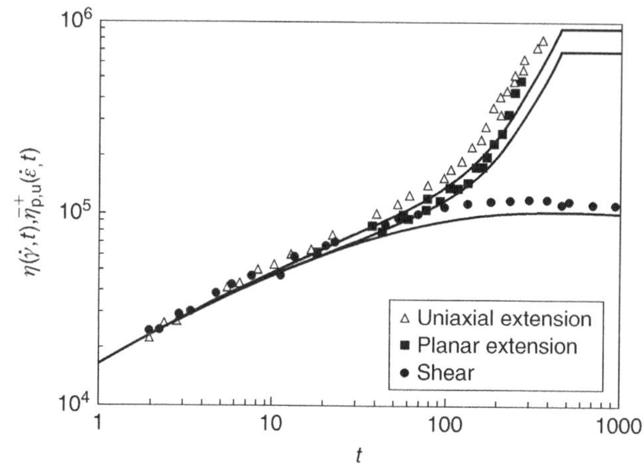


**FIGURE 3.7** Transient uniaxial elongation, planar extension and shear viscosity of IUPAC-X LDPE at strain rate of 0.01 s. Points, experimental measurements; lines, predictions of the Pompom model. Reprinted with permission from Ref. 45.

### 3.4.4 Theoretical Description of Extensional Strain Hardening in Branched Polymers

The elongational viscosity growth curves are usually given as log-log plots. However, when a semi-log plot is used, these curves may become straight lines at high strains. They can be approximated, then, by exponential functions of the strain: $\eta_E^+ \approx C_1 \exp\{k_1\varepsilon\}$. The exponential form of the function $\eta_E^+(\varepsilon)$ suggests the use of a specific formalism for describing the nonlinear viscoelastic response during the elongational viscosity growth; this is outlined below as reported by Gotsis et al. [17] and Tsenoglou et al. [47].

#### *3.4.4.1 Lodge's Rubber-Like Liquid Model with a Damping Function*

The stress growth during uniaxial elongation of polymeric melts can be described by several viscoelastic models. One of the most successful for this purpose is the Rubber-Like Liquid model of Lodge [48]. Assuming separability of time and strain, this model can be modified to account for nonlinear, strain-thinning effects by the incorporation of a damping function to the viscoelastic memory:

$$\sigma = \int_{-\infty}^{t} \mu(t-t')h(\lambda)\mathrm{F}(t-t')dt' \tag{3.13}$$

In this equation F is the Finger deformation tensor, $h(\lambda)$ is the damping function, $\lambda$ is the stretch as a function of present and past time $(\lambda(t,t'))$, and $\mu$ is the memory function. The Finger tensor in the case of uniaxial elongation (constant strain rate, $\dot{\varepsilon}$) is given by [41]:

$$\mathrm{F}(t-t') = \begin{bmatrix} \exp\left(2\dot{\varepsilon}(t-t')\right) & 0 & 0 \\ 0 & \exp\left(-\dot{\varepsilon}(t-t')\right) & 0 \\ 0 & 0 & \exp\left(-\dot{\varepsilon}(t-t')\right) \end{bmatrix}. \tag{3.14}$$

The memory function, $\mu$, is a material function of time and depends on the relaxation spectrum of the melt. A widely used form of $\mu$ comes from the multi-mode Maxwell model:

$$\mu(t) = -\frac{dG(t)}{dt} = \sum_{i=1}^{k} \frac{G_i(t)}{\tau_i} = \sum_{i=1}^{k} \frac{G_i(0)}{\tau_i} e^{-t/\tau_i}, \tag{3.15}$$

where $(\tau_i, G_i(0))$ are the sets of the relaxation times and moduli evaluated by fitting the dynamic shear moduli of the melts at the same temperature.

The damping function, $h(\lambda)$, physically signifies the extent of the stress loss due to the reduction of the entanglement density and segmental orientation following the deformation of a given magnitude, $\lambda$. Several forms for $h$ have been proposed and tried with variable degrees of success [e.g., 49, 50].

Tsenoglou have generalized this theory and modeled the nonlinear viscoelasticity under a general extensional flow defined by the velocity field $U_i = e_i X_i$, where $i = 1, 2, 3$, and $e_1 + e_2 + e_3 = 0$. The rate of strain tensor, $\underline{\underline{D}}$, for an incompressible fluid in such a flow, is

$$\underline{\underline{D}} = \begin{bmatrix} e_1 & 0 & 0 \\ 0 & e_2 & 0 \\ 0 & 0 & e_3 \end{bmatrix} = \begin{bmatrix} \dot{\varepsilon}_1 & 0 & 0 \\ 0 & m\dot{\varepsilon} & 0 \\ 0 & 0 & -(m+1)\dot{\varepsilon} \end{bmatrix} \tag{3.16}$$

Here, $e_1 \geq e_2 \geq e_3$ and $-0.5 \leq m \leq 1$. For uniaxial (i.e., axisymmetric) extension $m = -0.5$, for planar extension (pure shear) $m = 0$, and for biaxial compression, $m = 1$.

In a general extensional flow the relative Hencky (or logarithmic) strain, $\varepsilon$, and the principal stretch ratio, $\lambda$, are related:

$$\lambda = \exp(\varepsilon) = \exp\left(\int_{t'}^{t} \dot{\varepsilon}(t'')dt''\right).$$

The finger tensor, $F(t')$, which expresses deformation at present time $t$ relative to the configuration at past time $t'$, is in such flows:

$$\underline{\underline{F}}(t') = \begin{bmatrix} \lambda^2 & 0 & 0 \\ 0 & \lambda^{2m} & 0 \\ 0 & 0 & \lambda^{-2(m+1)} \end{bmatrix} \tag{3.17}$$

with its first and second invariants respectively equal to $I_1 = \lambda^2 + \lambda^{2m} + \lambda^{-2(m+1)}$ and $I_2 = \lambda^{-2} + \lambda^{-2m} + \lambda^{2(m+1)}$, and its eigenvalues equal to the square of the principal stretch ratios, $\lambda$, $\lambda^m$ and $\lambda^{-(m+1)}$.

A simple form for the damping function, $h(\lambda)$, that accommodates the essential phenomenology, i.e., the controlled exponential growth of $\eta_E^+$, and LCB effects of variable intensity, is the following:

$$h(\lambda) = \lambda^{-\beta}. \tag{3.18}$$

Here, $\beta(\geq 0)$ is an adjustable parameter that should depend on the branching number: Increasing $B_n$ leads to better network connectivity, improved resistance to strain-induced network destruction and, therefore, to less stress damping and smaller $\beta$ values.

Combining (3.17) and (3.18) with (3.13) we get:

$$\underline{\underline{\sigma}} = \int_{-\infty}^{t} \mu(t-t') \begin{bmatrix} \lambda(t,t')^{2-\beta} & 0 & 0 \\ 0 & \lambda(t,t')^{2m-\beta} & 0 \\ 0 & 0 & \lambda(t,t')^{-2(m+1)-\beta} \end{bmatrix} dt' \tag{3.19}$$

The differential form of this equation is often more useful than the integral version. For a series of relaxation modes, $(G_i, \tau_i)$, this becomes:

$$\dot{\underline{\underline{\sigma}}} - \underline{\nabla v}^T \cdot \underline{\underline{\sigma}} - \underline{\underline{\sigma}} \cdot \underline{\nabla v} + \left(\beta + \sum_{i=1}^{k} \frac{1}{\tau_i}\right) \underline{\underline{\sigma}} = \left(\sum_{i=1}^{k} \frac{G_i(0)}{\tau_i}\right) \underline{\underline{\delta}} \tag{3.20}$$

where $\underline{\underline{\delta}}$ is the unit tensor and $\underline{v}$ is the velocity vector. Notice that the combination of the first three terms is the upper convected time derivative:

$$\underline{\underline{\sigma}}' \equiv \dot{\underline{\underline{\sigma}}} - \underline{\nabla v}^T \cdot \underline{\underline{\sigma}} - \underline{\underline{\sigma}} \cdot \underline{\nabla v}$$

Both extension geometry and molecular structure affect the extent of melt strain damping. Deformations of similar magnitude but of different $m$ may differ in their ability of network connectivity destruction, either through contact extinction or loss of segmental orientation. The intensity of strain hardening follows, in general, the order of: uniaxial > planar > biaxial [51]. The extent of strain damping follows the reverse order. Tsenoglou et al. [47] suggest the following relation between $\beta$ and $m$

$$\beta = \beta_u \left(2 - \exp\left\{-\sqrt{\frac{m+0.5}{0.6}}\right\}\right) \tag{3.21}$$

The parameter $\beta_u$ depends on the molecular structure. Values of $\beta_u \geq 1$ correspond to linear polymers, devoid of branching. For $\beta = 0$ there is no stress damping and one recovers Lodge's original Rubber-Like Liquid model, which does not include strain-induced network destruction.

When the relaxation modulus of the melt takes the form of (3.15), the first and second normal stress difference evolution following extensional flow startup at $t = 0$, where $\lambda_t \equiv \exp(\dot{\varepsilon}t)$, as predicted by the theory above are

$$N_1^+(t, \dot{\varepsilon}) = \sigma_{11} - \sigma_{22} = \sum_{i=1}^{k} \left\{ \frac{(2-\beta)\dot{\varepsilon}\tau_i \lambda_t^{2-\beta} G_i(t) - G_i(0)}{(2-\beta)\dot{\varepsilon}\tau_i - 1} - \frac{(2m-\beta)\dot{\varepsilon}\tau_i \lambda_t^{2m-\beta} G_i(t) - G_i(0)}{(2m-\beta)\dot{\varepsilon}\tau_i - 1} \right\} \tag{3.22}$$

$$N_2^+(t,\dot{\varepsilon}) = \sigma_{22} - \sigma_{33} = \sum_{i=1}^{k}\left\{\frac{(2-\beta)\dot{\varepsilon}\tau_i\lambda_t^{2-\beta}G_i(t)-G_i(0)}{(2-\beta)\dot{\varepsilon}\tau_i-1} - \frac{(2m+\beta+2)\dot{\varepsilon}\tau_i\lambda_t^{-2-2m-\beta}G_i(t)+G_i(0)}{(2m+\beta+2)\dot{\varepsilon}\tau_i+1}\right\}. \quad (3.23)$$

Depending on the value of $\beta$ $(0 \leq \beta \leq 2)$ compared to the product of (deformation rate) $\times$ (relaxation time of the fluid), either strain hardening or stress plasticity may be predicted. The condition for plastic flow to occur is

$$\dot{\varepsilon} < \min_{1 \leq i \leq k}\left\{\frac{1}{(2-\beta)\tau_i}\right\}.$$

Then, the viscosities reach a steady state value equal to

$$\begin{aligned} \eta_1(\dot{\varepsilon}) &= \frac{N_1}{\dot{\varepsilon}} = 2(1-m)\sum_{i=1}^{k}\frac{\tau_i G_i(0)}{[1-(2-\beta)\dot{\varepsilon}\tau_i][1+(2m+\beta+2)\dot{\varepsilon}\tau_i]} \\ \eta_2(\dot{\varepsilon}) &= \frac{N_2}{\dot{\varepsilon}} = 2(2+m)\sum_{i=1}^{k}\frac{\tau_i G_i(0)}{[1-(2-\beta)\dot{\varepsilon}\tau_i][1-(2m-\beta)\dot{\varepsilon}\tau_i]} \end{aligned} \quad (3.24)$$

In the region of very low deformation rates, the Trouton viscosities assume the form

$$\begin{aligned} \eta_1^+(t,\dot{\varepsilon}\to 0) &= 2(1-m)\sum_{i=1}^{k}\tau_i(G_i(0)-G_i(t)) \\ \eta_2^+(t,\dot{\varepsilon}\to 0) &= 2(2+m)\sum_{i=1}^{k}\tau_i(G_i(0)-G_i(t)) \end{aligned} \quad (3.25)$$

Strain hardening of the viscosities is predicted to occur at critical times equal to

$$\begin{aligned} t_{SH1} &= \frac{1}{2(1-m)\dot{\varepsilon}}\ln\left[\frac{(2m-\beta)\sum_{i-1}^{k}G_i(t)(2m-n-1/\dot{\varepsilon}\tau_i)}{(2-\beta)\sum_{i=1}^{k}G_i(t)(2-\beta-1/\dot{\varepsilon}\tau_i)}\right], \\ t_{SH2} &= \frac{1}{2(1-m)\dot{\varepsilon}}\ln\left[\frac{(2m+\beta+)2\sum_{i=1}^{k}G_i(t)(2m+\beta+2+1/\dot{\varepsilon}\tau_i)}{(2-\beta)\sum_{i=1}^{k}G_i(t)(2-\beta-1/\dot{\varepsilon}\tau_i)}\right], \end{aligned} \quad (3.26)$$

provided that the expression bracketed inside the logarithmic term exceeds unity, i.e., on condition that $-0.5 \leq m < 0$.

***3.4.4.2 The Pompom Model*** A very successful theoretical model for the description of the rheology of branched molecules was developed by Mcleish and Larson [52]. The Pompom molecule, which approximates a long-chain branched macromolecule, consists of a crossbar with a certain length (therefore, with a characteristic molecular weight, $M_b$, relaxation time, $\tau_b$, and a molecular weight concentration, $\phi_b$, in the material), both ends of which are connected with the same number (functionality), $q$, of dangling arms (pompoms), each arm with molecular weight $M_a$. The orientation contribution to the stress is calculated by using the tube model of Doi and Edwards (DE) [53]. The following equations provide the set of constitutive equations for the Pompom model, that have to be solved in flow calculations

The stress:

$$\sigma = \frac{15}{4} G_0 \phi_b \left( \phi_b \lambda^2(t) + \frac{2qs_c}{2qs_a + s_b} \right) \mathbf{S}(t) \qquad \text{with} \quad \phi_b = \frac{s_b}{2qs_a + s_b} \tag{3.27}$$

Evolution of the orientation:

$$S(t) = \int_{-\infty}^{t} \frac{1}{\tau_b(t')} \exp\left\{ -\int_{t'}^{t} \frac{dt'}{\tau_b(t)} \right\} \frac{\left\langle \frac{\mathbf{E}\cdot\mathbf{uE}\cdot\mathbf{u}}{|\mathbf{E}(t,t')\cdot\mathbf{u}|} \right\rangle}{|\mathbf{E}(t,t')\cdot\mathbf{u}|} dt' \tag{3.28}$$

Evolution of the backbone stretch:

$$\frac{\partial}{\partial t}\lambda(t) = \lambda(t)(\nabla \mathbf{v} : \mathbf{S}) + \frac{1}{\tau_s}(1 - \lambda(t)) \quad \text{for} \quad \lambda < q \tag{3.29}$$

Evolution of the arm-withdrawal measure:

$$\frac{\partial s_c}{\partial t} = \left( q\frac{s_b}{2} + s_c \right) \nabla \mathbf{v} : \mathbf{S} - \frac{1}{2t_a(x_c)} \tag{3.30}$$

Backbone orientation time scale:

$$\tau_b = \frac{4}{\pi^2} s_b^2 \phi_b \tau_a(x_c(t)) q \qquad x_c = \frac{s_c}{s_a} \tag{3.31}$$

Backbone stretch time scale:

$$\tau_s = s_b \tau_a(0) q \tag{3.32}$$

Arms relaxation time scales:

$$\tau_a = \tau_0 \exp\left\{ \frac{15}{4} s_a \left( \frac{(1-x)^2}{2} - (1-\phi_b)\frac{(1-x)^3}{3} \right) \right\} \tag{3.33}$$

where $s_a = M_a/M_e$ and $s_b = M_b/M_e$ and $M_e$ is the entanglement molecular weight. Obviously the degree of branching, $B_n$, is related to $s_a$ and $s_b$ and their ratio.

McLeish and Larson [52] and Rubio and Wagner [54] have also given a simplified version in the form of a set of differential equations. The Pompom model gives strain hardening behavior in uniaxial and planar elongational flows.

***3.4.4.3 The Molecular Stress Function Theory*** Another useful model that describes well the elongational flow properties of branched polymers has been proposed by Wagner et al. [46]. It is based on a modification of the reptation model [53] and follows the formalism of the response functions. The elongational stress growth of entangled polymer melts in this model becomes

$$\sigma = -p\mathbf{I} + \int \mu(t-t')f^2\mathbf{S}(t-t')dt', \tag{3.34}$$

where $\mathbf{S}$ is the second-order orientation tensor describing the average orientation of the primitive paths in the reptation model [55]. The molecular stress function, $f$, is effectively a damping function that gives the contribution of the relative stretching of the macromolecular chain. When the melt is stretched, the function $f$ accounts for the change of the diameter of the reptation tube from the (initial) value $a_0$ to $a$

$$f^2 = \left(\frac{a_0}{a}\right)^2 = \beta' \exp \frac{1}{\beta'} \langle \ln u' \rangle_0 + (1-\beta'). \tag{3.35}$$

The parameter $\beta'$ represents the effect of the branching topology on the orientational free energy.

Equation (3.34) is solved for a proper form of $\mathbf{S}(t)$ and $\mu(t)$ in the same manner as in the previous section to evaluate the stresses and the elongational viscosity.

### 3.4.5 Different Elongational Flow Techniques

Uniaxial elongation is a difficult to implement technique. All available measuring setups can be used only for materials that have a rather high melt strength and viscosity. For this reason there have been numerous attempts to measure $\eta_E$ indirectly by methods that are simpler to implement, but where the rate may be variable or where the flow is not purely extensional. Two such techniques are the entry flow method and Fibre Spinning-Rheotens® [3].

This section compares three different elongational flows, used to measure the elongational viscosity of three commercial polyolefins with different molecular structures: a HDPE, a LLDPE, and a LDPE. All three have the same melt flow index of 2.2 g/10 min (at 190 °C) but different polydispersity. The flows that are examined are the uniaxial elongation, the entry flow and the fibre spinning flow.

**3.4.5.1 *Entry Flow*** The entry flow method has a shear flow component in addition to extension. Certain assumptions for the form of the functions of the shear and the extensional viscosities (models) of the fluid should be made a priori and the parameters of the models can be evaluated from the measured values of $\Delta P_{entr}$. This disadvantage may be compensated by the easiness and the wide range of deformation rates at which this technique can be applied.

Based on the work of Binding [56, 57], assuming that the shear and extensional viscosities, and the primary normal stress difference coefficient have power-law forms of the shear rate or the extension rate ($\eta = m\dot{\gamma}^{n-1}$, $\eta_E = k\dot{\varepsilon}^{t-1}$ and $\Psi_1 = p\dot{\gamma}^{q-1}$) Gotsis and Ke [3] related $\Delta P_{entr}$ to the flow rate, $Q$, the barrel radius, $R_o$, the barrel: capillary radii ratio, $\alpha$, and the entrance angle, $\theta$,

$$\begin{aligned}\Delta P_{entr} &+ \frac{2np(3n-1)}{3(2n+q+1)(3n+q+2)}\left(\frac{3n+1}{n}\right)^{q+2} \\ &\times \left(\frac{Q}{\pi R_0^3}\right)^{q+1}\left(\alpha^{3(q+1)}-1\right) \\ &- \frac{2m}{3n\tan\theta}\left(\frac{3n+1}{n}\right)^{n}\left(\frac{Q}{\pi R_0^3}\right)^{n}(\alpha^{3n}-1) \\ &= \frac{2kI_{nt}}{3t}\left(\frac{3n+1}{n+1}\right)^{t+1}\left(\frac{Q\tan\theta}{\pi R_0^3}\right)^{t}(\alpha^{3t}-1)\end{aligned} \tag{3.36}$$

With $I_{nt} = \int_0^1 \left|\frac{3hn+1}{n}\rho^{1+\frac{1}{n}}-2\right|^{t+1}\rho d\rho$

The parameters $m$, $n$, $p$, and $q$ are found from shear flow measurements. Plotting, then, the logarithm of the left-hand side term of (3.3) vs. the logarithm of $Q$, the slope gives the value of $t$ and then $I_{nt}$. From the intercept one finds $k$.

This estimate can be quantitatively related to the value of the extensional viscosity of the polymer averaged over the strain accumulated in the entrance region:

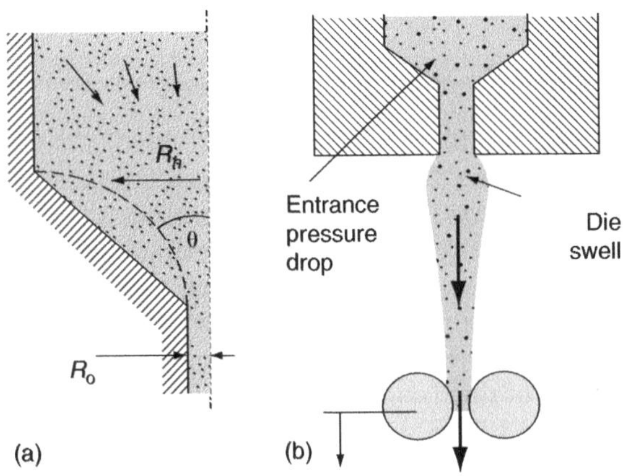


**FIGURE 3.8** (a) The entry flow method and (b) The fiber spinning method.

$$\dot{\varepsilon} = \frac{6V \ln\alpha}{R_0}, \quad \varepsilon_{entr} = 2 \ln \alpha, \tag{3.37}$$

where $V$ is the piston velocity.

The technique is supposed to give an average steady state value for $\eta_E$.

Hence comes the problem of the comparison of this value to the true ones that are obtained from uniaxial elongation. This problem is more serious for materials that cannot reach a steady state in uniaxial elongation.

***3.4.5.2 Fiber-Spinning Flow*** The Rheotens method imitates an industrial process and is relatively simple to apply. Again the problem exists here of the rheological relevance of the obtained information and its relation to the actual elongational viscosity of the melt.

In the fiber spinning flow, the force is constant along the filament and the fluid does not experience much shearing during the stretching. However, the stress and the extension rate are not constant along the filament. The equations that describe this flow are the following:

$$\begin{aligned} &Q = \pi R_o^2 v \qquad v_f = 2\pi\nu' R_w \qquad \lambda = \frac{v_f}{\alpha^2 v} \\ &\varepsilon = \ln \lambda \qquad \dot{\varepsilon} = \frac{v\alpha^2}{H} \lambda^{z/L} \ln \lambda \\ &\sigma = \frac{F\alpha^2 \lambda^{z/L}}{\pi R_o^2} \qquad \eta_E = \frac{\sigma(L)}{\dot{\varepsilon}(L)} \end{aligned} \tag{3.38}$$

where $v_f$ is the linear velocity at the wheels of the Rheotens, $R_w$ is the wheel radius, $H$ the length of the filament, $\nu'$ is the wheel rotational frequency and $F$ is the drawing force. The viscosity is evaluated at the pinching wheels. The total accumulated strain, which affects the value of the extensional viscosity that is measured at the wheels, is the sum of the elongational strain contributions of the entrance region and of the stretching of the filament.

Wagner et al. [58] developed a method that uses master curves extracted from the Rheotens curves. The filament velocity along the spin line was measured and found to increase linearly with the spin line length $x$ below a critical draw ratio $V_p$, while for higher draw ratios the velocity increased over proportionally with $x$. With this observation, for $V < V_p$, the strain rate is constant: $\dot{\varepsilon} = (v_0/L)(V - V_S)$, where $V$ and $V_s$ are the draw ratios at the pinch rollers and the (extrapolated) starting point of the filament respectively. The corresponding elongational viscosity for $V < V_p$ is

$$\eta = \frac{F_p V L}{A_0 v_0 (V_p - V_s)} \tag{3.39}$$

For the nonlinear regime, $V > V_p$, the spinline velocity was assumed to be described by a power-law dependence (with power-law index $n < 1$) of the melt tension on the extension rate. The viscosity then becomes

$$\eta = \frac{F_p V_p L}{A_0 v_0 (V_p - V_s)} \left(\frac{V}{V_p}\right)^{(n-1)/n} \tag{3.40}$$

***3.4.5.3 Comparison*** The shear viscosity curves of the three materials at 190 °C are shown in Figure 3.9. The same figure shows the extensional viscosities as functions of the strain. Since the melts have similar shear viscosity

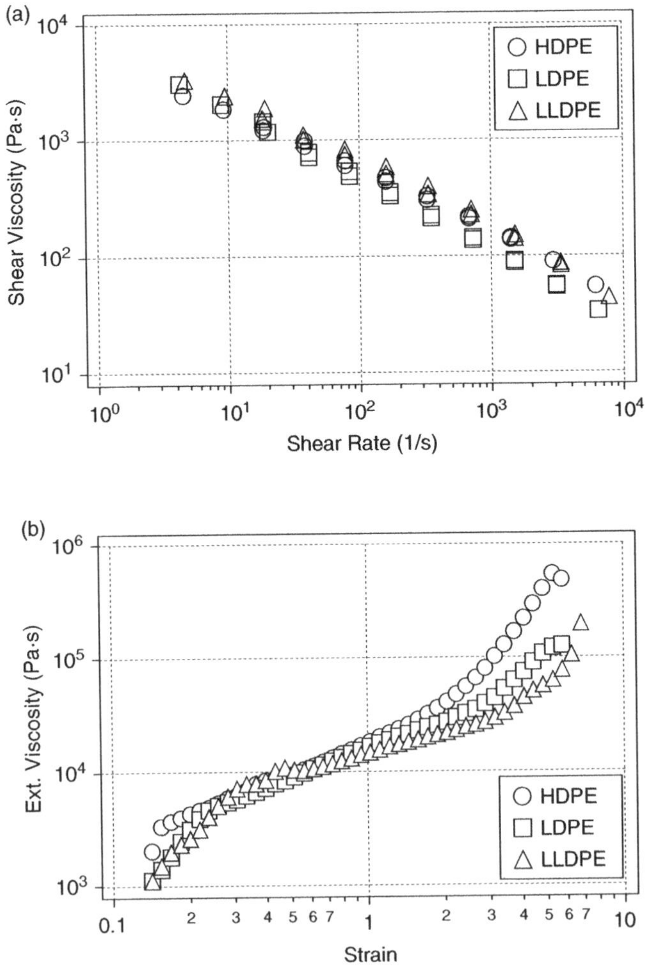


**FIGURE 3.9** (a) Steady state shear curves of LDPE, HDPE, and LLDPE samples, at 190°C. (b) Comparison of the elongational viscosity growth of the three materials at an elongational rate of 1 $S^{-1}$. Reprinted with permission from Ref. 3.

curves at this temperature, the differences in their elongational behavior are the result of their differences in branching and MWD. Eventually all three melts show strain hardening but their difference is greater at higher strains, where LDPE grows faster.

In all experiments that involve stretching flows the highly branched LDPE shows the highest resistance. This is expected from a branched macromolecule. It is surprising, however, that the LLDPE shows even less elasticity (as this is manifested by the value of the extensional viscosity, by its strain hardening, and by dynamic mechanical measurements) than the linear HDPE. The explanation could be a wide molecular weight distribution of this particular grade of HDPE. The eventual (delayed) strain hardening of this LLDPE is probably due to some degree of LCB, as was discussed earlier.

The extensional viscosity calculated from the data of $\Delta P_{entr}$ is shown (as lines) in Figure 3.10a. They correspond to 5 s.u. accumulated in the entry region (contraction ratio of 12) and are compared with the extensional viscosity that was measured in uniaxial elongation and averaged also over the first 5 s.u. (points). The agreement is reasonable given the approximations used and the averaging procedure.

Figure 3.10b shows the extensional viscosity of HDPE calculated using the Rheotens method (3.39) and compared to the values of the viscosity measured in uniaxial elongation and plotted at constant accumulated strain vs. extension rate. One universal value for the degree of strain relaxation for all rates seems not possible to be extracted from this graph.

The apparent elongational viscosity of the IUPAC LDPE sample A18 estimated from the Rheotens experiments using the method of Wagner et al. [58] is shown in Figure 3.11. The maximum of the transient elongational viscosity corresponds to the critical tension $\sigma_p$. At higher extension rates the apparent elongational viscosity decreases according to the power law model. The abrupt change of slope at the maxima is due to the change of the model at this point. With increasing die exit velocity $v_0$, the elongational viscosity curves are shifted to smaller viscosities and larger extension rates. This is presumably due to the effect of the rheological prehistory.

In conclusion, the estimates for the elongational viscosity of polymer melts given by the entry flow method correspond to the values of the viscosity measured in uniaxial elongation averaged over strain equal to what is accumulated in the entrance region. The Rheotens method can also be used for the same purpose but the relevant strain in this case includes, in addition to what is accumulated during the stretching of the filament, also part of the strain accumulated in the entrance region of the capillary. The latter method is less reliable for such estimates because the effect of the strain relaxation in the intermediate flows is difficult to quantify.

### 3.4.6 Window of Instability in the Shear Flow of LDPE

Waddon and Keller [59] found that in a narrow temperature interval above its melting point the flow resistance of the LDPE melt may drop significantly at

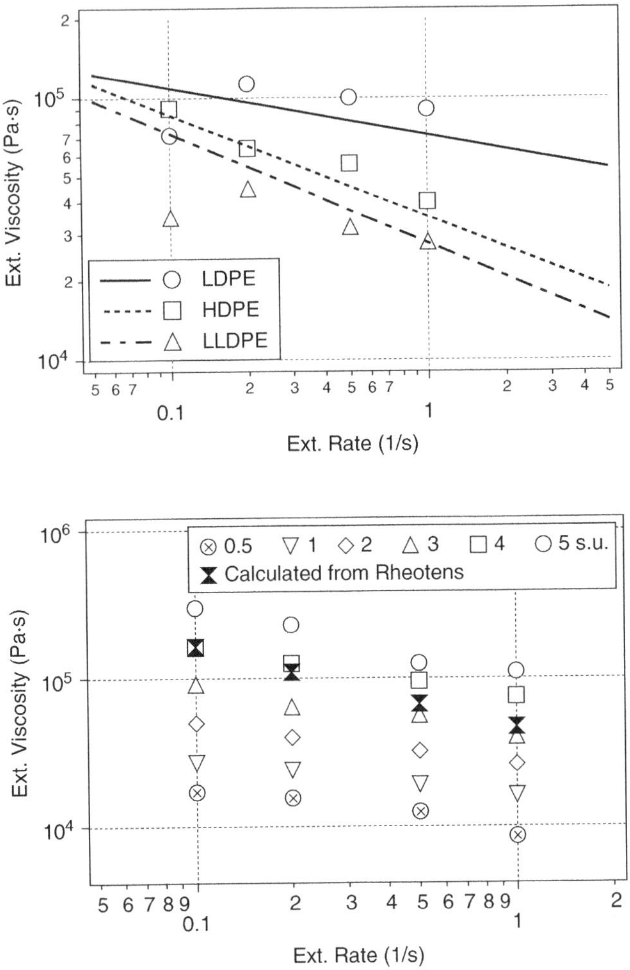


**FIGURE 3.10** (a) Comparison of the entry flow method and the uniaxial elongation. (b) Elongational viscosity of HDPE at constant strain compared with the calculated values from the Rheotens data. Reprinted with permission from Ref. 3.

certain shear rates, displaying a sharp minimum at a sharply defined temperature within this interval, while at the same time extrudate distortions—in the case that such are present when outside the temperature window—are absent. These effects are conspicuous even by visual inspection. Such abrupt singularities point to some kind of structure formation. Further, the effect is of potential practical interest for the extrusion of, e.g., filaments, rods, and pipes, and it is important as a source of instabilities within processing equipment.

Within a narrow temperature interval of 148–152°C during extrusion of LDPE melts Waddon and Keller [59] observed a reduction in the extrusion

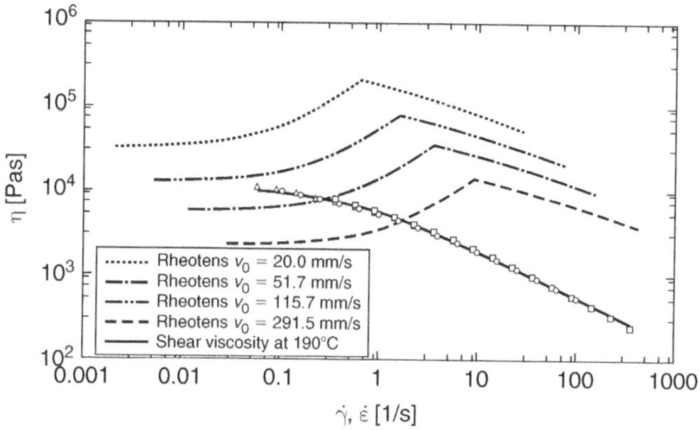


**FIGURE 3.11** Apparent elongational viscosity $\eta(\dot{\varepsilon})$ of IUPAC LDPE A18 for different die exit velocities $v_0$ in the Rheotens experiment; comparison to shear velocity $\eta(\dot{\gamma})$. Reprinted with permission from Ref. 58.

pressure, hence the flow resistance, with a sharp minimum at around 150°C. They also noticed the associated disappearance of flow instabilities and extrudate distortions. The effect was assigned to a flow-induced phase transformation. In addition, waddon established the reversibility of the effect with temperature.

The sharp change with temperature seems to be the result of a thermodynamic phase change from the isotropic melt to a mesophasic hexagonal (*h*) phase (the mobile hexagonal phase of PE) and the associated structure formation [60]. As opposed to the normal orthorhombic crystal form of the PE, the hexagonal phase is metastable and hence cannot be observed under ambient conditions. At elevated pressures, it becomes stable and it was found to be mobile in molecular terms displaying features that are reminiscent of the liquid crystal state of polymeric matter, i.e. low viscosity. This results in the decrease of the pressure in extrusion, or an apparent decrease of the viscosity.

Figure 3.12b shows the flow curve of the LDPE within the temperature window of instability. Following an initial rising (normal) portion there is a sharp discontinuity: The stress drops suddenly at a critical value of $\dot{\gamma}$, leveling off to a lower plateau value before rising again steadily. The extent of the plateau region depends on the length of the capillary used to measure the viscosity: A longer capillary produces a more extended plateau region. Since the stress in the material cannot fall in absolute terms upon increasing shear rate (physically unacceptable), then this will lead to instability in the flow, which, in practical terms, will mean that the melt will start slipping at the walls or on an internal surface.

If the flow takes place at temperatures within the instability window, then the viscosity becomes a nonunique function of the strain rate (Fig. 3.12b). The phenomenon is important because it leads to instabilities in the processing

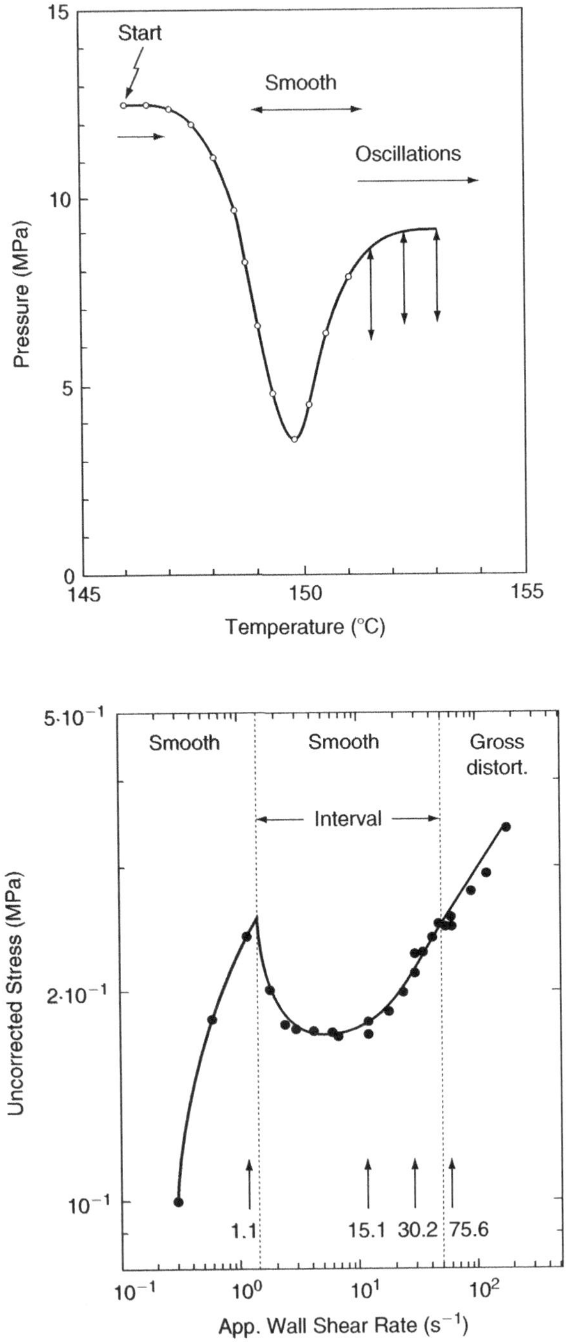


**FIGURE 3.12** (a) Pressure vs. temperature trace recorded during heating at a fixed extrusion rate showing the pressure minimum associated with an extrusion window. (b) Stress vs. strain rate curve at the temperature where the minimum of the curve in (a) is observed. Reprinted with permission from Ref. 60.

equipment but it also affects melt fracture. Extruding within the window gives smooth extrudates and lower viscosities. On the other hand, the instabilities usually manifest themselves as slip on the walls of the processing machinery. The sudden lower stresses may lead to failure in processing equipment where this is not accounted for. An example is the possible failure of the lubricating action of the melt in gear pumps, if they operate at these temperatures, with possible grave consequences (seizure of the axles).

Stress-induced crystallization may also lead to a sudden increase of the viscosity during extrusion, as it was observed by van der Vegt and Smit [61]. In this case, the crystallization was induced by the elongational flow in the contraction before the extrusion die and it was not a peculiarity of the polymer.

## 3.5 LONG-CHAIN BRANCHED POLYPROPYLENE

Commercial iPP is produced via Ziegler-Natta or metallocene catalysis. These processes lead to iPP with highly linear chains and relatively narrow molecular weight distributions. As a consequence, traditional commercial iPP has relatively low MS and, thus, rather poor processing characteristics in processes where the type of flow is predominantly elongational. The latter include foaming, thermoforming, extrusion coating, and blow molding. To use iPP in these forming processes, one needs to modify the polymer and enhance its MS. Broad or bimodal molecular weight distributions can result in some strain hardening and enhanced melt strength. The most efficient way to enhance the melt strength of polymers with linear chains, however, is by grafting LCB on the chain backbone. It has been known from practise that the situation improves significantly even in the presence of limited long-chain branching [9].

A lot of effort has been put in producing high melt strength PP by the polymer industry in the last years. Several commercial grades are available, mostly produced by grafting long-chain branches on iPP linear backbone, either by electron beam irradiation [62] or in the melt by using peroxides with relative low decomposition temperature [5]. These methods produce LCB-PP with broadened molecular weight distribution and complex branch structures. The direct synthesis of long-chain branched PP has also been tried. Metallocene catalysis has been reported, either directly [63] or via the addition of premade PP macromonomers [64]; conjugated diene monomers [65] have been used; metallocene-mediated polymerization of PP has been attempted in the presence of T-reagent *p*-(3-Butenyl)styrene [66, 67]. A good review of methods to synthesize LCB–PP is available [67]. Other methods involve the modification by peroxide reactions in the solid state in the presence or not of a co-agent [68, 69]. The free radical reactivity and the solubility of the reagents in iPP seem to be the most important factors for the efficiency of the modification reaction.

### 3.5.1 Shear Flow Rheology of Sparsely Branched PP

***3.5.1.1 Shear Stresses*** The zero-shear melt viscosities of branched polymers depart markedly from the 3.4-power dependence on molecular weight expected for linear polymers. This leads one to try to use the departure of the shear rheology from the norm as a possibly quantitative indication of the branched structure of the polymer. For example, Schreiber [70] proposed interpreting the ratio $(\eta_0)_{lin}/(\eta_0)_{br}$ simply as an index covariant with the amount of long-chain branching.[1]

Such an index could also be useful during, e.g., reactive modification of PP. Monitoring the degree of branching is essential in optimizing a long-branch grafting modification or, conversely, in preventing accidental cross-linking, which in other cases is an undesirable consequence of process-induced degradation. This can be done in principle by conventional GPC methods with considerable difficulty. Since even a modest reorganization of the molecular architecture can alter the viscoelastic behavior significantly [38], monitoring the rheology, instead, seems to be an appealing and inexpensive alternative. Branching progression, for example, translates into a dramatic increase in elasticity as manifested by changes in the zero shear recoverable compliance, $J_e^0$. This is difficult to monitor given the inability of experimentally reaching the necessary $G'(\omega)/\omega^2$ plateau in low-frequency dynamic mechanical measurements and the extreme sensitivity of $J_e^0$ to molecular weight polydispersity. Monitoring and interpreting the relatively modest, yet readily measurable increase of the zero shear rate viscosity that accompanies branching is a more appealing alternative.

In this part a theory is presented from [18] that correlates the behavior in simple shear flow with the microstructure of a melt that contains LCB. The theory assumes that linear polymer chains of initial (weight average) molecular mass $M_L$ have been broken through scission and some of their fragments have cross-linked with neighbors to form new molecules. The result is the description of the melt as a blend consisting of chains with the initial linear architecture and of branched chains.

For low degree of branching of less than one branching point per molecule on average, most of the linear fragments will end up as parts of larger branched polymers. A fraction of the initially linear chains, after breaking (on the average) in the middle, will cross-link with mostly unbroken chains. If cross-linking occurs not too far from a chain middle point, this results in structures resembling three-armed stars of average arm-molecular mass $M_a \approx M_L/2$ and, therefore, of total molecular mass $M_B = 3M_a \approx 3M_L/2$. The "blend" $(BL)$ consists of a fraction $(1-x)$ of linear $M_L$ chains and a fraction $x$ of $1.5M_L$ three-armed stars.

[1] $(\eta_0)_{br}$ denotes the Newtonian limiting melt viscosity observed at a given temperature for a sample of interest, and $(\eta_0)_{lin}$ is the corresponding viscosity for a linear polymer of equal molecular weight, after both have been corrected for small effects due to short branches

Estimates of the impact of the microstructure on the melt fluidity may be made based on fundamental precepts of molecular rheology of the entangled state. The molecular weight dependence of the viscosity of the linear ($L$) component is

$$\eta_L = \eta_C \left(\frac{M_L}{M_C}\right)^{3.5}, \qquad (3.41)$$

where $M_C \approx 2M_e$, with $M_e$ the molecular weight between two successive entanglements. For PP, $M_e$ is roughly equal to 5600. $\eta_C$ is the melt viscosity at the entanglement crossover, where $M_L = M_C$.

The zero-shear viscosity, $\eta_B$, and the relaxation time of a melt of entangled star-like molecules, vary (in essence) exponentially with the molecular weight of each of the protruding arms:

$$\frac{\eta_B}{\eta_C} = \sqrt{\frac{M_a}{M_e}} \exp\left\{\alpha\left(\frac{M_a}{M_e} - 1\right)\right\} \qquad (3.42)$$

Since $M_C \approx 2M_e$ and $M_a \approx M_L/2 \approx M_B/3$, this is equivalent to

$$\frac{\eta_B}{\eta_C} = \sqrt{\frac{M_L}{M_C}} \exp\left\{\alpha\left(\frac{M_L}{M_C} - 1\right)\right\} \qquad (3.43)$$

According to the molecular theory, the coefficient $\alpha$ is independent of the branching point functionality, and equal to 15/8; comparison with experiment, however, indicates a smaller value, $\alpha \approx 0.4316-0.6015$. This is most likely due to unaccounted constraint release interference, since the 15/8 value was derived for tethered chain relaxation in an environment of immobile obstacles, while in a melt that environment is quite fluid.

The viscosity of the $BL$ blend, $\eta_{BL}$, is assumed to be subject to logarithmic additivity; i.e., $\eta_{BL} \approx (\eta_B)^x (\eta_L)^{1-x}$:

$$\eta_{BL} \approx \eta_C \exp\left\{x\alpha\left[\frac{M_L}{M_C}\right]\right\}\left(\frac{M_L}{M_C}\right)^{3.5-3x} \qquad (3.44)$$

Since $\eta_{BL}/\eta_L \approx (\eta_B/\eta_L)^x$ and comparing (3.41) and (3.44) we can derive an expression for the degree of branching, $B_n = x$, from the measured viscosity ratio of the melt-strengthened ($BL$) over the initial ($L$) states

$$B_n = \frac{\ln\left\{\dfrac{\eta_{BL}}{\eta_L}\right\}}{\alpha\left[\left(\dfrac{M_L}{M_C}\right) - 1\right] - 3\ln\left\{\dfrac{M_L}{M_C}\right\}} \qquad (3.45)$$

If weight average molecular mass data for the initial linear polymer are not available, this may be approximately calculated, provided that the viscosity is known for another linear size, $L$, at the same temperature

$$M_L = M_{L'}\left(\frac{\eta_L}{\eta_{L'}}\right)^{1/3.5} = M_C\left(\frac{\eta_L}{\eta_C}\right)^{1/3.5}, \tag{3.46}$$

and (3.45) may be restated as follows

$$B_n = \frac{\ln\left\{\frac{\eta_{BL}}{\eta_L}\right\}}{\alpha\left[\left(\frac{\eta_L}{\eta_C}\right)^{1/3.5} - 1\right] - \frac{6}{7}\ln\left\{\frac{\eta_L}{\eta_C}\right\}} \tag{3.47}$$

Verification of the above equation has been given [18] by comparing with viscosity data obtained from LCB–PP samples made by modifying linear iPP using bulky peroxydicarbonates (PODIC) for radical initiation in reactive extrusion [5]. The degree of branching $B_n$ in the modified melt was controlled by increasing the added amount of ethyl-hexyl PODIC (Fig. 3.13a) or by varying the PODIC species used: PODICs with larger side groups generally result in more polypropylene branching (Fig. 3.13b). The zero shear viscosity was determined by measuring the dynamic shear viscosity and extrapolating to zero frequency using the Cross Viscosity model

$$\eta(\dot{\gamma}) = \frac{\eta_0}{1 + (\lambda\dot{\gamma})^n}. \tag{3.48}$$

Similar to LDPE, Figure 3.13 shows that the viscosity of LCB–PP increases with the degree of branching at low $\omega$, yet this increase disappears, or even reverses, at high rates of shear. Figure 3.14 compares the branching indices $B_n$ calculated via (3.45) with those measured using SEC by Lagendijk et al. [5], plotted versus the corresponding zero shear viscosity, $\eta_0$. Considering that for the given linear polypropylene precursor, $M_L$ was 410 kg/mol and $M_C$ = 11.2 kg/mol, the extracted value for parameter $\alpha$ was equal to 0.42. It is not surprising and it is consistent with other observations that this value is smaller than the theoretical, $\alpha = 15/8$. These results verify that measurable changes of the shear rheology upon introduction of long-chain branches on a linear PP can be used as a characterization tool for the molecular structure that results from this modification.

The method above works only for polymers whose branches are the same as their backbone (same chain flexibility), when there is no significant change of the molecular weight on modification, and for sparsely branched polymers (i.e. for $B_n < 1$). It should be noted here that, on increasing $B_n$ above 1, $\eta_0$ may go through a maximum and start decreasing again [37], something that has been seen for LDPE, which usually has a lower zero shear viscosity than HDPE with equivalent $M_w$, and for the highly branched commercial HMS–PP resins (e.g., Profax®).

To use this method to get an estimate of $B_n$ from the value of the zero shear viscosity, one needs the equivalent value of the linear polymer that has the same molecular weight. If the branched structure has different molecular weight, the application of this method is not straightforward. One way to circumvent the differences in molecular weight is to normalize $\eta_0$ by $M_w^{3.4}$ for the comparison with the linear [69], assuming that

$$B_n \propto \frac{\eta_0}{M_w^{3.4}}. \tag{3.49}$$

This, however, is sufficient only for qualitative comparisons of samples.

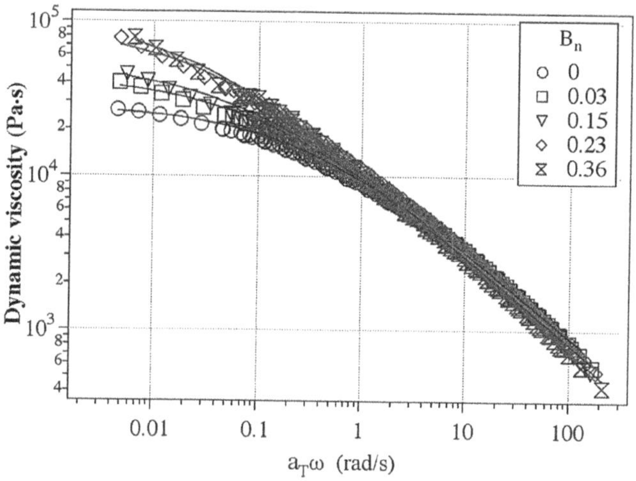


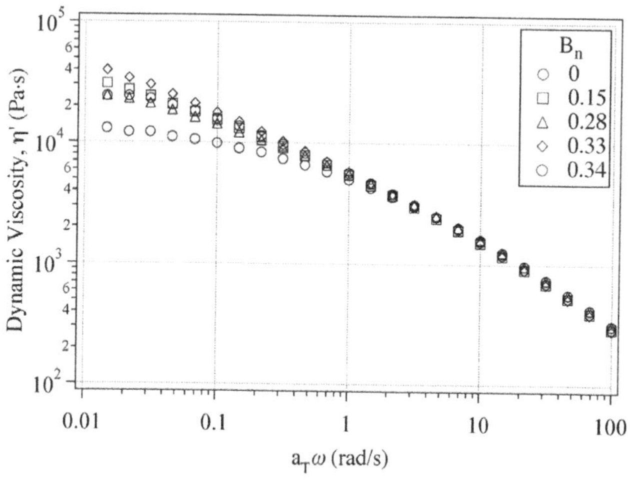


**FIGURE 3.13** Dynamic complex viscosity of a linear PP precursor and its branched derivatives, produced by using (a) increasing amounts of a peroxydicarbonate initiator or (b) the same amount of different peroxydicarbonates. Data from Ref. 18, shifted at 190 °C.

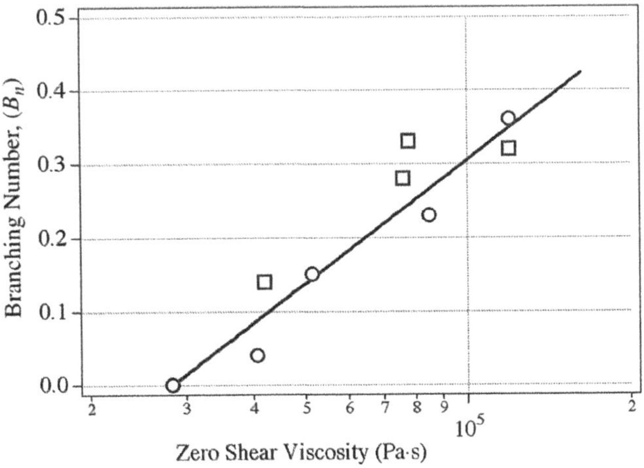


**FIGURE 3.14** Degree of branching vs. the zero shear viscosity of branched samples of PP. The symbols represent experimental results of data in Figure 3.13a (o) and Figure 3.13b (□). Data from Ref. 18. The line represents the predictions of (3.45) for $\alpha = 0.42$.

A more detailed method of estimating LCB by comparing the zero shear viscosity of the linear and the branched polymer has been proposed by Janzen and Colby [37]. This method works even when the molecular weight changes. An average molecular weight between branches, $M_b$, is introduced to characterize LCB and the zero shear viscosity is related to $M_w$ and $M_b$

$$\eta_0 = AM_w\left[1 + \left(\frac{M_b}{M_c}\right)^{2.4}\right]\left(\frac{M_w}{M_b}\right)^s \tag{3.50}$$

where $A$ is a parameter that can be evaluated if $\eta_0$ and $M_w$ of the linear polymer are known. The exponent $s$ is given for sparsely branched polymers by

$$s = \max\left[1, 1.5 + 1.125B\ln\left(\frac{M_b}{90M_{Kuhn}}\right)\right]. \tag{3.51}$$

$B$ is a constant. The Kuhn length, $M_{\mathrm{Kuhn}}$ for polypropylene is 187.8 g/mol [67].

For a branched polymer: $M_b < M_w$. The ratio $M_b/M_w$ should decrease as the degree of branching increases, because more of the molecular weight corresponds then to the parts of chain between the branch points or to the branches themselves. The method has been tested with LDPE and branched polyesters [37] and polypropylenes [67].

Still another attempt to estimate $B_n$ from shear rheology data was made by Shroff and Mavridis [71]. A long-chain branching index (LCBI) was defined for sparsely branched PEs.

$$LCBI = \frac{\eta_0^{1/a_3}}{[\eta]_B} \frac{1}{k_3^{1/a_3}} - 1, \tag{3.52}$$

where the parameters $k_3$ and $a_3$ are evaluated from the relation between the zero-shear viscosity and the intrinsic viscosity of linear polymers, $[\eta]_L$

$$\eta_0 = k_3 [\eta]_L^{a_3} \tag{3.53}$$

For a series of Exact® (Exxon) metallocene polyethylenes this becomes

$$LCBI = \frac{\eta_0^{0.179}}{[\eta]} \frac{1}{4.8} - 1.$$

For the rheological data of Raju et al. [27] the expression becomes

$$LCBI = \frac{\eta_0^{0.196}}{[\eta]} \frac{1}{4.7} - 1.$$

Other such indices include the Dow rheology index (DRI) based on the parameters $\eta_0$ and $\tau_0 = 1/\omega_0$ of the Cross model used to fit the dynamic viscosity

$$DRI = 0.1\left(3.65 \times 10^6 \frac{\tau_0}{\eta_0} - 1\right). \tag{3.54}$$

DRI is a measure of the rheological polydispersity, and as such, it depends on both MWD and LCB. It should be used as a measure of $B_n$ only when comparing samples with the same (narrow) MWD.

**3.5.1.2 Primary Normal Stress Difference** There are few reports on $N_1$ measured in cone-and-plate instruments. Yamaguchi and Wagner [19] have compared the melt of a HMS–PP Profax® (MFR = 3.0, produced by Himont) with that of a linear PP that had higher molecular weight and similar polydispersity index. The shear viscosity of the linear polymer is higher than the one of the branched polymer, as should be expected from the higher molecular weight of the corresponding sample. The primary normal stress difference of the HMS–PP, however, is significantly higher than the one of the linear PP, reflecting its higher melt elasticity. The stress ratio $(N_1/\tau_{12})$ of the branched PP in these samples was 4 times higher than the linear polymer, indicating that its elasticity at these shear rates was four times that of the linear polymer.

### 3.5.2 Dynamic Rheology

Figure 3.15 displays curves of the dynamic moduli ($G'(\omega)$ and $G''(\omega)$) for a linear iPP and several samples that were prepared by this iPP using peroxides and co-agents to graft long branches on the backbone [5, 69]. The molecular weight of the branched samples was not significantly higher than the one of the linear iPP.

The linear iPP shows the typical curves for the moduli of a viscoelastic melt with low elasticity. At low frequencies, $G'$ is lower than $G''$. The inverse of the crossover frequency ($\omega_{cr}$), where $G'$ crosses over $G''$ is the characteristic relaxation

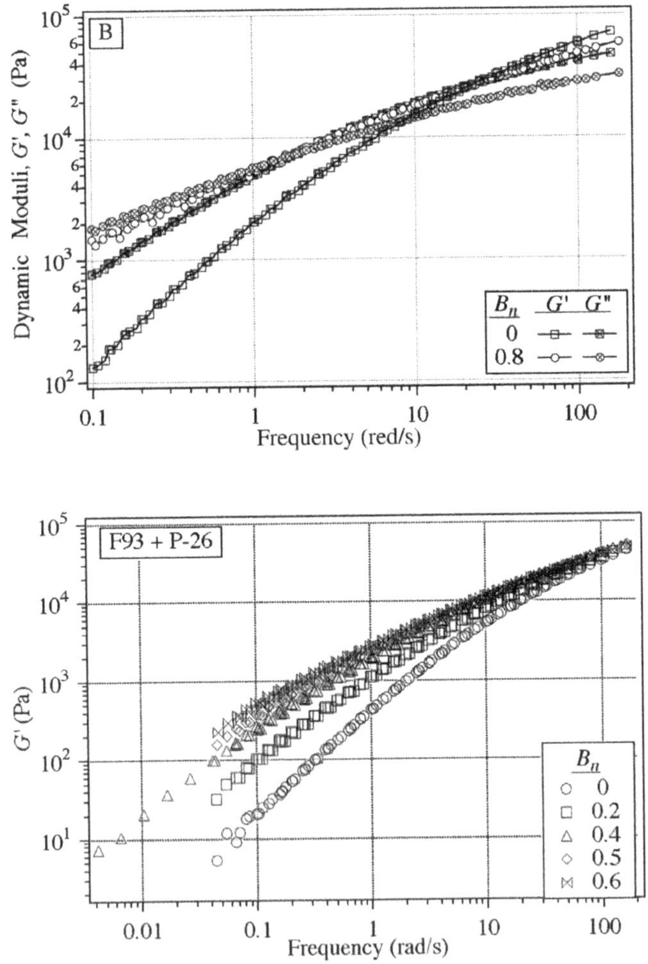


**FIGURE 3.15** Dynamic moduli of melts of linear iPP or this polymer modified with different amounts of PODIC to get LCB. The iPP precursors are: *B*, Borealis HC1000, *F93*, BP/Amoco, Fortilene F9300; and *P-26*, dimyristyl peroxydicarbonate used as initiator for the modification reactions. Reprinted with permission from Ref. 17.

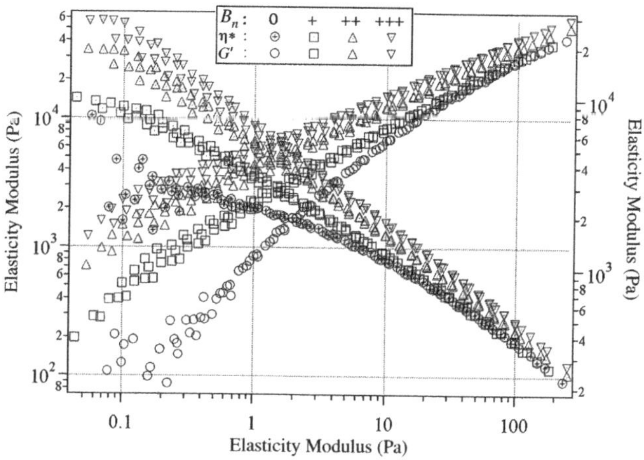


**FIGURE 3.16** Elasticity modulus, $G'(\omega)$, and dynamic complex viscosity, $\eta^*(\omega)$, curves for iPP modified to have increasing values of $B_n$ (shifted at 200°C). Reprinted with permission from Ref. 69.

time for the polymer, $t_{cr} = 0.02$ s. Similar to LDPE, however, the branched PP samples show higher values for $G'$ than the original iPP (Fig. 3.16). The values of $G'$ at low frequencies increase with $B_n$, while they are less affected at higher frequencies, where they approach an almost $B_n$-independent rubber plateau. This is because the high-frequency dynamic response reflects small segment molecular dynamics. The terminal region moves to lower frequencies with increasing $B_n$.

The loss modulus, $G''$, is affected less by the addition of branches on iPP. The value of the crossover frequency decreases with $B_n$ (or $t_{cr} = 1/\omega_{cr}$ increases). Longer $t_{cr}$ (lower $\omega_{cr}$) indicates higher elasticity and, indirectly, more branching. The branched polymers have more elastic melts than the equivalent linear iPP.

Figure 3.16 shows the complex viscosity vs. frequency curve for some modified iPP. The form of this curve changes in the branched samples and the shadow of two inflection points (around 0.5 and 15 rad/s) starts appearing in the logarithmic plots. As explained before, these are indications of increasing degree of branching.

The plot of loss angle $\delta(\omega)$ for branched iPP samples [9] is shown in Figure 3.17. The linear iPP displays a monotonically decreasing curve of $\delta(\omega)$. For the branched samples the curve changes in position and in form. There is an inflection in the curve of the loss angle, which hints to the development of a plateau at frequencies around 1 rad/s. The magnitude and breadth of this plateau should depend on $B_n$ [25]. This is also implied in Figure 3.17, where the inflection is stronger at higher LCB content. This plot, however, indicates only that there is an increase of the elasticity of the melt and no information can be extracted as to the type of the branching or, even, the origin of this enhanced elasticity. Quantitative conclusions from these curves, therefore, cannot be drawn.

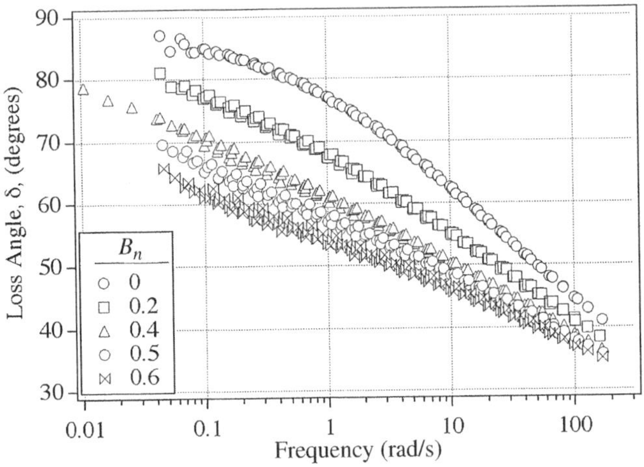


**FIGURE 3.17** Loss angle vs. frequency at 190°C for PP melts with various degrees of LCB. Samples as in Figure 3.15b. Reprinted with permission from Ref. 17.

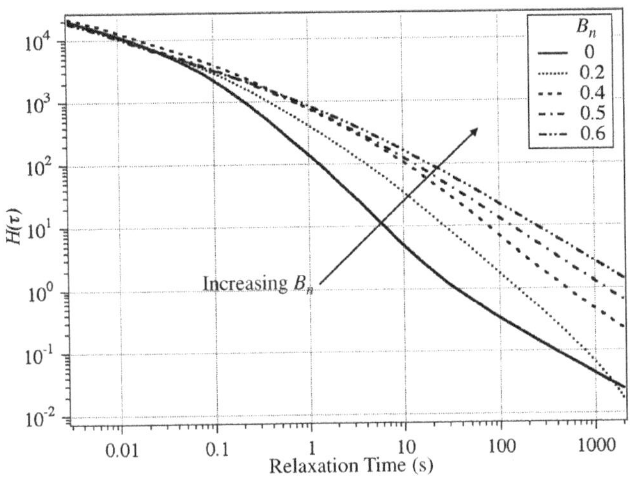


**FIGURE 3.18** Relaxation spectra of linear and branched PP melts and their dependence on the branching number, $B_n$. Data from Ref. 17.

The relaxation spectra for the linear polypropylene and samples with increasing values of $B_n$ produced by modifying the linear polymer are shown in Figure 3.18. Following the trends of the dynamic moduli, the spectra become broader as $B_n$ increases: $H(\tau)$ increases with $B_n$ at long relaxation times, while there is little change at short relaxation times. This is another indication that the melt elasticity increases with $B_n$.

While the activation energy for flow, $E_a$, increases with the degree of LCB ($B_n$) in polyethylenes [14], $E_a$ remains constant in linear and sparsely branched PP samples at around 40 kJ/mol, and there is no definitive trend of reduction or increase of this value with varying $B_n$ [5]. Therefore, when the degree of branching is small, as it is the case in sparsely branched polypropylenes, it has no perceptible influence on the activation energy of flow. The value of $E_a$ of the highly branched commercial Profax® polypropylene is somewhat higher. Blends of Profax® and linear iPP have been found to show an activation energy for flow with a value around that of pure Profax [5]. The study of the effect of branching on the activation energy of flow is complicated by the fact that branched PP is not a thermorheologically simple fluid.

### 3.5.3 Elongational Flow Rheology of LCB–PP Melts

The elongational viscosity of linear iPP melt are essentially nonstrain hardening. The extensional viscosity of the branched PP melts, on the other hand, initially show a simple monotonic growth as a function of the strain with a decreasing first derivative that compares well with the theory of linear viscoelasticity. Beyond a deformation of around 1 s.u., however, the curves show an increase in their slope as the viscosity growth accelerates; this is the region of strain hardening. At even higher strains the branched melts break more or less by elastic fracture. However, the force measured at those high strains is very low, and perhaps below the sensitivity of most rheometers.

The theories of Section 3.4.4 are now used for the description of $\eta_E^+$ of LCB-PP. The stretch in uniaxial elongational flow at constant strain rate is: $\lambda = \exp\{\varepsilon\} = \exp\{\dot{\varepsilon}t\}$. The damping function of (3.18) then becomes

$$h(\lambda) = \exp[-\beta\dot{\varepsilon}(t - t')]. \tag{3.55}$$

With this $h(\lambda)$ Lodge's Rubber-Like Liquid model with a damping function gives the following for the viscosity growth in uniaxial elongation

$$\begin{aligned}\eta_E^+ &= \frac{\sigma_{11} - \sigma_{22}}{\dot{\varepsilon}} \\ &= 1\dot{\varepsilon}\int_{-\infty}^{t} \mu(t-t')\exp\{-\beta\dot{\varepsilon}(t-t')\}\big(\exp(2\dot{\varepsilon}(t-t')) - \exp(-\dot{\varepsilon}(t-t'))\big)dt'.\end{aligned} \tag{3.56}$$

Values of $\beta > 2$ are not physically acceptable because they lead to a decreasing $\eta_E^+(\varepsilon)$ at high $\varepsilon$. For values in $0 \le \beta \le 2$ we have

$\beta = 0$: There is no damping and the original Lodge Rubber-Like Liquid model is recovered. Strain hardening is then predicted for strain rates higher than the inverse of the longest relaxation time used in $\mu$ and no steady state.

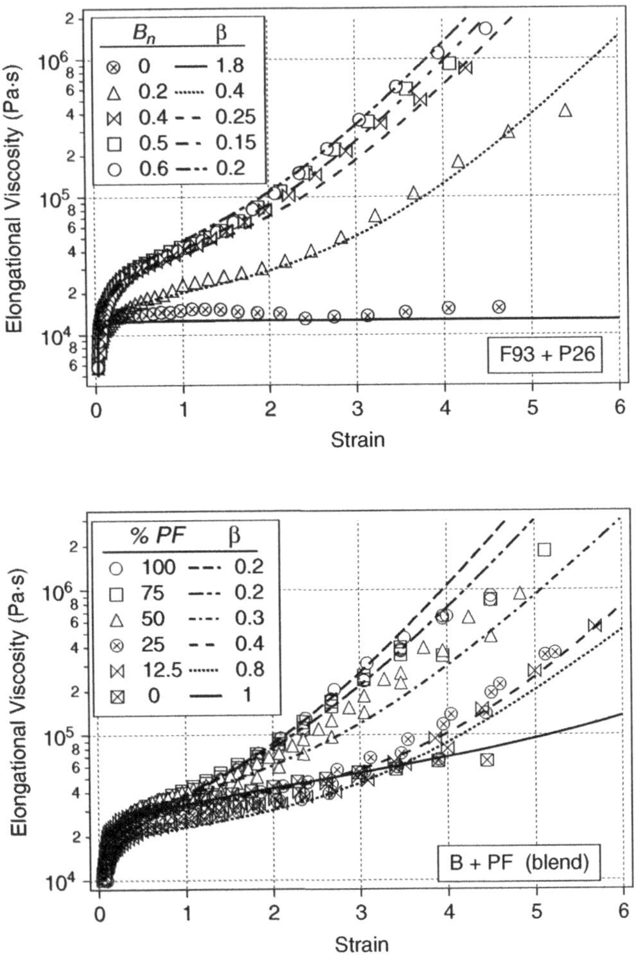


**FIGURE 3.19** Elongational viscosity growth at extension rate around 0.1 $s^{-1}$ and at 190°C for samples (a) resulting from peroxide modification of a linear iPP to add controlled degrees of branching (as in Fig. 3.15), and (b) blends of a branched (*PF*, Profax-PF-814 by Basell) and a linear iPP. The fits of (3.56) and the corresponding values of the strain hardening parameter $\beta$ are also shown. Reprinted with permission from Ref. 17.

The slope of the $\eta_E^+(\varepsilon)$ curve at high strains is 2 in a semi-log plot, similar to the one characterizing the stretching of a generalized-Hookean spring.

$\beta = 2$: There is complete damping and no strain hardening. The elongational viscosity does not accelerate; it merely increases monotonically and reaches a steady state value asymptotically.

$0 < \beta < 2$: The degree of damping increases as $\beta$ decreases (Fig. 3.19).

The fit of this model for some sparsely branched PP melts is shown in Figure 3.19. The relaxation data used in $\mu(\tau)$ were evaluated from the dynamic moduli measurements of each melt [17]. It can be seen in Figure 3.19 that the model of Lodge with this damping function (3.55) and (3.56) does a relative good job in describing the growth of the elongational viscosity of the linear and the branched samples, thus proving to be adequate to model strain hardening for sparsely branched PP melts. The strain sensitivity exponent $\beta$ of the damping function seems to decrease when $B_n$ increases. There is, therefore, a relation between the branching number and this parameter. At the highest values of $B_n$, $\beta$ comes close to zero and we recover the original model of Lodge.

Figure 3.19 also shows the elongational viscosity growth and the corresponding fits of (3.56) for the melts of blends of a linear iPP and a commercial branched HMS–PP (produced by electron beam irradiation) [9]. Even though the molecular weight, its distribution, and the degree of branching change among the samples in this case, there is still a monotonic increase of the degree of strain hardening with the amount of branches present; this translates to a corresponding decrease of $\beta$ from a value of 1 for the pure linear iPP to a value of 0.2 for the blends containing 75% HMS–PP or higher.

Examining now the prediction of the molecular stress function theory for PP we see that the linear PP melt does not show any strain hardening in the range of strains that can be achieved, and the value of $\beta'$ in (3.34) and (3.35) is around 2. As $B_n$ increases, however, the value of $\beta'$ decreases abruptly, and for branching numbers $> 0.5$ the appropriate value of $\beta'$ is 1/2. Because of this abrupt change with $B_n$, the parameter $\beta'$ is not a convenient measure of branching in polypropylene.

**3.5.3.1 *The Estimation of $B_n$ From Elongational Rheology Data*** The best quantitative way to estimate $B_n$ is by directly measuring this strain hardening. The whole $\eta_E^+(\varepsilon)$ curves can be fitted by viscoelastic models that implement a damping function and the estimated model parameters are used to infer the value of $B_n$.

The application of this method for the model of (3.55) is shown in Figure 3.20 for PP samples of known $B_n$. This figure shows that the parameter $\beta$ decreases monotonically with the increase of $B_n$. For $B_n \leq 0.8$, therefore, one can use the value of $\beta$ as a measure of the degree of branching. An empirical relationship between $\beta$ and the degree of branching, $B_n$, which describes the systematic data of Gotsis et al. [17] shown in Figure 3.20 is

$$\beta_u \approx 2\exp(-a\sqrt[3]{B_n}) \tag{3.57}$$

with $a = 2.9$. For a series of blends of commercially branched PP and a linear PP prepared so that $0 \leq B_n \leq 5$, it was found that $a = 1.5$. This indicates that, $a$ is not a universal constant but depends on the MWD breadth and the details of branching. Furthermore, if one would extrapolate the data in Figure 3.20, then

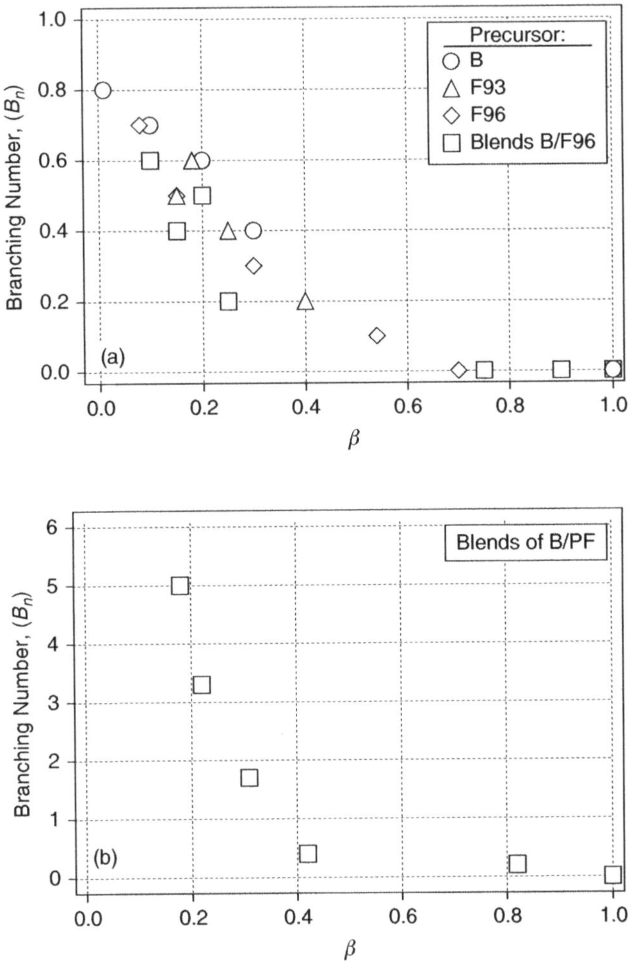


**FIGURE 3.20** Branching number, $B_n$, as a function of the strain sensitivity parameter $\beta$ of the damping function (3.55). (a) Data from P26 peroxy modified linear iPP precursors as in Figure 3.15. *F96*, BP/Amoco Fortilene 96000. The modified blends of B/F96 differ because they have both a broader MWD and branches. (b) Data from blends of linear and commercially branched chains. Reprinted with permission from Ref. 17.

all melts that have >0.8 long chain branches per molecule would be able to fit the original Lodge model without the need of a damping function ($\beta = 0$) and would behave as generalized-Hookean solids even at large deformations. That is, the method is not selective enough for large values of the degree of Branching.

A simpler way to characterize the strain hardening of a melt is to compare the value of the elongational viscosity at two different strains: one before the point where the upward change of the slope of the viscosity growth curve takes place

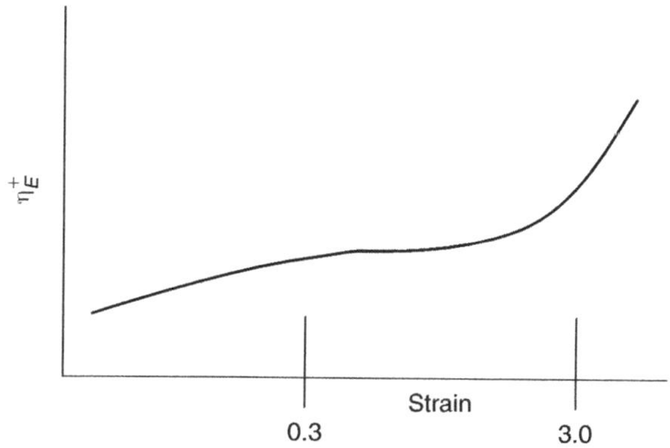


**FIGURE 3.21** A strain hardening elongational viscosity growth curve as it has been seen for long-chain branched PP, and the definition of the strain hardening index. This is a log-log plot. Notice that steady-state elongational viscosity is not always reached for fluids of this type before they break. Reprinted with permission from Ref. 9.

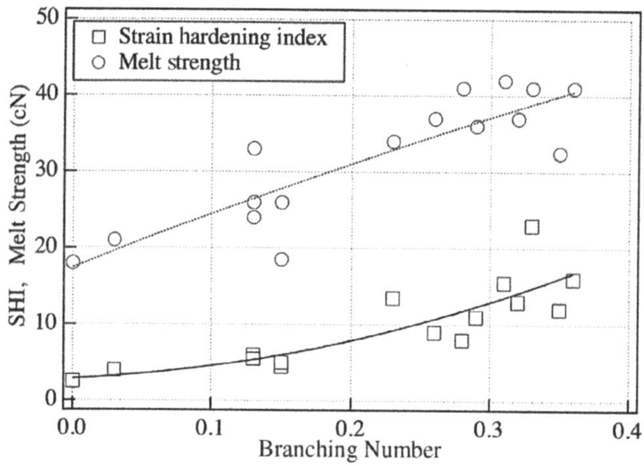


**FIGURE 3.22** The strain hardening index and the melt strength measured at 190°C for branched iPP samples as a function of the number of branches. Reprinted with permission from Ref. 9.

and one after that point. Gotsis et al. [17] chose the strains of 0.3 and 3 s.u. (Fig. 3.21) for these. The ratio of the values of the elongational viscosity at these two strains is defined as the strain hardening index (SHI):

$$SHI = \frac{\sigma_E(\varepsilon = 3)}{\sigma_E(\varepsilon = 0.3)}$$

The relation of SHI and MS with the branching number, $B_n$, is shown in Figure 3.22 for some branched PP samples [5]. When the chain has more branches the melt becomes more strain hardening and its strength increases.

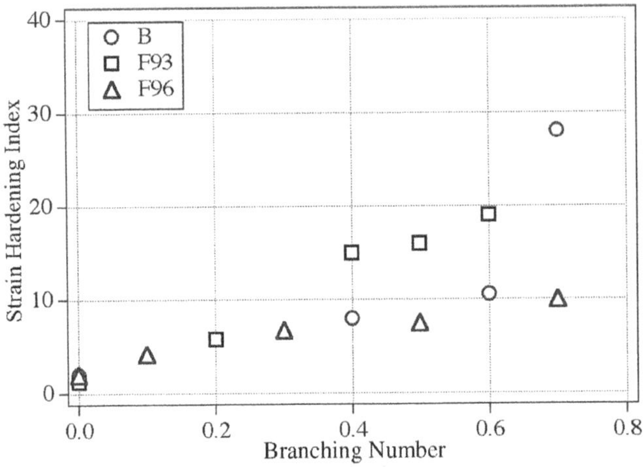


**FIGURE 3.23** Strain hardening vs. the branching number of samples prepared using different precursors. Data from Ref. 9.

There is, therefore, a close relation between the processability of PP in thermoforming, the strain hardening behavior of its melt in uniaxial elongational flow and the number of long-chain branches on its chain.

An application of this method is shown in Figure 3.23 for several (branched) PP samples resulting from modification of three different linear precursors (polypropylene resins B, F96, and F93 [17]). Even though the modification results in approximately the same number of branches per chain on the two different precursors, F93 and F96, the effect of these branches on the strain hardening of the melts is different. The samples resulting from precursor F96 show consistently lower values for SHI than the ones resulting from F93, which has lower molecular weight than F96.

Obviously the relations between $B_n$ and MS or SHI are not so simple or absolute. Nor can they be generalized easily to all polymers. As discussed already, the melt strength and the degree of strain hardening of the elongational viscosity also depend on other parameters, such as the molecular weight and the molecular weight distribution of the modified polymer, the length of the branches, the distribution of these lengths, the tacticity of the branches and the distribution of the branches on main chains with different lengths [72]. For this reason no attempt should be made to fit the curves in Figure 3.22 by an empirical equation. To fully understand the effect of branches on the rheology and the processing properties of the PP melts, systematic measurements are needed on samples with well characterised molecular structure, that will include variations on all mentioned parameters.

### 3.5.4 Shear Modification of the Branched Structure

When forming processes involving high shearing flows are applied to the melts of branched polymers, then the branched structure may be affected. This is known in LDPE as shear modification and results in a reduction of the melt elasticity [73–75]. The phenomenon has also been observed in branched polypropylene and found to be stronger than for LDPE, with the elasticity slower to recover by annealing [19].

Shearing of highly branched PP (HMS–PP) in a batch mixer at moderate shear rates for 4 minutes may result in 25% reduction of the melt strength. This decrease is reversible by annealing: It may take 1 h at 190°C to recover to the unsheared value. The reduction of the melt elasticity depends on the accumulated shear strain. The times needed for the recovery by annealing depend also on the accumulated strain.

Since the damage is not irreversible, it seems that the process of shear modification corresponds to changes in the conformation of the branches around the backbone. Münstedt [76] suggested that the long-chain branches get aligned along the backbone and thus offer less chance for entanglements. In the formalism of the reptation theory, this means that the backbone chain drags the branches into its tube [77] during shear. Upon annealing, the branches come out of the backbone tube, i.e., they dealign, to increase their entropy. The recovery process is slower for the branched PPs (relative to LDPE) because most of the branches of the HMS–PP are very long, while the (greater in number) branches in LDPE have a wider length distribution, the shorter ones relaxing much faster than the longer ones [19].

The practical conclusion of all this is that the engineer should account for possible reduction in the melt elasticity in the processing steps preceding the forming step where this elasticity is desired. High preshearing action should be avoided if possible. Therefore, it is advisable, e.g., that a starved extruder and a gear pump be used to feed the film die, or a twin screw extruder be used for foaming, etc.

## 3.6 PROCESSABILITY AND LCB

When linear polymers are modified to obtain a few long-chain branches, the molecular weight of the polymers increases only slightly but the shear viscosity of their melts can change considerably [9]. That is, their MFI decreases sharply. This increase of the shear viscosity seems to be related with the change of the Melt Strength, as it is shown in Figure 3.24. However, the increase of the force measured in the Rheotens experiment for these samples is so large that it cannot be explained by the increase of the shear viscosity that results either from the slight increase of the average molecular weight, or from the addition of LCB [9]. The strong enhancement of the MS with long-chain branching is mainly due to the increase of the strain hardening behavior of the melt elongational viscosity.

The combination of a high melt strength and a low shear viscosity would be a desired aspect for thermoforming processing. The ratio of the melt strength of the polymer over its shear viscosity expresses this combination and can be used to normalize and quantify the improvement of the thermoforming

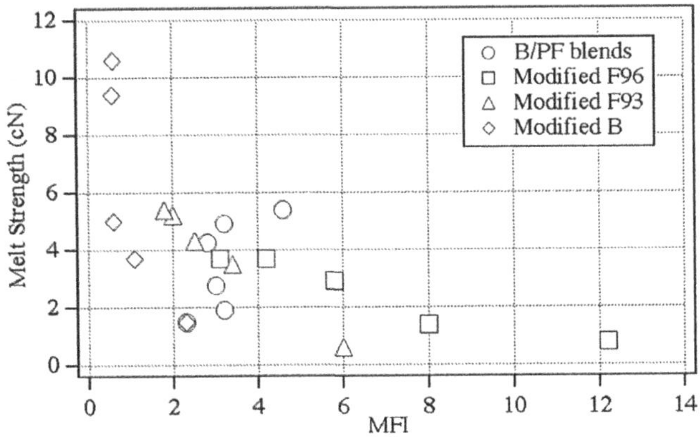


**FIGURE 3.24** Melt strength vs. MFI for linear and branched melts as in Figure 3.20. Reprinted with permission from Ref. [9].

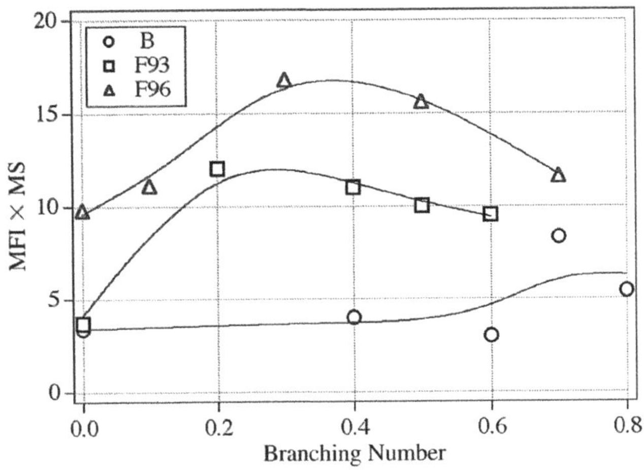


**FIGURE 3.25** Normalized melt strength (PITA) as a function of the branching number for the samples described in Figure 3.20. Data from Ref. [9].

processing properties of PP due to LCB. Because of the dependence of the (non-Newtonian) viscosity of the thermoplastics on shear rate, a convenient standard measure of the viscosity for comparisons in engineering processes is the inverse of the value of the MFI. The product of MFI and melt strength, then, can be used as an index for the processability of the PP melt in thermoforming applications. A higher value of this processability index for thermoforming applications (PITA) indicates a melt with lower viscosity and higher melt strength. This index is shown in Figure 3.25 for branched samples resulting from the modification of three linear PP precursors [9]. The figure shows that there may be a maximum in the value of PITA within a range of branching

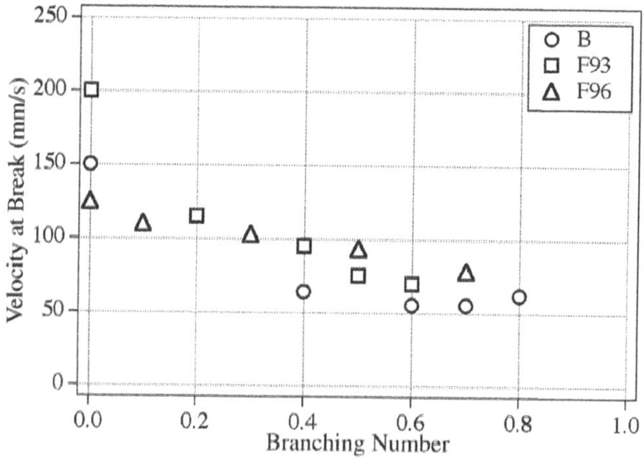


**FIGURE 3.26** Velocity at break at the Rheotens experiment as a function of the branching number for the samples described in Figure 3.20. Data from Ref. 9.

numbers. The position of this maximum depends on the molecular characteristics of the precursor polymer.

The maximum velocity that the molten thread can be drawn before it breaks at a Rheotens instrument is shown in Figure 3.26. The velocity at break decreases in general with the amount of added branches. This effect decreases when the MW of the precursor decreases or its MWD becomes broader [17]. The changes of the velocity at break that accompany the modifications could partly explain the form of the curves in Figure 3.25. It seems that the samples with too many branches break too early in the melt drawing experiment, before they can reach their potential for high elongational stresses that their branched structure would suggest. The maxima in Figure 3.25, therefore, correspond to the optimum combinations of enhanced elongational viscosity and reasonable melt ductility.

One way to evaluate the relative effectiveness of the methods used to modify linear chains and add LCB is to divide the normalized melt strength shown in Figure 3.25 by the normalized melt strength of the equivalent linear polymer. This also accounts for the differences in the melt properties of the precursors. This is illustrated in Figure 3.27. The molecular weight and the MWD of the precursor polymer affects the obtained improvement of the processing properties, at least in the case when peroxides are used to induce the addition of the branches. For example, Figure 3.27 indicates that the best choice from the three linear precursors for this modification method is the low viscosity/narrow molecular weight F-93. To obtain the optimum thermoforming properties, this should be modified using 1 to 2 mmol dimyristyl PODIC/100 g PP [9].

In another example, the processability index is used to compare the efficiency of blends, made from linear iPP and commercial highly branched PP (Figure 3.28), with the modified linear PPs above. In terms of performance in

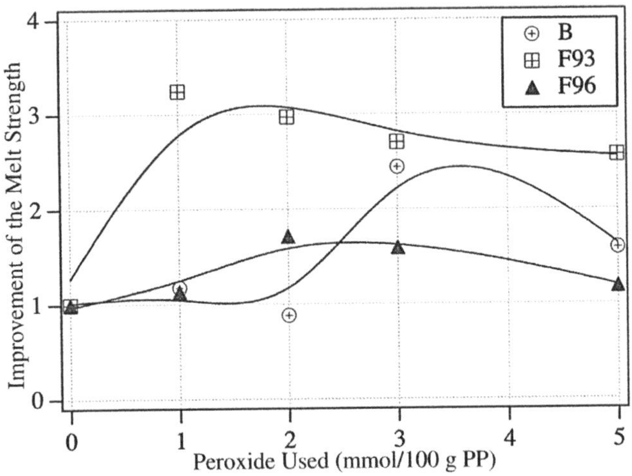


**FIGURE 3.27** Effectiveness of the modification for the improvement of the thermoforming properties of PP samples resulting from three different linear precursors. Data from Ref. 9

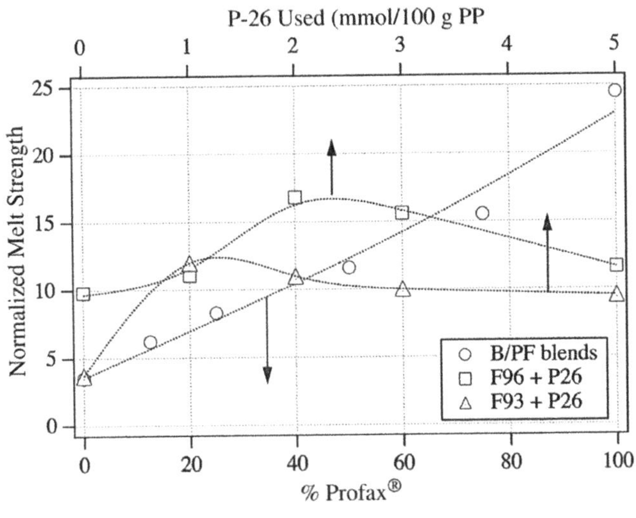


**FIGURE 3.28** Comparison of the normalized melt strength of blends of linear and branched polymers with that of the modified linear polymers. Data from Ref. 9.

thermoforming processing, a blend of 65% Profax® HMS-PP in linear iPP (Borealis-HC1000) will be equivalent to the product of the modification of F96 iPP using 3.2 mmol Perkadox-26 peroxide per 100 g PP. A blend of 42% Profax in linear iPP will be equivalent to F93 iPP modified with 2.1 mmol peroxide/100 g PP. At lower fractions of HMS–PP the combination of the values of the viscosity and the MS of the blends is worse than the one of the peroxide modified polypropylenes. Above 42% and 65% HMS–PP, respectively, the blends have superior thermoforming processing properties in comparison to the modified linear polymers. So what is the answer to the practical question, "Should we modify a linear PP or should we buy a highly branched PP to blend with a linear iPP to use it in a thermoforming process"? To answer this question, one also has to keep in mind that the cost of commercial HMS–PP is currently higher than that of modifying a commercial linear iPP.

When the normalized melt strength is compared with the values of *SHI* for the several PP samples (Fig. 3.23), then it is obvious that the strain hardening of the elongational viscosity enhances PITA. To be able to use PP efficiently in thermoforming or foaming processes, where high melt strength is required, strain hardening melts are needed. The relation between the strain sensitivity parameter $\beta$ and $B_n$ could, in principle, give an indication of how many branches would be needed for a certain degree of strain hardening and the corresponding PITA enhancement.

The fast drop of $\beta$ with the degree of branching means that one does not always need a very high degree of modification in order to achieve optimum melt strength behavior. The optimum value for the relative melt strength may be achieved at rather low values of branching, where $\beta$ has already reached a value close to 0. Very high branching levels, on the other hand, could decrease the maximum extensibility of the melt during stretching. The result could be that the thermoforming processing properties of the melt and its ability to produce high quality foam may show an optimum at a certain degree of branching as we will see in a following section.

### 3.6.1 Application in the Production of PP Foam

One of the major reasons that strain hardening and enhanced melt strength are desired for common PP is the possibility of using this polymer in the production of high-quality foam. The PP that is used now for foaming is mainly HMS–PP. Linear PP cannot produce foam with low-density and uniform cell structure because of its low melt strength. We saw that LCB enhances the MS of PP, but an optimum for the value of the degree of branching may exist, beyond which the quality of foam is reduced.

An example of the applicability of iPP modified for improved MS in the manufacturing of PP foam has been presented by Gotsis et al. [9]. A linear PP precursor and its branched samples were used and the foaming was conducted in a laboratory twin-screw extruder with nitrogen gas injected in the barrel at a point where the material had already melted. The foam expanded at the cylindrical exit die of the extruder and was collected by a conveyor belt. Details on the process can be found in the work of van de Wetering [78]. The resulting foam samples were compared in terms of foam density and cell size distribution. The conditions were not optimized since the research did not focus on the study of the foaming process but on the differences in the foaming behavior of samples with different degrees of long-chain branching.

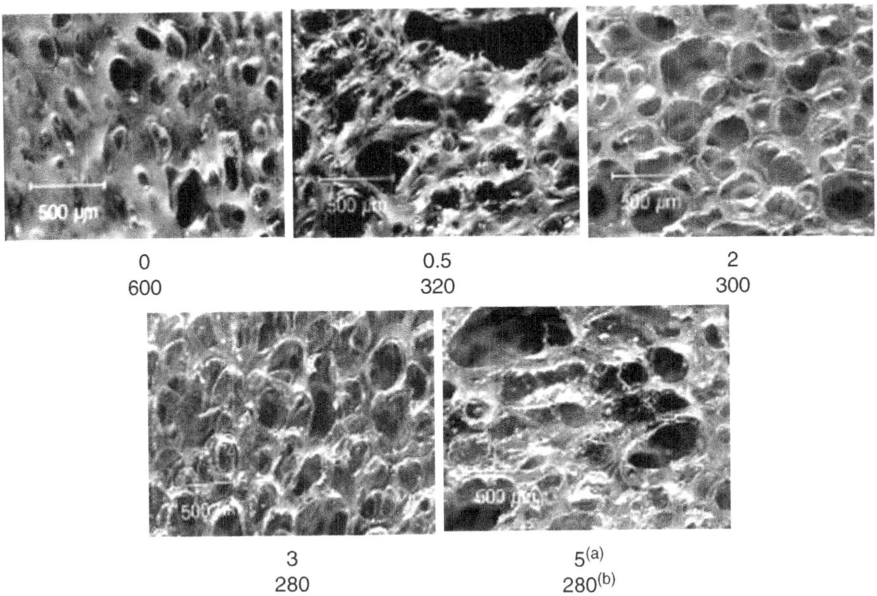


**FIGURE 3.29** Scanning electron micrographs of the foam that was made using (initially linear) PP modified by different amounts of peroxide: (a) in mmol peroxide/100 g PP; (b) the density of the produced foam in kg/m$^3$. Reprinted with permission from Ref. 9.

The density of PP at room temperature is around 900 kg/m$^3$. A foamed PP should have at most half this density, preferably much less. When the unmodified linear PP is used to make foam, a density of 600 kg/m$^3$ is often obtained, and thus linear PP is considered not foamable under the processing conditions described here. However, when the linear iPP was modified to get sparse LCB, then the resulting foam density fell to around 300 kg/m$^3$ [78]. Recent reports by Borealis® bring the density of foam produced by branched PP down to 45 kg/m$^3$!

Except the foam density, other important parameters for the quality of the produced foam are the cell size and the cell size distribution. The scanning electron micrographs in Figure 3.29 show the differences in the appearance of the foam cells that result by using PP with different levels of branching. It can be argued that the fourth case, the polymer with $B_n = 0.3$ shows the best picture from the lot. This is verified in Figure 3.29, which shows the cell size distribution in the foams. One can see here that the sample with $B_n = 0.3$ shows a minimum in the average cell size and the narrowest cell size distribution.

Figure 3.30 indicates that very high degrees of branching may not be necessary for the optimization of the material properties that control the easiness and quality of foaming or, in general, thermoforming processing. It seems that there is an optimum at an intermediate level of modification. As it was indicated above, a possible reason for this is that long-chain branching increases

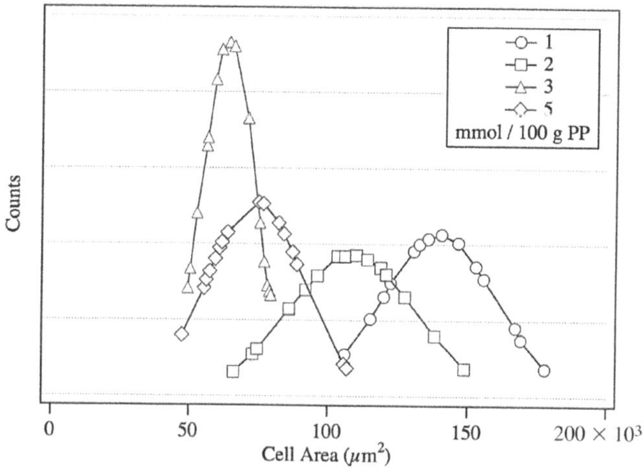


**FIGURE 3.30** Cell size distributions in the foams shown in Figure 3.29. Reprinted with permission from Ref. 9.

the melt strength and strain hardening but it may also decrease the strain at break of the melt. A good compromise between these properties should give the optimum.

### 3.6.2 Some Conclusions on the Rheology of LCB Polyolefins

Most commercial polymerization processes for iPP produce linear chains. If desired, LCB is added afterward to the linear precursor. The degree of LCB that is added on PP is usually much lower than that of LDPE. However, even low values of $B_n$ affect the rheology of the PP melt considerably. The branches are long (of the same order of magnitude as the main chain) and attach preferentially on the longer chains in the distribution. This results in a greater probability of entanglements and higher melt elasticity relative to polyethylene with similar degree of branching.

The zero shear viscosity increases with $B_n$ and the shear thinning becomes more severe. Most available reports do not show high enough values for $B_n$ in PP samples, such that the zero shear viscosity goes over the predicted maximum in a $\eta_0(B_n)$ curve and starts decreasing.

The elasticity of the melt also increases with $B_n$. This increase is reflected on the change of the relaxation spectrum, which becomes flatter at long relaxation times. This results in higher values of the storage modulus and lower values of the loss angle: The $\delta(\omega)$ curve obtains two deflection points and shifts to lower values as $B_n$ increases. The loss modulus also increases slightly reflecting the small change of the shear viscosity.

The most spectacular change of the rheology of the PP melt on addition of long-chain branches occurs during elongational flow. Then the elongational viscosity of the melt becomes strain hardening. *SHI*, an index that characterizes the degree of this strain hardening, increases with the increase of the number of branches per molecule.

The main reason for the addition of LCB on PP is the enhancement of the melt strength. The MS increases strongly with $B_n$, as it is related to the degree of strain hardening. Even when divided by the viscosity, the ratio, which expresses the processability of the melt in thermoforming processes, relates well with $B_n$ and the quality of the product, e.g. the foam produced from branched samples.

Some long-chain branching, therefore is beneficial for the processes that rely on elongational flows. However, an optimum in the degree of branching may exist for the application of the polymer in specific processes. A very large number of branches per chain may reduce the processability of PP because the branches may reduce the strain at break of the melt.

## 3.7 ACKNOWLEDGMENTS

Sincere thanks to Professor Christos Tsenoglou for his critical review of this chapter. A large part of what has been described here is based on his theoretical work on the rheology of branched polymers.

## REFERENCES

1. M. Gahleitner, *Prog. Polym. Sci.*, 26, 895–944 (2001).
2. C.B. Gell, W.W. Graessley, V. Efstratiadis, et al., *J. Polym. Sci. B: Polym. Phys.*, 35, 1943–1954 (1997).
3. A.D. Gotsis and S.F. Ke, *SPE ANTEC Tech. Papers*, 45, 1156–1161 (1999).
4. A. Ghijsels, J.J.S.M Ente, and J. Raadsen, *Intern. Polym. Process.*, 5, 284–286 (1990).
5. R.P. Lagendijk, A.H. Hogt, A. Buijtenhuijs, and A.D. Gotsis, *Polymer*, 42 (25),10035–10043 (2001).
6. A. Ghijsels and J. De Clippelier, *Intern. Polym. Process.*, 9, 252–257 (1994).
7. A. Ghijsels, C.H.C. Massardier, and R.M. Bradley, *Intern. Polym. Process.*, 12, 147–154 (1997).
8. V.V. De Maio and D. Dong, *SPE ANTEC Tech. Papers*, 43, 1512–1516 (1997).
9. A.D. Gotsis, B.L.F. Zeevenhoven, and A.H. Hogt, *Polym. Eng. Sci.*, 44, 973–982 (2004).
10. F.N. Cogswell, *Polymer Melt Rheology, Woodhead Publishing*, Cambridge, UK 1981.
11. T. Takahashi, J. Takimoto, and K. Koyama, *J. Appl. Polym. Sci.*, 72, 961–969 (1999).
12. L. Berger and J Meissner, *Rheol. Acta*, 31, 63–74 (1992).
13. J. Vega, M. Aguilar, J. Peon, et al., *e-Polymers*, 46, 1–35 (2002).
14. J.F. Vega, A. Santamaria, A. Muñoz-Escalera, and P. Lafuente, *Macromolecules*, 31, 3639–3648 (1998).

15. D. Yan, W.J. Wang, and S. Zhu, *Polymer*, 40, 1737–1744 (1999).
16. L.J. Kasehagen and C.W. Macosko, *J. Rheol.*, 42, 1303–132 (1998).
17. A.D. Gotsis, B.L.F. Zeevenhoven, and C. Tsenoglou, *J. Rheol*, 48, 895–914 (2004).
18. C. Tsenoglou and A.D. Gotsis, *Macromolecules*, 34, 4685–4687 (2001).
19. M. Yamaguchi and M.H. Wagner, *Polymer*, 47, 3629–3635 (2006).
20. J.M. Dealy and K.F. Wissbrun, *Melt Rheology and Its Role in Plastic Processing* Van Nostrand Reinhold, New York 1990.
21. S.E. Bin Wadud and D.G. Baird, *J. Rheol.*, 44, 1151–1167 (2000).
22. C. Gabriel and H. Münstedt, *J. Rheol.*, 47, 619–630 (2003).
23. R. Hingmann and B.L. Marczinke, *J. Rheol.*, 38, 573–587 (1994).
24. S. Kurzbeck, F. Öster, H. Münstedt, et al., *J. Rheol.*, 43, 359–373 (1999).
25. P.M. Wood-Adams, J.M. Dealy, A.W. deGroot, and O.D. Redwine, *Macromolecules*, 33, 7489–7499 (2000).
26. A. Muñoz-Escalona, P. Lafuente, J.F. Vega, and A. Santamaría, *Polym. Eng. Sci.*, 39, 2292–2303 (1999).
27. V.R. Raju, G.G. Smith, G. Martin, et al., *J. Polym. Sci. B: Polym. Phys.*, 17, 1183 (1979).
28. S.T. Milner, *J. Rheol.*, 40, 303 (1996).
29. Paula Wood-Adams and Stèphane Costeux, *Macromolecules*, 34, 6281–6290 (2001).
30. Helmut Münstedt, Stefan Kurzbec, and Lothar Egersdörfer, *Rheol. Acta*, 37, 21–27 (1998).
31. C. Gabriel and H. Münstedt, *Rheol. Acta*, 41, 232–244 (2002).
32. G.D. Wignall, R.G. Alamo, J.D. Londono, et al., *Macromolecules*, 29, 5332–5335 (1996).
33. M.J. Hill, P.J. Barham, and J. van Ruiten, *Polymer*, 34(14), 2975–2980 (1993).
34. M.J. Hill, R.L Morgan, and P.J. Barham. *Polymer*, 38(12), 3003–3009 (1997).
35. A. Santamaria, *Mat. Chem. Phys.*, 12, 1–28 (1985).
36. P.M. Wood-Adams and J.M. Dealy, *Macromolecules*, 33, 7481–7488 (2000).
37. J. Janzen and R.H. Colby, *J.Mol.Struct.*, 485/486, 569–584 (1999).
38. W.W. Graessley, *Acc. Chem. Res.*, 10, 332–339 (1977).
39. Cèsar A. García-Franco, Srivatsan Srinivas, et al., *Polymer*, 38(12), 3003–3009 (1997).
40. H.H. Winter and F. Chambon, *J. Rheol.*, 30, 367(1986).
41. R.B. Bird, R.C. Armstrong, and O. Hassager, *Dynamics of Polymer Liquids*, vol. 1, John Willey & Sons, New York 1977.
42. I.J. Chen and D.C. Bogue, *Trans. Soc. Rheol.*, 16, 59–78 (1972).
43. Florian J. Stadler, Joachim Kaschta, and Helmut Münstedt, *Macromolecules*, 41, 1328–1333 (2007).
44. W.W. Graessley, *Macromolecules*, 15, 1164–1167 (1982).
45. N.J Inkson, T.C.B. McLeish, O.G. Harlen, and D.J. Groves, *J. Rheol.*, 43, 873–896 (1999).
46. M.H. Wagner, H. Bastian, P. Hachmann, et al., *Rheol. Acta*, 39, 97–109 (2000).

47. C.J. Tsenoglou, E. Vogiatzis, and A.D. Gotsis, *J. Non-Newt. Fluid Mech.*, 138, 33–43 (2006).
48. A.S. Lodge, *Elastic Liquids, Academic Press*, London 1964.
49. M.H. Wagner, *Rheol. Acta*, 15, 135–142 (1976).
50. R.G. Larson, *Constitutive Equations for Polymer Melts and Solutions,* Butterworth-Heinemann, Boston 1988.
51. A. Nishioka, T. Takahashi, Y. Masubuchi, et al. *J. Non-Newt. Fluid Mech*, 89, 287 (2000).
52. T.C.B. McLeish and R.G. Larson, *J. Rheol.*, 42, 81–112 (1998).
53. M. Doi and S.F. Edwards, *The Theory of Polymer Dynamics*, Oxford Univ. Press, Oxford, UK 1988.
54. M. Doi and S.F. Edwards, *The Theory of Polymer Dynamics*, Claventis Press, Oxford 1988.
55. M.H. Wagner and J. Schaeffer, *Rheol. Acta*, 33, 506–516 (1994).
56. D.M. Binding, *J. Non. Newt. Fluid Mech.*, 27, 173–189 (1988).
57. D.M. Binding, *J. Non. Newt. Fluid Mech.*, 41, 27–42 (1991).
58. M.H. Wagner, A. Bernnat, and V. Schulze. *J. Rheol*, 42, 917–928 (1998).
59. A.J. Waddon and A. Keller, *J. Polym. Sci. Part B: Polym. Phy*. 28, 1063(1990).
60. J.W.H. Kolnaar and A. Keller, *J.Non-Newt.Fluid Mech.*, 67, 213–240 (1996).
61. A.K. van der Vegt and P.P.A. Smit, *Adv. Polym. Sci. Technol., Soc. Chem. Ind.*, 26, 313 (1967).
62. M. Rätzsch, *Pure Appl. Chem.*, A36(11), 1759–1769 (1999).
63. W. Weng, W. Hu, A.H. Dekmerzian, and C.J. Ruff, *Macromolecules*, 35, 3838–3843 (2002).
64. A.E. Cherian, E.B. Lobkovsky, and G.W Coates, *Macromolecules*, 38, 6259–6268 (2005).
65. Z. Ye, F. Alobaidi, and S. Zhu, *Ind. Eng. Chem. Res.*, 43, 2860–2870 (2004).
66. L. Bing and T.C. Chung, *Macromolecules*, 32, 8678 (1999).
67. J.A. Langston, R.H. Colby, T.C.M. Chung, et al., *Macromolecules*, 40, 2712–2720 (2007).
68. M. Rätzsch, M. Arnold, E. Borsig, et al., *Pr. Polym. Sci.*, 27(7), 1195–1282 (1998).
69. E. Borsig, M. van Duin, A.D. Gotsis, and F. Picchioni, *Eur. Polym. J.*, 44, 200–212 (2008).
70. H.P. Schreiber and E.B. Bagley, *J. Polym. Sci.*, 58, 29–48 (1962).
71. R.N. Shroff and H. Mavridis, *Macromolecules*, 34(21), 7362–7367 (2001).
72. A. Malmberg, C. Gabriel, T. Steffl, et al., *Macromolecules*, 35, 1038–1048 (2002).
73. M. Yamaguchi and C.G. Gogos, *Adv. Polym. Tech.*, 20, 261(2001).
74. M. Yamaguchi, D.B. Todd, and C.G. Gogos, *Adv. Polym. Tech.*, 22, 179 (2003).
75. M. Yamaguchi and M. Takahashi, *Polymer*, 42, 8663 (2001).
76. H. Münstedt, *Colloid Polym. Sci.*, 259, 966 (1981).
77. P.J.R. Leblans and C. Bastiaansen, *Macromolecules*, 22, 3312 (1989)
78. E. van de Wetering, "Foam Extrusion of Peroxide-Modified Polypropylene," graduation report, Hoogeschool Enschede, Dept. of Chemical Technology, Enschede, the Netherlands 2001.

# CHAPTER 4

# STRUCTURE AND RHEOLOGY OF FIBER SUSPENSIONS

AARON P. R. EBERLE,[1,2] KEVIN ORTMAN,[2] and DONALD G. BAIRD[2]
[1] NIST Center for Neutron Research, National Institute of Standards and Technology, Gaithersburg, MD, USA
[2] Department of Chemical Engineering, Virginia Polytechnic Institute and State University, Blacksburg, VA, USA

## CONTENTS



*Applied Polymer Rheology: Polymeric Fluids with Industrial Applications*, First Edition.
Edited by Marianna Kontopoulou.

## 4.1 INTRODUCTION

Polymer composites, protective coatings, membranes, ceramics, and structural materials are examples of a diverse class of fiber composites whose precursors are multicomponent fluids consisting of various disperse solids and a continuous phase, the suspending medium. Control of the fiber microstructure and bulk flow properties of the suspensions is vital to the optimization of most manufacturing processes. Hence the rheological behavior and its connection to the suspension microstructure is of technical importance.

In this chapter we review the rheological properties and the connection between rheology and microstructure of fiber suspensions. This does not include nanocomposites or natural fiber composite materials, which are covered in detail in separate chapters. Still, this field is extensive and so we limit the discussion to the major topics and general behaviors of bicomponent fluids consisting of discontinuous and homogeneously dispersed fibers in a suspending medium. This chapter includes the current theoretical treatment of both short and long fibers, and Newtonian and non-Newtonian suspending media. In addition, the chapter discusses the experimental rheological behavior and its connection to the fiber orientation distribution. Additional detail regarding the experimental rheology can be found in the works of Eberle et al. [1], Zirnsak et al. [2], and Larson [3], while the theory can be found in Chung and Kwon [4], Advani and Tucker [5], and Petrie [6].

The chapter is organized to first give the reader a brief background in specific areas deemed necessary to the concepts discussed here, which include suspension classification and experiments. Then the rheology of rigid fiber suspensions is discussed, followed by flexible fibers. The separation of rigid and flexible fiber suspensions reflects the dramatic differences in both the experimental behavior and approaches to modeling.

## 4.2 BACKGROUND

The topics discussed in this chapter are complex, and the reader is assumed to have a basic understanding of rheology. To minimize confusion the rheological material functions will be defined following that of the official nomenclature for the Society of Rheology as they are discussed in detail elsewhere [7]. On the other hand, the description of fiber suspensions and the instrumentation used to characterize their behavior directly relates to the interpretation of the rheology. Hence an accurate classification of materials and rheometric devices is needed to delineate the contributing factors to the rheological behavior from both an experimental and numerical perspective. In the subsequent Sections we address this by classifying fiber suspensions by their concentration and length as it relates to flexibility. An important aspect of this work is relating the rheological behavior to the microstructure, which is not a trivial task in fiber suspensions where large diameters and opaque matrices prevent the use of scattering techniques. As a result the current approaches to orientation analysis are included. This is followed by a review of the different rheometers and rheometer geometries commonly used, including a discussion of their strengths and weaknesses with respect to obtaining accurate measurements.

Fibers longer than about 1 μm or so will settle as a result of gravity unless the density of the fluid is matched to that of the particle. An approximation for the sedimentation time, $t_s$, of a cylinder in a viscous fluid was proposed by Chaouche and Koch [8] in which $t_s = 8\eta_s L/\Delta\rho g d^2[\ln(2a_r) - 0.72]$ where $\eta_s$ is the suspending medium viscosity, $L$ is the fiber length, $\Delta\rho$ is the difference in densities, $g$ the acceration due to gravity, $d$ is the fiber diameter, and $a_r$ is the fiber aspect ratio defined as $a_r = L/d$. From this simple expression it is shown that $t_s$ is directly proportional to $\eta_s$ which conveniently minimizes the effect of sedimentation in most composite fluids discussed here. Therefore, the reader can assume that the contribution of sedimentation to changes in the fiber microstructure is negligible unless otherwise specified.

Brownian motion refers to the random movement of any sufficiently small particle as a result of the momentum transfer from suspending medium molecules. The relative effect that Brownian motion may have on orientation, of anisotropic particles, in a dynamic system can be estimated using the rotary

Peclet number, $\mathrm{Pe} \equiv \dot{\gamma}/D_{ro}$ where $\dot{\gamma}$ is the shear rate, and $D_{ro}$ rotary diffusivity, which defines the ratio of the thermal energy in the system to the resistance to rotation. Doi and Edwards [9] estimate $D_{ro} = 3k_B T[\ln(a_r) - \Omega]/\pi\eta_s L^3$, where $k_B$ is Boltzmann's constant, $T$ is temperature, and $\alpha$ is a constant in which $\Omega = 0.8$ is a good estimate [10]. Using the scaling argument that when $\mathrm{Pe} \gg 1$ Brownian effects can be neglected, which in general is true for particles whose longest dimension is $\gtrsim 10\ \mu\mathrm{m}$ [3]. In the following sections the fiber suspensions can be considered non-Brownian, unless stated otherwise, and as such, the numerical analysis of these systems will not include any Brownian terms. The reader is referred to the works of Doi and Edwards [9] on the dynamics of Brownian rods if interested in such topics.

### 4.2.1 Fiber Suspension Classification

As discussed later, the dynamic behavior of fiber can be highly dependent on the concentration, and fiber characteristics such as length and flexibility. In this Section we seek to separate fiber suspensions based on their degree of interaction and flexibility with respect to the applied shear stress. While concentration, length, and flexibility lead to specific classifications, polydispersity is characterized using common statistical averages such as the number average, $L_n$, and weight average, $L_w$, and as such will not be discussed.

***4.2.1.1 Concentration*** Fiber suspensions are typically classified into three concentration regimes: dilute, semidilute, and concentrated, which are based on their volume fraction, $\phi = \pi n L^3/4a_r^2$ where $n$ is the number of fibers per unit volume. The dilute regime is such that the fibers within the suspension are free to both rotate and translate without hydrodynamic interaction or direct contact. Theoretically, this occurs when the average distance between the center of mass of two fibers is greater than $L$, leading to the constraint of $n < 1/L^3$ or $\phi < a_r^{-2}$. The transition to semidilute occurs just above the dilute upper limit. Here hydrodynamic interaction is the predominant phenomenon with little fiber contact. However, the suspension orientation state is not subject to geometric constraints, which can constrict orientation states. Interaction between fibers becomes theoretically possible for homogeneously dispersed fibers when $n > 1/L^3$ but when the mean spacing between fibers, $S_\mathrm{m}$, is on the order of the fiber diameter, physical contact becomes an increasingly significant phenomenon. Therefore, the upper limit is subject to the constraint $S_\mathrm{m} \gg d$ [9]. The mean spacing between fibers is a function of the orientation state of the fibers within the suspension. For a random orientation state the mean spacing is estimated as $S_m \cong 1/(nL^2)$ and for a suspension whose fibers are completely aligned the mean spacing is of the order $S_m \cong (nL)^{-1/2}$ [11]. This leads to two upper limits: $n \ll 1/(dL^2)$ for random and $n \ll 1/(d^2L)$ for aligned orientation.

The concentrated regime is where $n \geq 1/(d^2L)$ or $\phi \geq a_r^{-1}$. In this range, the dynamic properties of the fibers can be severely affected by fiber–fiber interactions and can lead to solid-like behavior [12]. It is interesting to note that most fiber composites of industrial interest typically have fiber concentrations of $\phi > 0.1$ and fall within the concentrated regime. The concentration regimes follow the theories for molecular rods, as, for example, described in Doi and Edwards [9]. In addition to the three regimes defined above, molecular theories define a critical concentration in which molecules will preferentially align and become nematic similar to a phase transition. However, it has yet to be proven that fiber suspensions will also go through this transition.

***4.2.1.2 Flexibility*** Fiber suspensions continuously evolve as the demand for high performance, lightweight, and cost-effective parts drive the development of new materials. Glass, carbon, nylon, Kevlar®, and Nomex® are examples of fibers that have been used to reinforce polymers leading to a wide range of fiber composites with varying properties. While it is reasonable to assume that the rheology of fiber suspensions, regardless of composition, should follow defined trends, in reality, it depends on the flexibility of the fiber with respect to the applied shear stress. This will be discussed in detail later for both the experimental rheology and, perhaps more important, in the section describing the approaches to simulating fiber suspensions.

Within this text the term *flexibility* will be used to describe how stiff a fiber behaves in the presence of flow. In this context, a fiber exhibiting a larger degree of flexibility is easier to bend within a specified flow field. The flexibility of a fiber commonly leads to two classes of materials, short and long. Short fiber suspensions are defined as those that remain rigid within a specified shear field, while long fiber suspensions consist of fibers that flex or bend. Switzer and Klingenberg [13] characterized the flexibility of a fiber using an effective stiffness ($S^{\text{eff}}$) dimensionless number defined as

$$S^{\text{eff}} = \frac{E_{Y}\pi}{4\eta_{s}\dot{\gamma}a_{r}^{4}} \tag{4.1}$$

where $E_Y$ is the Young's modulus, and $\eta_s$ is the suspending medium viscosity. From this expression, the flexibility of a fiber can be seen to increase with aspect ratio for a material with a given Young's modulus.

### 4.2.2 Microstructure Analysis

Direct measurement of fiber orientation is a key development used to establish structure–property relationships and test advances in constitutive theory. However, the measurement of fiber orientation within a suspension in its molten or consolidated state is a complex task, especially since the orientation can vary considerably through the thickness of the sample. Current methods used to quantify the orientation include transmission optical microscopy [14], contact microradiography [15], scanning acoustic microscopy [16], scanning electron microscopy [17], nuclear magnetic resonance (NMR) imaging [18], X-ray tomographical imaging [19], and reflection microscopy [20]. Many of these techniques offer different strengths but have similar weaknesses with image resolution and effective sampling size. Transmission optical microscopy, contact microradiography, and scanning acoustic microscopy provide only two-dimensional images, typically with poor resolution [21]. An automated approach has yet to be developed that incorporates the imaging ability of scanning electron microscopy, NMR imaging, and X-ray tomography [22], which reduces the sample size that can be measured. Reflection microscopy has become the most popular and widely used method. Though it is not clear when the ideas for this method were developed, it is known that its foundation stems from simple geometric analysis of ellipsoids and was directly used to quantify fiber orientation by Darlington and McGinley [15] as early as 1975. This method suffered from the same insufficiencies in being able to process large sample sections to acquire accurate orientation distributions until the advent of economical, high-speed, image analysis systems [23, 24].

In reflection microscopy a section of the sample is cut, often normal to the flow direction, encapsulated in an epoxy resin and polished to a smooth surface. The epoxy prevents the degradation of the fibers at the edge of the sample. Contrast between the matrix and the fiber is produced by oxygen plasma etching; this process creates a rough surface around the fibers, which appear dark in reflected light [22]. The three-dimensional orientation is then calculated from the ellipse that appears at the intersection of a plane.

The orientation of each fiber can be described using a unit vector, **u**, along the backbone of the fiber. The zenith and azimuthal angles $\theta$, $\varphi$, respectively, defining the components of **u** for a each fiber can be calculated based on the ratio of the minor $m$ and major $M$ axis and depends on which plane is being processed (4.2); the other angle can be measured directly from the image [5]. Figure 4.1a depicts the elliptical foot print in the $x_1 - x_2$ plane and the angle $\varphi$ measured directly from the image.

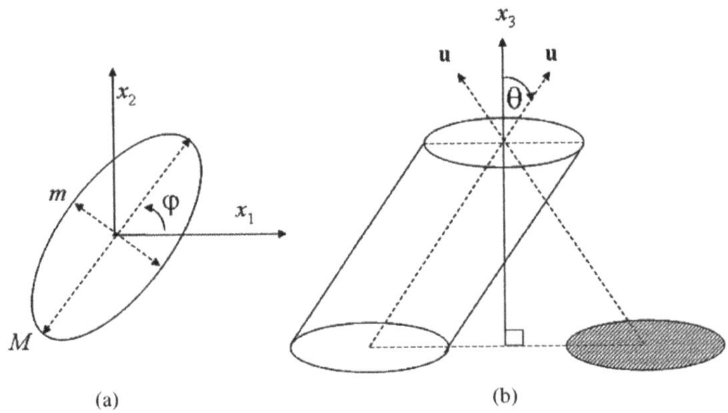


**FIGURE 4.1** Fiber orientation angles $\varphi$, and $\theta$ determined from the elliptical projection [22]. Reprinted with permission from Ref. 26.

$$\cos\theta|_{1-2} = \frac{m}{M}, \cos\varphi|_{2-3} = \frac{m}{M} \tag{4.2}$$

The subscript 1–2 or 2–3 refers to the plane being processed. The limitation of calculating the angle based on (4.2) is that the calculated angle is always between 0 and $\pi$ causing an inherent ambiguity. For example, in the $x_1 - x_2$ plane it is impossible to distinguish between a fiber that is oriented $(\theta_1, \varphi_1)$ and $(\pi - \theta_1, \varphi_1)$ because their cross-sections are identical. Figure 4.1b depicts two possible unit vectors **u** for the same elliptical footprint. Recently, Lee et al. [26] and Eberle et al. [25] used confocal laser microscopy to view the footprints of fiber cross-sections at two planes. This reduced the ambiguity of the technique.

Additional limitations to this method include the many assumptions that are made regarding the distribution of fiber lengths and fiber orientations when correction factors for counting bias are applied [27, 28]. Also, the technique is limited to short fiber composites due to increased and incalculable errors associated with fibers that are not perfectly straight [21]. In the case of long fibers, no dedicated assessment protocol has been identified in literature [29]. Techniques such as X-ray tomography have the ability to capture the 3D spatial distribution, but currently there is no analysis software capable of quantifying the orientation.

### 4.2.3 Rheometry Flow Field and Boundary Effects

Throughout the literature many different shear rheometers have been used to measure the rheological material functions of fiber suspensions. This includes torsional, capillary, sliding plate, concentric cylinder, falling ball, and Couette. A complete review and mathematical description of the flow field and design equations for each geometry can be found in Macosko [30]. By far the most widely used for polymer composites are the torsional, and capillary rheometers. Traditionally, these types of rheometers were designed to measure the rheology of homogeneous fluids in which case small gaps and capillaries were part of the design criteria to minimize sample volume and velocity field inhomogeneities. For fiber suspensions this represents possible sources of error through fiber-boundary interactions and flow field complications associated with curvilinear streamlines, inhomogeneous flow fields, and particle migration during testing. Subsequently, we briefly discuss both torsional and capillary rheometers in relation to measuring the rheology of fiber suspensions for the primary purpose of informing the reader of possible sources of error that are not easily corrected for.

Torsional rheometers are commonly used to measure the low shear rate $\dot{\gamma} < 10\ \mathrm{s}^{-1}$, steady-state, and transient rheology. While many fixture designs and geometries are available the cone-and-plate or parallel-disk fixtures are the most commonly used. The cone-and-plate geometry provides a homogeneous shear field but has a small gap of $< 0.1$ mm near the center and does not allow for gap control. Boundary interactions can occur when the rheometer gap is small compared to the characteristic length of the filler [31]. For a fiber, the characteristic length can significantly change, depending on its orientation. For instance, a fiber whose orientation is in the shear direction has a characteristic length that is equal to the length of the fiber, while a fiber that is aligned in the flow direction has a characteristic length equal to the diameter of the fiber. To minimize fiber–boundary interaction Blankeney [32] and Attanasio et al. [33] suggested that the sample thickness should be greater than three times the characteristic length of the filler. Boundary interactions are most probable in the case of a fiber tumbling in the shear plane and suppress and broaden the overshoot behavior of the transient material functions [34]. To prevent this Eberle et al. [35] suggested the use of a novel sample geometry in the cone-and-plate geometry where the center of a premolded sample disk was removed, called a *donut sample*, effectively eliminating the region most likely to cause complications due to boundary interaction.

Parallel-disks allow for gap control but the shear rate varies from the center of the plate to the outer edge resulting in an inhomogeneous shear field. Eberle et al. [35] found that that inhomogeneous shear field induces excessive fiber–fiber interaction in concentrated fiber suspensions and can severely affect the magnitude of the stress growth overshoot peak and width of the overshoot. A comparison of the shear stress growth coefficient, $\eta^+$, measured using the parallel-disk, and cone-and-plate geometry with and without the donut sample can be seen in Figure 4.2. $\eta^+$ measured using the parallel-disk has an exaggerated overshoot and steady-state plateau compared to the cone-and-plate measurements. $\eta^+$ measured using the cone-and-plate with unmodified sample

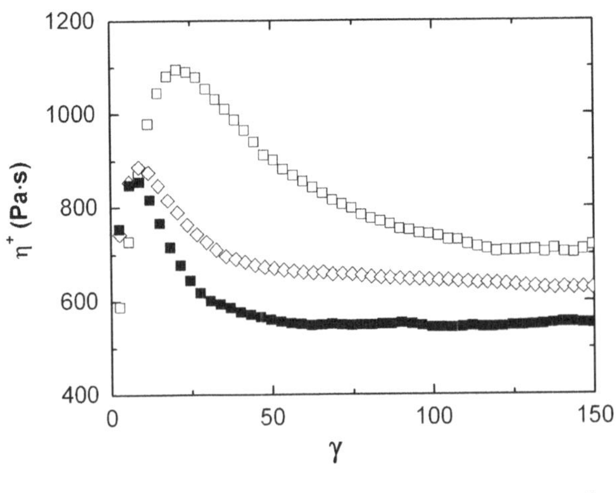


**FIGURE 4.2** Stress growth vs. strain in startup of flow measurements for 30 wt % short glass fiber filled polybutylene terephthalate at $\dot{\gamma} = \dot{\gamma}_R = 1\ s^{-1}$. Shear stress growth coefficient, $\eta^+$, measured using the parallel-disk (□), the cone-and-plate with unmodified sample (■), and the cone-and-plate with donut sample (◇). Reprinted with permission from Ref. 35.

has a broadened overshoot and enhanced steady state plateau. These behaviors are even more exaggerated in the transient normal stresses.

The geometries described above all have limitations to their effective use in rheological measurements of fiber suspensions, especially in measuring the transient material functions. Ideally, any rheological measurement should be performed in a "sufficiently large" device so that the boundary and flow field effects are minimized. In the future it is expected that sliding plate rheometers, such as that developed by Giacomin et al. [36], will be more widely used; though these too suffer from finite absolute strain limitations.

Capillary rheometers are used to investigate the high shear rate rheology $\dot{\gamma} > 10\ s^{-1}$. In this case the suspension is forced through an abrupt contraction and the pressure drop is measured across the contraction and capillary length. Studies have shown that particle migration can occur away from the channel wall across streamlines, in laminar flow, due to small inertia and wall effects termed *tubular pinch* [37, 38]. The degree of particle migration in fiber suspensions is ill-defined. However, if migration does occur away from the boundaries, the regions of highest shear rates close to the capillary wall could contain a far less concentration of fibers, which results in a stress response similar to the pure matrix.

## 4.3 RIGID FIBERS

### 4.3.1 Dilute Suspensions: Fiber-Orientation Kinematics

The orientation of a single fiber can be described with a unit vector **u** along the major axis as shown in Figure 4.3, where $\varphi$ and $\theta$ are the azimuthal and zenith angles, respectively. The components of **u** in spherical coordinates are

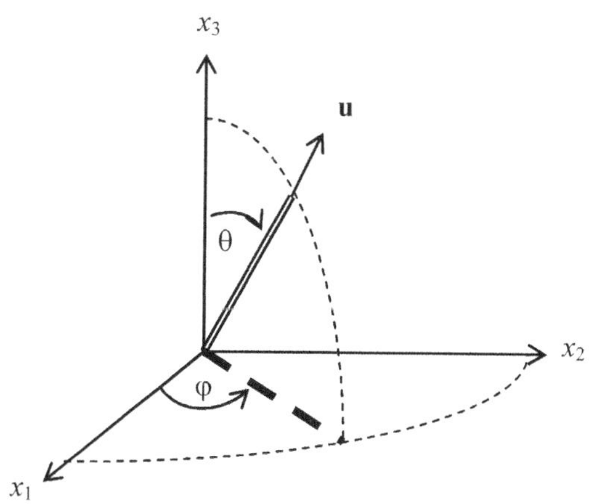


**FIGURE 4.3** Unit vector **u** describing the orientation state of a fiber in spherical coordinates. Reprinted with permission from Ref. 26.

$$u_1 = \sin\theta\cos\varphi \tag{4.3}$$

$$u_2 = \sin\theta\sin\varphi \tag{4.4}$$

$$u_3 = \cos\theta \tag{4.5}$$

In the following section when discussing simple shear flow ($v_1 = \dot{\gamma}\ y$ and $v_2 = v_3 = 0$) we define $x_1$ as the flow direction, $x_2$ the direction of the velocity gradient, and $x_3$ the neutral or vorticity direction.

The first theoretical work that is easily extendable to rigid fibers is that of Jeffery [39]. Jeffery extended Einstein's [40] approach to solving the equations of motion for the flow of a Newtonian fluid around a spherical particle to that of a neutrally buoyant ellipsoidal particle in the absence of Brownian motion. Following Jeffery's analysis, in spherical coordinates, the rotation of an ellipsoid subject to simple shear flow is governed by the following differential equations:

$$\frac{\partial\varphi}{\partial t} = \frac{\dot{\gamma}}{a_r^2 + 1}(a_r^2\cos^2\varphi + \sin^2\varphi) \tag{4.6}$$

$$\frac{\partial\theta}{\partial t} = \lambda\frac{\dot{\gamma}}{4}\sin 2\theta\sin 2\varphi \tag{4.7}$$

where $\lambda$ is a shape constant given by

$$\lambda = \left(\frac{a_r^2 - 1}{a_r^2 + 1}\right) \tag{4.8}$$

The analytical solutions to the system of differential equations (4.6) and (4.7) are

$$\tan\varphi = a_r\tan\left(\frac{\dot{\gamma}t}{a_r + a_r^{-1}} + k\right),\ \tan k = \left(\frac{1}{a_r\tan\varphi_0}\right) \tag{4.9}$$

$$\tan\theta = \frac{Ca_r}{\sqrt{a_r^2\sin^2\varphi + \cos^2\varphi}},\ C = \frac{\tan\theta_o}{\sqrt{a_r^2\sin^2 k + \cos^2 k}} \tag{4.10}$$

where $C$ and $k$ are constants that may be obtained from initial orientation of the ellipsoid, $\varphi_o$ and $\theta_o$. Both $\varphi(t)$ and $\theta(t)$ are functions of time and are periodic resulting in a 3D orientation that never reaches a steady state under dynamic conditions termed a Jeffery orbit. The relative time lapse during rotation depends on the particle's initial orientation, aspect ratio, and shear rate seen by the particle. High $a_r$ particles spend the majority of time aligned parallel to the direction of flow [6]. The period of rotation, $T$, required for a rotation of $2\pi$ about the particle center of mass is:

$$T = \frac{2\pi}{\dot{\gamma}}\left(a_r + \frac{1}{a_r}\right) \tag{4.11}$$

Predictions for the analytical solution to Jeffery's analysis that clearly show the periodicity of the particles orientation can be seen in Figure 4.4a. which is a graphical representation of (4.9) and (4.10) for an ellipsoidal particle with an $a_r = 5$ and initial orientation conditions $\varphi_o = 175°$ and $\theta_o = 45°$.

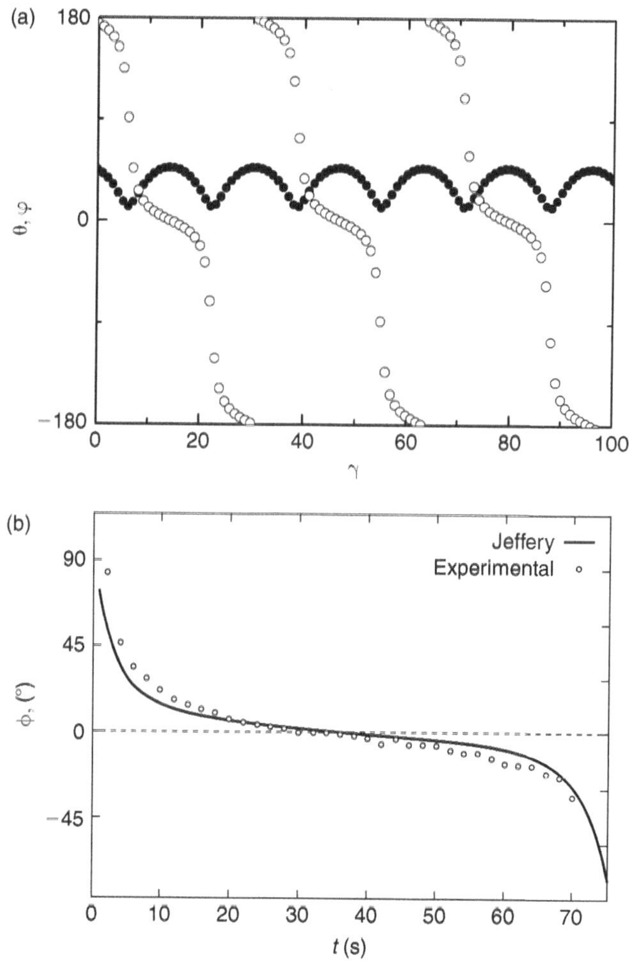


**FIGURE 4.4** (a) Zenith and azimuthal angles defining the orientation of an ellipsoidal particle in simple shear flow as predicted by the analytical solution to the Jeffery's equation of motion, (5.9) and (5.10). The solution is for an ellipsoid with $a_r = 5$ with initial conditions of $\varphi_o = 175°$, $\theta_o = 45°$. Reprinted with permission from Ref. 41. (b) Evolution of a nylon fiber compared to predictions of Jeffery's equation. Reprinted with permission from Ref. 42.

The above analysis pertains to ellipsoidal particles but the theory can be extended to non-ellipsoidal particles by using an effective aspect ratio ($a_{re}$) that can be determined by comparing the rate of rotation and $a_r$ of a particle [43]. Work has been done both theoretically [44, 45] and empirically [31, 46] to establish relations between the particle shape and $a_{re}$. Cox [44] found $a_{\rm re}$ for blunt ended bodies to be represented by

$$a_{re} = \frac{1.24a_r}{(\ln a_r)^{1/2}} \tag{4.12}$$

This work suggests that for blunt ended particles $a_{re} < a_{\rm r}$ and, therefore, they have a slower period of rotation than particles with a rounded end. An exception to this trend would be for $a_{\rm r} < 4.8$, in which case (4.12) predicts $a_{re} > a_{\rm r}$.

Experimentally, Jeffery orbits have been observed in various systems containing high aspect ratio particles [42, 47–49]. The transient shear response of a dilute suspension of rigid fibers in a Newtonian suspending medium under simple shear flow conditions will rotate around a vorticity axis unless acted upon by another force, i.e., direct contact and/or hydrodynamic interactions between other particles or boundaries [47]. This behavior can be seen in Figure 4.4b for a single nylon fiber, $a_{\rm re} = 27.5$, suspended in a Newtonian fluid subject to simple shear flow. Though the predictions using Jeffery's equation seem to agree well with experimental results, Moses et al. [42] state the model tends to overpredict the rate of fiber reorientation.

In the limit that $a_{\rm r} \to \infty$, then $\lambda \to 1$ and the solution to (4.9) and (4.10) can be simplified, in which the equations are a function of strain, $\gamma = \dot{\gamma}t$, only:

$$\tan\theta = \tan\theta_{\rm o}\sqrt{\gamma^2 \sin^2\varphi_{\rm o} + 2\gamma \sin\varphi_{\rm o}\cos\varphi_{\rm o} + 1} \tag{4.13}$$

$$\tan\varphi = \frac{1}{\cot\varphi_{\rm o} + \gamma} \tag{4.14}$$

In this case the solution is no longer periodic, or simply, the period of rotation is infinite. As a result, a particle will rotate from its initial orientation to align itself in the flow direction. This can be seen in Figure 4.5, which graphically depicts the solutions to (4.13) and (4.14) for various initial conditions. For a more in-depth discussion of the motion of a fiber predicted by Jeffery's analysis, the reader is referred to Barbosa and Bibbó [41].

Equations (4.6) and (4.7) can be generalized for any flow field in terms of $\mathbf{u}$ leading to [48–50],

$$\dot{\mathbf{u}} = \frac{D\mathbf{u}}{Dt} = \mathbf{W}\cdot\mathbf{u} + \lambda[\mathbf{D}\cdot\mathbf{u} - \mathbf{u}(\mathbf{u}\cdot\mathbf{D}\cdot\mathbf{u})] \tag{4.15}$$

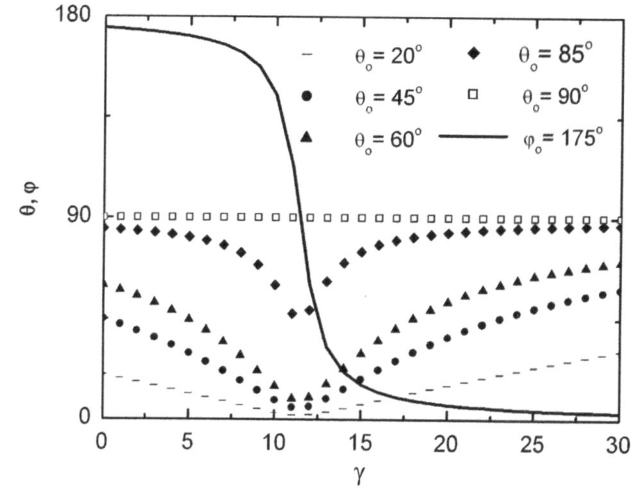


**FIGURE 4.5** Evolution of orientation for an ellipsoidal particle with infinitely large $a_r$ subject to simple shear flow as predicted by (5.13) and (5.14). Predictions were subject to the initial conditions $\varphi_o = 175^\circ$, $\theta_o = 20^\circ$, $45^\circ$, $60^\circ$, and $90^\circ$. Reprinted with permission from Ref. 41.

where $\dot{\mathbf{u}}$ is the substantial time derivative of the unit vector $\mathbf{u}$, $\mathbf{W} = [(\nabla \boldsymbol{v})^t - \nabla \boldsymbol{v}]/2$ is the vorticity, and $\mathbf{D} = [\nabla \boldsymbol{v} + (\nabla \boldsymbol{v})^t]/2$ is the rate of strain tensor and $\nabla \boldsymbol{v} = \partial v_j/\partial x_i$. Because the particles are assumed rigid and inextensible, $\mathbf{u}(\mathbf{u} \cdot \mathbf{D} \cdot \mathbf{u})$ is subtracted, which is the term associated with stretch along the primary axis of a flexible particle or molecule. Equation (4.15) can be simplified further, in limiting case that $\lambda \rightarrow 1$ and by noting the quantity $(\mathbf{u} \cdot \mathbf{W} \cdot \mathbf{u})$ is zero due to the antisymmetric nature of $\mathbf{W}$, as [6, 51]

$$\dot{\mathbf{u}} = \nabla \mathbf{v} \cdot \mathbf{u} - \mathbf{u}(\mathbf{u} \cdot \nabla \mathbf{v} \cdot \mathbf{u}) \tag{4.16}$$

We now consider the case of an ideal suspension, which consists of a large number of fibers that are identical in size and shape and homogeneously dispersed. To predict the motion of such a suspension using (4.15) or (4.16) one would have to track the movement of each fiber from its initial condition, resulting in an overwhelming amount of computational effort. Fortunately the fiber orientation state can be easily represented using a distribution function, $\psi(\theta, \varphi, t)$, that describes the probability of a rod at time $t$ being in the range $\theta, \varphi$ to $\theta + d\theta, \varphi + d\varphi$. In certain cases, it is convenient to represent the distribution function in terms of the unit vector $\mathbf{u}$ as $\psi(\mathbf{u}, t)$. The distribution function is normalized so that the integral over the unit sphere is equal to unity:

$$\int_0^{2\pi}\int_0^{\pi} \psi(\theta, \varphi, t) \sin\theta d\theta d\varphi = \int \psi(\mathbf{u}, t) d\mathbf{u} = 1 \tag{4.17}$$

$\psi$ (**u**, $t$) must also be symmetric, i.e., $\psi$ (**u**) = (−**u**). This is a result of rod symmetry in that a rod whose orientation is **u** is indistinguishable from one whose orientation is −**u** [5, 6]. Finally the distribution function must be able to describe the evolution of fiber orientation under dynamic conditions, termed the continuity condition [51]. To satisfy this condition one must employ a convection-diffusion type of equation. The equation used by Doi [52] and commonly referred to by others is the Smoluchowski equation or a modified Fokker-Planck equation [53]. The Smoluchowski equation is a macroscopic description of a multiparticle system. For a suspension of non-interacting fibers subject to an external flow field and neglecting Brownian motion the Smoluchowski equation can be written in the simple form:

$$\frac{D\psi}{Dt} = -\frac{\partial}{\partial u} \cdot (\dot{\mathbf{u}}\psi) \tag{4.18}$$

Equation (4.18) is valid for any $\dot{\mathbf{u}}$. For dilute suspensions we can substitute (4.16) into (4.18):

$$\frac{D\psi}{Dt} = -\frac{\partial}{\partial \mathbf{u}} \cdot [(\nabla \mathbf{v} \cdot \mathbf{u})\psi - \mathbf{u}(\mathbf{u} \cdot \nabla \mathbf{v} \cdot \mathbf{u})\psi] \tag{4.19}$$

Using the distribution function for predicting the particle orientation in complex flows is cumbersome because the amount of information computed is so large. For this reason, most theoretical work has been formulated around having to calculate the distribution function explicitly. Typically, this is accomplished with orientation tensors [5]. Using the distribution function, an infinite number of even-order tensors can be formulated that define an averaged orientational state of the system, often referred to as structure tensors. The structure tensors of interest with respect to modeling fiber suspensions are the second- and fourth-order tensors defined as:

$$\mathbf{A}(t) = \int \mathbf{uu}\psi(\mathbf{u}, t)d\mathbf{u} \tag{4.20}$$

$$\mathbf{A}_4(t) = \int \mathbf{uuuu}\psi(\mathbf{u}, t)d\mathbf{u} \tag{4.21}$$

The trace of **A** is always equal to 1 and for a completely random orientation state **A** = 1/3 **I**, where **I** is the unity tensor. In the limit that all the fibers are perfectly aligned in the $x_1$ direction the only nonzero component is $A_{11} = 1$. The fourth-order tensor arises in the theoretical analysis for both the particle motion and in calculating the contribution of the hydrodynamic interaction of the fibers to the extra stress. For a more complete description of orientation tensors and their use in representing fiber suspensions the reader is referred to Advani and Tucker [5].

To obtain an evolution equation in a form containing **A**, when the distribution function is unknown, one must take the material derivative of (4.20) and substitute in the governing equation for $\psi$ and $\dot{\mathbf{u}}$, (4.18), on the right-hand side. Integration by parts, with some simplification, leads to

$$\frac{D\mathbf{A}}{Dt} = \mathbf{A} \cdot \nabla\mathbf{v} + \lambda[(\mathbf{A} \cdot \nabla\mathbf{v})^t - 2\mathbf{D} : \mathbf{A}_4] \tag{4.22}$$

To solve this equation a closure approximation is need to express $\mathbf{A_4}$ in terms of **A**. Equation (4.22) is also commonly written in the following form [5]

$$\frac{D\mathbf{A}}{Dt} = (\mathbf{W} \cdot \mathbf{A} - \mathbf{A} \cdot \mathbf{W}) + \lambda(\mathbf{D} \cdot \mathbf{A} + \mathbf{A} \cdot \mathbf{D} - 2\mathbf{D} : \mathbf{A}_4) \tag{4.23}$$

The fourth-order tensor $\mathbf{A_4}$ has been the subject of much research because to complete the analysis one must use a closure or decoupling approximation to express the fourth-order tensor in terms of the second-order tensor. A closure approximation is some function that approximates a higher-order orientation tensor with lower-order orientation tensors and/or the invariants of lower-order orientation tensors. In the context of this chapter the closure approximation is vital in establishing an equation of change for the average orientation state of the system and in calculating the extra stress contribution as a result of the hydrodynamic drag discussed later. In addition the higher-order orientation tensors (tensors greater than second-order) arise in any continuum model that describes the mechanical or rheological properties of a two-phase system containing particles whose orientation can be anisotropic. Numerous closure approximations have been suggested and tested, which are beyond the scope of this text. The reader is directed to the works of Hand [54], Hinch and Leal [55], Barthés-Biesel and Acrivos [56], Advani an Tucker [57], Chung and Kwon [58], and Jack and Smith [59] for in-depth reviews.

### 4.3.2 Dilute Suspensions: Rheology

***4.3.2.1 Theory*** A general expression for the total stress in a dilute suspension of high aspect ratio non-Brownian particles can be derived from the theories of Hand [54] and Giesekus [48] and is commonly referred to as the Lipscomb model [60, 61]

$$\sigma = -P\mathbf{I} + 2\eta_s\mathbf{D} + 2c_1\phi\eta_s\mathbf{D} + 2\phi\eta_s N\mathbf{D} : \mathbf{A}_4 \tag{4.24}$$

where $\boldsymbol{\sigma}$ is the total stress, $\eta_s$ is the suspending medium viscosity, $c_1$ is a constant, and $N$ is a dimensionless parameter that represents the coupling between hydrodynamic stress contribution and the fiber orientation. The third term on the right side of the equation is the viscosity enhancement as a result of the fiber and is similar to the enhancement term for a dilute suspension of spheres proposed by Einstein [40]. Lipscomb et al. [60] gives $c_1$ to be equal to 2. The fourth term on the right side of the equation is the contribution to stress from the hydrodynamic drag of the fluid over the fiber. As in the equation for evolution of fiber orientation, a closure approximation is needed in (4.24) to express $\mathbf{A_4}$ in terms of $\mathbf{A}$. Using (4.24) it is straightforward to show that the shear stress growth coefficient, $\eta^+$, and the first normal stress growth function, $N_1{}^+$, are

$$\eta^+ = \sigma_{12}/\dot{\gamma} = \eta_s[1 + \phi(c_1 + 2NA_{1212})] \tag{4.25}$$

$$N_1^+ = 2\phi\eta_s\dot{\gamma}N(A_{1112} - A_{2212}) \tag{4.26}$$

where the fourth-order tensor components are a function of time and can be predicted with the analysis presented in Section 4.3.1. pertaining to the evolution of fiber orientation.

For dilute suspensions Lipscomb et al. [60] gives $N$ to be a function of fiber aspect ratio,

$$N = \frac{a_r^2}{2\ln a_r} \tag{4.27}$$

Other works for dilute suspensions such as Batchelor [11] give $N$ to be a function of the number of fibers per unit volume, $c$, and/or fiber length, $L$, and aspect ratio.

$$N = \frac{a_r^2}{3\ln(2a_r)}f(\varepsilon),\ f(\varepsilon) = \frac{1 + 0.64\varepsilon}{1 - 1.5\varepsilon} + 1.659\varepsilon^2,\ \varepsilon = [\ln(2a_r)]^{-1} \tag{4.28}$$

#### *4.3.2.2 Experimental Characterization*

*4.3.2.2.1 Viscosity* It is clear from fiber suspension theory that the bulk extra stress of a fiber suspension subject to shear flow is intimately linked to the fiber orientation. In the case of dilute suspensions where fibers will continue to rotate around the vorticity axis an oscillating stress response that is a function of the orientation state of the particle has been experimentally observed. Figure 4.6 depicts the oscillating shear stress upon startup for a dilute suspension of fibers in a Newtonian fluid by way of the specific viscosity normalized by the fiber concentration $\eta_{sp}/\phi$, vs. time, normalized by the period of fiber rotation. The decay of the stress oscillation is believed to be a result of several interaction, including boundary, particle–particle, hydrodynamic or slight aspect ratio variations [63]. In most reported rheological measurements of dilute suspensions stress oscillations decay to a steady state, which exhibits an enhanced

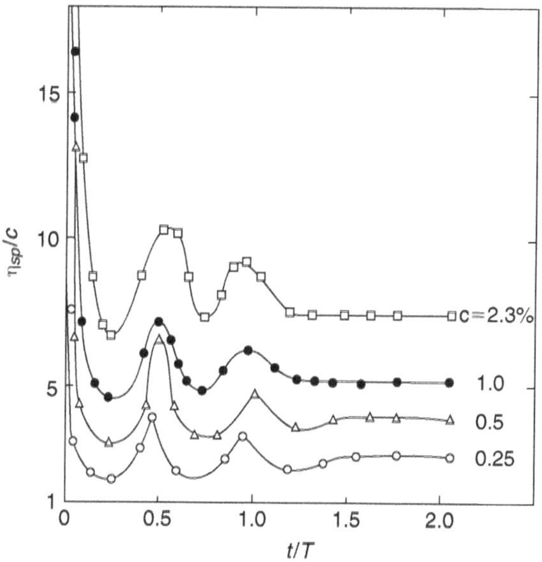


**FIGURE 4.6** Start-up of shear flow experiment depicted through the specific viscosity ($\eta_{sp} = \eta_r - 1$), normalized by the glass fiber volume fraction, vs. time, normalized by the period of oscillation $T$. Tests were performed with a dilute suspension of rigid rods of varying concentration and constant aspect ratio, $a_r = 5.2$, in a Newtonian suspending medium. The rods were initially aligned parallel to the direction of velocity gradient with the use of an electric field. All tests were performed on a torsional cone-and-plate rheometer at $\dot{\gamma} = 2.51\ s^{-1}$. Reprinted with permission from Ref. 62.

viscosity compared to the suspending medium and follows similar trends to the theory.

Shear thinning behavior of dilute fiber suspensions in Newtonian fluids has been observed. Ganani and Powell [64] who reviewed much of the early literature found that for suspensions containing fibers with aspect ratios in the range of 35–45, a weak shear thinning behavior was noticed at low shear rates followed by a Newtonian plateau $\dot{\gamma} > 10\ s^{-1}$. For fibers with $a_r > 100$, the authors found a strong shear rate dependence over a broad shear rate range of $0.1-100\ s^{-1}$. When considering the theory, (4.25), the source of this non-linear behavior becomes more evident as the only parameter with a shear rate dependence is the fiber orientation. Polydispersity, initial orientation, inhomogeneities in the system, and hydrodynamics can all play an important role resulting in a distribution of orientation states that can vary with shear rate.

The literature pertaining to dilute suspensions in non-Newtonian fluids is limited because in most cases the stresses are dominated by the suspending medium and difficult to measure. However, what is available suggests the viscosity increase is proportional to the fiber concentration. In highly viscoelastic fluids stress oscillations are rare and if present will dampen after one oscillation period. This is a result of competing viscoelastic and hydrodynamic effects [65, 66].

*4.3.2.2.2 Normal Stress* Dilute fiber suspensions in Newtonian suspending media can exhibit a nonzero first normal stress function, $N_1$, which displays a similar connection to the fiber orientation as the shear stress. An oscillating fiber orientation will manifest in the form of a transient and oscillating $N_1^+$. If steady state in the fiber orientation is reached, $N_1$ has been found to be linearly proportional to the shear rate and suspending medium viscosity. These trends have been shown experimentally and coincide with theory, as in (4.26). Nonzero normal stresses in fiber suspensions with Newtonian suspending media arise when the fiber orientation has a component out of the plane of shear and is not a result of enhanced elasticity. This becomes more apparent when considering dilute suspensions in non-Newtonian suspending media. A small amount of fiber or dilute suspensions can display a decrease in $N_1$ compared to the neat fluid before rising in proportionality to $\phi$ [67].

*4.3.2.2.3 Small-Amplitude Oscillatory* In small-amplitude oscillatory measurements of dilute fiber suspensions in Newtonian suspending media only the real or viscous component of the complex viscosity and complex modulus exist [68]. This reaffirms that rigid fibers do not contribute to elasticity. In contrast, the normal stresses that can arise are directly related to the fiber orientation component out of the shear plane. Small-amplitude oscillatory measurements will be discussed in more detail in the section pertaining to nondilute suspensions.

### 4.3.3 Nondilute Suspensions: Fiber-Orientation Kinematics

***4.3.3.1 Semidilute Suspensions*** In semidilute suspensions the oscillations present in the dilute suspensions are less common. This is a direct result of the contributing factors that can dampen the oscillations being more prevalent as the fiber concentration is increased. The common approach to predict this behavior relies on Jeffery's equation for the fiber orientation with the assumption that the fiber is infinitely long [11], or $\lambda = 1$, in which case,

$$\frac{DA}{Dt} = (\mathbf{W}\cdot\mathbf{A}-\mathbf{A}\cdot\mathbf{W}) + (\mathbf{D}\cdot\mathbf{A}+\mathbf{A}\cdot\mathbf{D}-2\mathbf{D}:\mathbf{A}_4) \tag{4.29}$$

As discussed in Section 4.3.1, the predictions are no longer periodic but transient in that the fibers will rotate from their initial orientation state to align in the flow direction.

***4.3.3.2 Concentrated Suspensions*** In the first attempts at simulating the fiber orientation of concentrated suspensions in mold filling, Jeffery's equation for infinitely long fibers, (4.29), was used [69]. Comparison between fiber orientation measurements of injection-molded parts and simulation results suggested Jeffery's equation overpredicts the degree of alignment and the shear strain needed to align the fibers. As a result Folgar and Tucker [70] modified Jeffery's theory to include a phenomenological term that prevents full alignment of fiber orientation, termed the Folgar-Tucker (F-T) model. The F-T model can be written in terms of **A** as follows [5]

$$\frac{D\mathbf{A}}{Dt} = (\mathbf{W}\cdot\mathbf{A}-\mathbf{A}\cdot\mathbf{W}) + (\mathbf{D}\cdot\mathbf{A}+\mathbf{A}\cdot\mathbf{D}-2\mathbf{D}:\mathbf{A}_4) + 2C_I\dot{\gamma}(\mathbf{I}-3\mathbf{A}) \qquad (4.30)$$

where $C_I$ is a phenomenological parameter. The last term on the right-hand side of the equation is similar to the isotropic diffusivity term in theories for Brownian rods [9]. The F-T model allows for the control of the steady-state fiber orientation through the magnitude of $C_I$, but the rate of fiber reorientation is dominated by the flow field for the case of small $C_I$, which is typically in the range of 0.016–0.0001 [71]. Currently, there is no theoretical approach for calculating the interaction coefficient, $C_I$, in the F-T model and is determined by fitting predictions to experimental results, which can be time intensive to produce [72]. Bay [71] developed an empirical expression for concentrated suspensions that is a function of the fiber volume fraction and aspect ratio as

$$C_I = 0.0184\exp(-0.7148\phi a_r) \qquad (4.31)$$

Equation (4.31) predicts that $C_I$ decreases for increasing $\phi a_r$ and represents fiber screening. Phan-Thien et al. [73] proposed a model in which $C_I$ increases with increasing $\phi a_r$ as

$$C_I = m_1[1.0-\exp(-m_2\phi a_r)] \qquad (4.32)$$

where $m_1$ and $m_2$ are fit parameters, which they found to be, $m_1 = 0.03$, $m_2 = 0.224$.

The F-T model improves the predictions of the steady-state fiber orientation but has little effect on the strain at which the steady-state orientation occurs. In an attempt to control the rate of fiber reorientation, Huynh [74] and Sepehr et al. [75] both included a term to reduce the rate of fiber orientation termed the strain reduction factor by Huynh and the slip coefficient by Sepehr. The slip coefficient, α, can be added to the F-T model as follows

$$\frac{D\mathbf{A}}{Dt} = \alpha[(\mathbf{W}\cdot\mathbf{A}-\mathbf{A}\cdot\mathbf{W})+(\mathbf{D}\cdot\mathbf{A}+\mathbf{A}\cdot\mathbf{D}-2\mathbf{D}:\mathbf{A}_4) + 2C_I\dot{\gamma}(\mathbf{I}-3\mathbf{A})] \qquad (4.33)$$

and has a value between 0 and 1. However, the addition of the slip or strain reduction factor to the equations governing fiber motion results in a loss of objectivity of the equation. This becomes important when the coordinate frame is translated or rotated as might be required in complex flow simulations of mold filling. However, the physical aspects of the predictions, in the case of simple shear flow, are still acceptable.

Model predictions of fiber orientation using the strain reduction factor in combination with the F-T model show an improved agreement with experimental fiber orientation in simple flow and complex flows in injection molded plaques. Figure 4.7 depicts the experimental fiber orientation in startup of simple shear flow with predictions of the fiber orientation using the F-T model with slip and two values of $C_I$. The value $C_I = 0.006$ was determined by fitting and $C_I = 0.0001$ represents small $C_I$. Predictions of the F-T model with the slip parameter seem to agree with the limited experimental data. However, there is evidence that suggests that the fiber component in the 2-direction increases at small strains to unpredicted values, which is attributed to fiber contact [25].

Wang et al. [76] suggested a model in which the rate of fiber orientation could be reduced without violating objectivity as

$$\frac{D\mathbf{A}}{Dt} = (\mathbf{W}\cdot\mathbf{A} - \mathbf{A}\cdot\mathbf{W}) + \{\mathbf{D}\cdot\mathbf{A} + \mathbf{A}\cdot\mathbf{D} - 2[\mathbf{A}_4 + \ldots(1-\alpha)(\mathbf{L}_4 - M_4 : \mathbf{A}_4)] : \mathbf{D}\} + 2\alpha C_I\dot{\gamma}(\mathbf{I} - 3\mathbf{A}) \tag{4.34}$$

where $\mathbf{L}_4$ and $\mathbf{M}_4$ are functions of $\mathbf{A}$. Equation (4.34) is nearly identical to the F-T model with the slip parameter. The difference is $\mathbf{A}_4$ is replaced by a reduced-strain

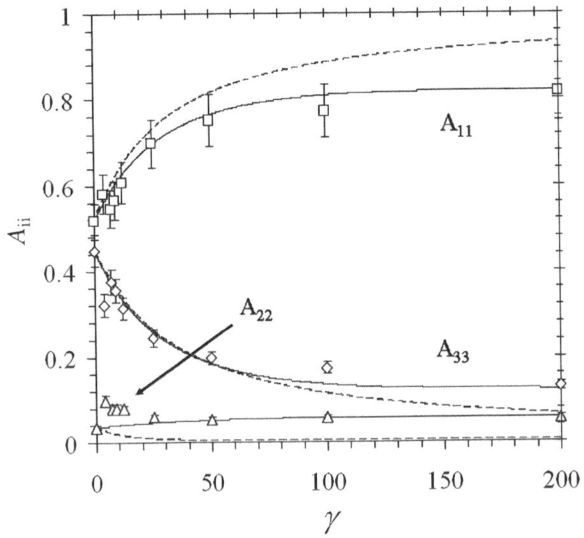


**FIGURE 4.7** Experimental and predicted fiber orientation represented through the $A_{ii}$ components in startup of simple shear flow at $\dot{\gamma} = 1\ \text{s}^{-1}$. The data points are for a concentrated short fiber suspension ($\phi = 0.1766$, $a_r = 28.2$) with a polybutylene terephthalate matrix. The lines represent the predictions of the Folgar-Tucker model with the addition of a slip term with $\alpha = 0.3$, $C_I = 0.0001$ (dashed line) and $C_I = 0.006$ (solid line). Reprinted with permission from Ref. 26.

closure approximation $[\mathbf{A}_4 + (1-\alpha)(\mathbf{L}_4 - \mathbf{M}_4 : \mathbf{A}_4)]$ and is the origin of the term *Reduced-Strain Closure* (RSC) model. The model gives similar predictions to the F-T model with slip while maintaining material objectivity. Note (4.34) still contains $\mathbf{A}_4$ which can be expressed in terms of $\mathbf{A}$ with any available closure.

### 4.3.4 Nondilute Suspensions: Rheology

***4.3.4.1 Semidilute Suspensions*** In the semidilute regime interparticle hydrodynamic interaction is the predominant phenomenon that can affect the dynamic behavior of the fiber microstructure and contribute to the extra stress. The functional form of the extra stress in this regime is the same as that in the dilute case, (4.24). However, $N$ is more complex and accounts for interparticle hydrodynamics. Dinh and Armstrong [51] followed the slender body theory of Batchelor to estimate the hydrodynamic drag as a function of interparticle spacing as,

$$N = \frac{a_r^2}{3\ln(2h/D)} \quad h = \begin{cases} (nL^2)^{-1} & \text{random} \\ (nL)^{-\frac{1}{2}} & \text{aligned} \end{cases} \tag{4.35}$$

where $h$ is the interparticle spacing given for a completely random and aligned fiber orientation, and $D$ is the fiber diameter. As with some semidilute theories, Dinh and Armstrong set $c_1$ equal to zero in (4.24).

Predictions for the Dinh and Armstrong model are well documented and will be reviewed here: first in the case of simple shear flow, then elongational flow. In startup of shear flow, the viscosity of the suspension is greater than the solvent viscosity by a factor of $nL^3$ and the normal stress difference coefficients are equal to zero. The transient shear viscosity and normal stress difference coefficients are a function of shear strain alone and not shear rate. At large strains $\eta^+$ approaches the viscosity of the suspending medium, $\eta^+ \lim_{\gamma\to\infty} \to \eta_s$. This is a direct result of the approximation of the hydrodynamic drag on the fiber by a line integral along the backbone of the fiber. The thickness is neglected in the calculation, resulting in no contribution to the extra stress, which becomes apparent when the fibers fully align in the principle flow direction. The transient first normal stress differences approach zero at large strains. The dimensionless elongational stress growth viscosity is a function of the Hencky strain alone and not the strain rate [54, 77, 78].

Shaqfeh and Fredrickson [79] proposed a theory for dilute and semidilute suspensions that was based on a diagrammatic representation of the multiscattering expansion for the averaged Green's function to obtain an expression for the hydrodynamic energy dissipation from viscous drag of the fluid on the particle. For an isotropic fiber orientation distribution they give

$$N = \frac{4a_r^2}{3}\left\{\frac{1}{\ln(1/\phi) + \ln\ln(1/\phi) + C}\right\} \tag{4.36}$$

where $C''$ is a constant given as $C'' = -0.66$, and 0.16 for random and aligned, respectively. Phan-Thien and Graham [80] proposed the following phenomenological expression

$$N = \frac{a_r^2(2-\phi/G_v)}{2[\ln(2a_r)-1.5](1-\phi/G_v)^2} \tag{4.37}$$

where $G_v = 0.53 - 0.013a_r$, for the range of $5 < a_r < 30$. The predictions of both the theories of Shaqfeh and Fredrickson and Phan-Thien and Graham are similar to that of the Dinh and Armstrong [51] model with the exception of the magnitude of the transients which are linked to the value of $N$.

***4.3.4.2 Concentrated Suspensions*** Currently there is no theory for concentrated suspensions that accounts for fiber contact. However, semidilute theory has been used to some extent to model their stress behavior, in which case $N$ is fit to the rheological material functions of a fluid instead of calculating $N$ from theory [71, 81, 82].

***4.3.4.3 Viscoelastic Matrices*** In the majority of fiber composites of industrial significance the matrix is polymeric and exhibits nontrivial viscoelastic behavior, which increases the complexity of modeling such suspensions. The first attempts ignore the fiber and treat the suspension as a homogeneous viscoelastic fluid [81]. Advances in the field have emphasized the importance of including the fibers to the stress formalism and current treatment includes the fiber and suspending medium by viewing the system as a bicomponent fluid [77, 78, 83].

### 4.3.5 Nondilute Suspensions: Experimental Characterization

***4.3.5.1 Steady-State Shear Rheology*** In addition to an enhanced viscosity, nondilute fiber suspensions in Newtonian suspending media commonly exhibit a pronounced shear thinning response. The degree of shear thinning can be a function of fiber concentration, length, and suspending medium viscosity. This behavior is attributed to the destruction of transient network structures of fibers at increasing shear rates [2].

Suspensions with non-Newtonian suspending media can exhibit various shear thinning behaviors. Multiple factors can contribute to the shear thinning, the most dominant of which being the fiber network, and the suspending medium shear thinning characteristics. In general, the shear viscosity of suspensions containing low concentrations of low aspect ratio fibers (in the dilute and semidilute regime) usually approach a Newtonian plateau at low shear rates, and in many instances show little change from the behavior of the neat matrix [34]. The shear viscosity of suspensions containing a high concentration of fibers or fibers with a large aspect ratio (in the concentrated regime), exhibits a more pronounced behavior. At low shear rates the shear viscosity can exhibit a Newtonian plateau or rise in an unbounded manner and exhibit yield-like characteristics [84]. At high shear rates, shear thinning can occur at a reduced shear rate, which can result in a shear viscosity similar to that of the neat suspending medium. However, not all concentrated samples display yield-like behavior. Guo et al. [85] reported the absence of yield-like behavior for a high volume fraction $\phi = 0.384$ suspension in low-density polyethylene. The source of yielding is most likely complex interactions between the suspending medium and the fiber surface, which may not be present in all suspensions.

The more pronounced effect of fibers on shear viscosity at low shear rates and decreased effects at high shear rates is a reflection of changes in the suspension's microstructure [34]. For non-Brownian fiber suspensions the fiber orientation distribution is irreversibly changed by flow and leads to a shear history dependent rheology [34]. Higher shear rates impart a higher degree of fiber orientation in the flow direction [84]. As a result, a sample that is presheared may not exhibit the same shear viscosity compared to a sample that has not been sheared or has an isotropic fiber orientation distribution [86].

It is well established that nondilute fiber suspensions in Newtonian suspending media will exhibit a steady state $N_1$. This can be seen in Figure 4.8, which depicts the $N_1$ dependence on suspending medium viscosity at similar fiber concentration and $a_r$. Experimentally, the magnitude of $N_1$ follows a linear trend with shear rate, suspending medium viscosity, and fiber concentration, which coincides with theory [2]. Delineating the source of $N_1$ in non-Newtonian suspending media is not straightforward. The contribution to $N_1$ from the elasticity in the suspending medium is suppressed by the fiber [2]. However, it is speculated that the contribution from the fiber is consistent with theory. The mechanisms that contribute to the suppression of the elasticity are poorly understood but manifest in important processing applications such as die swell suppression.

***4.3.5.2 Small-Amplitude Oscillatory Rheology*** The addition of fibers to a non-Newtonian fluid increases the magnitude of the complex viscosity, $|\eta^*|$; the degree of this increase depends on the concentration, aspect ratio and orientation distribution of the fibers [85, 87]. The dependence of $|\eta^*|$ on glass fiber concentration and aspect ratio is not obvious. Similar to the shear viscosity behavior at low shear rates, $|\eta^*|$ can exhibit an enhanced Newtonian plateau [85, 87] or rise in an unbound manner at low frequencies [88–90]. At high frequencies, $|\eta^*|$ can begin to merge onto that of the matrix [89] or follow a similar shear thinning curve of the matrix at a constant enhance value [85, 87]. Furthermore, $|\eta^*|$ increasingly deviates from the shear viscosity at increasing

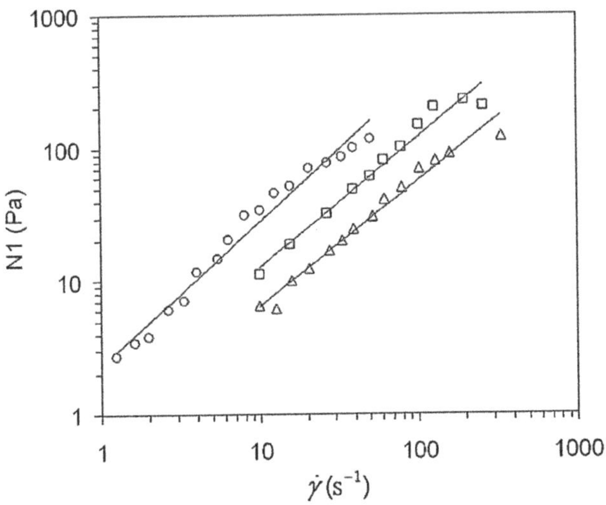


**FIGURE 4.8** $N_1$ vs. shear rate for various glass fiber suspensions with a constant fiber concentration ($\phi = 0.044$, $a_r = 276$) and different Newtonian suspending medium viscosities (Δ, 14; □, 16; ○; 120 Pa · s). All measurements were performed on a torsional rheometer with cone-and-plate geometry. Reprinted with permission from Ref. 2.

fiber concentrations for both Newtonian and non-Newtonian fiber suspensions [91–94]. This suggests that the Cox-Merz relationship [95] does not hold for fiber-filled systems.

It has been speculated that small-amplitude oscillatory shear in the linear strain region is too weak to induce fiber reorientation of the same magnitude as in steady shear flow [85]. This explains why there is a discrepancy between dynamic oscillatory and steady shear rheological measurements. However, there is experimental evidence that suggests that oscillatory flow can lead to irreversible orientation changes to some degree. Kim and Song [90] reported that for a suspension containing short glass fiber-filled ($\phi = 0.1766$, $a_r \sim 21.4$, concentrated regime) in a polybutylene terephthalate matrix exhibiting weak shear rate dependence, $|\eta^*|$ changed after repeated dynamic oscillatory tests on the same sample. The repeated tests are denoted as run 1–6 in Figure 4.9a. A newly loaded sample exhibited a yield-like behavior at low frequencies. After repeated tests on the same sample a Newtonian plateau developed, suggesting irreversible fiber reorientation.

Similar to the shear viscosity, the initial orientation of the sample can have an effect on $|\eta^*|$. A sample whose initial orientation is random will exhibit a larger $|\eta^*|$ than a sample whose fiber is oriented in the flow direction. Mobuchon et al. [87] reported a reduction of over 50% in $|\eta^*|$ after preshearing a concentrated short glass fiber-filled polypropylene ($\phi = 0.124$, $a_r \sim 21.4$, concentrated regime) (Fig. 4.9b). The reduction in $|\eta^*|$ was attributed to the alignment of the fibers in the flow direction during preshearing; the greater the preshear stress the greater the overall fiber alignment.

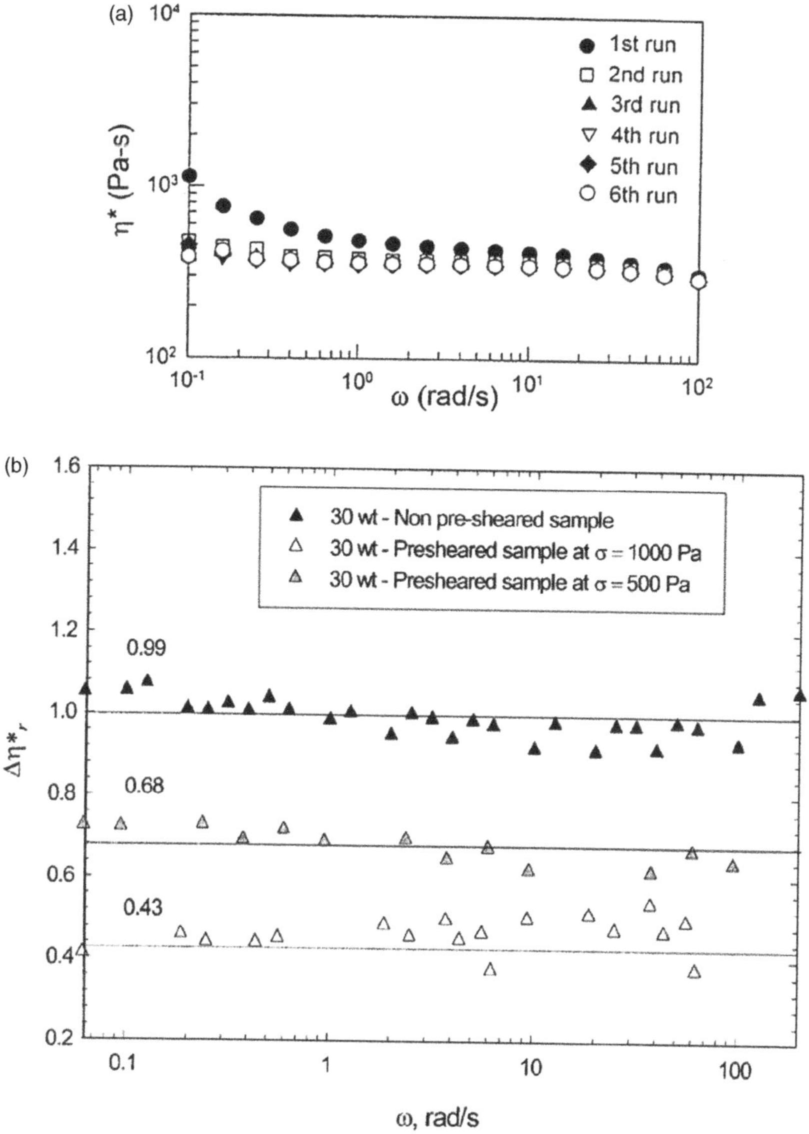


**FIGURE 4.9** (a) $|\eta^*|$ in small-amplitude ($\gamma_o = 0.15$) oscillatory shear for a short glass fiber-filled polybutylene terephthalate ($\phi = 0.1766$, $a_r \sim 21.4$, concentrated regime). Reprinted with permission from Ref. 90. (b) The effects of preshear on $|\Delta\eta_r^*|$ vs. frequency for a short glass fiber-filled polypropylene ($\phi = 0.124$, $a_r \sim 21.4$, concentrated regime). In this case $|\Delta\eta_r^*|$ is the magnitude of the complex viscosity of the presheared sample normalized by the no preshear sample values. Reprinted with permission from Ref. 82.

The dynamic functions of glass fiber suspensions in non-Newtonian fluids show a strain amplitude dependence that markedly increases with concentration and aspect ratio. Mutel and Kamal [96] tested PP containing short glass fibers at concentrations ranging from 10–40 wt % ($\phi = 0.0354 - 0.1805$). They found that the addition of short glass fibers to a PP caused increasing strain amplitude dependence on increasing concentration. On the contrary, the neat PP suspending medium exhibited linear strain dependence through a range of 5–50 strain %. Kim and Song [90] found that in a short glass fiber suspension, $|\eta^*|$ decreased at constant frequency as the strain amplitude was increased. This was attributed to larger strains increasing the average orientation of the fibers along the flow direction.

***4.3.5.3 Transient Responses*** Nondilute fiber suspensions can exhibit profound transient rheological material functions compared to their neat suspending media. In general, the behavior is characterized by a large and irreversible overshoot in both the shear stress growth $\sigma^+$, and the first normal stress growth $N_1^+$, functions. It has been well established through theory and experiment that the transient features are directly connected to the presence of fibers and their orientation within the suspension. For this reason, there has recently been a drive to establish unique material parameters for fiber suspension theory by fitting to the startup of flow and flow-reversal measurements [71, 82]. Subsequently, we review the key aspects of the transient rheology in relation to the fiber orientation. Certainly, this overview is incomplete and the reader is referred to additional sources [1, 71, 97, 98].

As discussed at the beginning of the chapter, the rheology, and especially the transient rheology of fiber suspensions can be highly dependent on the rheometer geometry used to characterize suspensions in steady shear. Measurements performed using parallel disks, which impose an inhomogeneous velocity gradient field from the center of the plate to the outer rim, lead to exaggerated transients in both $\sigma^+$ and $N_1^+$ [35]. In these instances the overshoot peak in $N_1^+$ can be orders of magnitude greater than the steady-state value and take over 150 strain units to reach a steady state. However, this transient response does not represent the true stress growth material function as a suspension would exhibit in a homogenous flow field [35].

The transient stresses of nondilute fiber suspensions in both Newtonian and non-Newtonian suspending media do not show an oscillatory response, as in dilute suspensions, but rise to a peak that decays to a steady state (stress overshoot) or simply decays to a steady state value with no apparent peak. Increasing fiber concentration and/or fiber aspect ratio increases the magnitude of the overshoot and the time needed to reach steady-state in a similar manner [84]. In general, $\sigma^+$ and $N_1^+$ approach a steady state in roughly 50 strain units. The overshoot is a result of an evolving microstructure, where, upon flow the fibers orient themselves toward the flow direction, a phenomenon that is irreversible. This behavior is seen in Figure 4.10a, which shows a stress growth measurement of a concentrated short glass fiber filled polybutylene

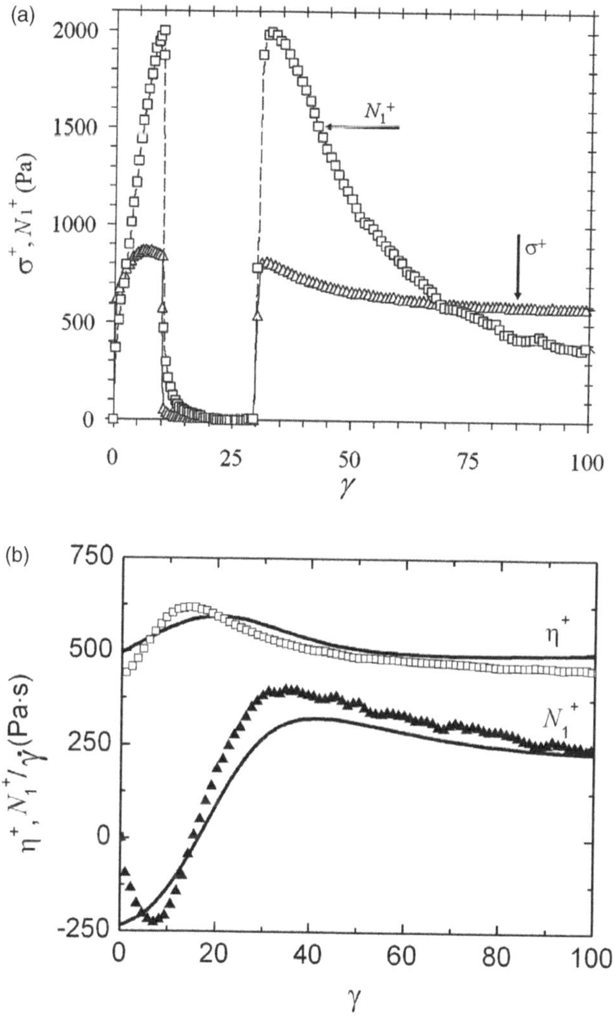


**FIGURE 4.10** Shear stress growth and first normal stress growth coefficients for (a) interrupted stress growth. Reprinted with permission from Ref. 67. (b) Flow reversal following steady shear of a short glass fiber-filled polybutylene terephthalate (30 wt %). The lines represent the fit of the Folgar-Tucker model with the addition of the slip term, combined with the Lipscomb stress equation. Reprinted with permission from Ref. 67.

terephthalate (PBT), which was interrupted at the peak of the $N_1^+$ overshoot. When the flow was interrupted the stresses decayed to zero following the relaxation time of the neat polymer, which is consistently observed in Newtonian and non-Newtonian suspending media [99]. However, when the flow was reapplied, the stresses immediately grew to their previous value, the overshoot peak, which in turn continued to decay toward a steady state. This behavior is consistent for both $\sigma^+$ and $N_1^+$.

A sample whose initial fiber orientation is aligned parallel to the direction of the velocity gradient will exhibit the greatest overshoot. In contrast, a sample whose initial fiber orientation is parallel to the flow direction will exhibit the lowest overshoot, with a randomly oriented sample falling between these two ideal cases [84]. Ramazani et al. [77] showed that preshearing a sample before the stress growth experiments removed the initial overshoot exhibited by randomly oriented fiber suspensions. The peak of the shear stress overshoot scales with strain at varying shear rates and typically occurs between 2 and 10 strain units. Current theory suggests that the peak in the shear stress overshoot corresponds to an average fiber orientation of 45° with respect to the flow direction [77, 84]. Experiments that connect the fiber microstructure to the rheological measurements have shown that the shear stress growth overshoot coincides with an overshoot in the $A_{12}$-component. Steady state occurs when the fiber orientation reaches a steady state, generally aligned in the flow direction, although complete alignment can be restricted due to fiber interaction [72].

Flow reversal experiments provide an avenue to condition a fiber suspension sample before testing. A flow reversal test begins with a sample being sheared in one direction for a predetermined time or strain. Subsequently, the flow is applied in the opposite direction either immediately or after some period of wait time. This effectively conditions the sample, which is shear history dependent, and increases the reproducibility of the measurements. Furthermore, it leads to rather unique values of $\sigma^+$ and $N_1{}^+$. The stress response to a flow reversal experiment can be seen in Figure 4.10b by way of the shear stress growth coefficient $\eta^+$ and $N_1{}^+$ for a short glass fiber-filled PBT. Upon flow reversal $\eta^+$ exhibits an overshoot, the peak of which occurs at an increased strain compared to the overshoot of a fresh sample. More interesting is the behavior of $N_1{}^+$, which displays an initial undershoot before rising to a small overshoot that decays toward a steady state. In flow reversal, the transients are believed to be a result of the fibers, which are not completely aligned in the flow direction, tumbling backward. It is interesting that current theory is able to predict the major features of this experiment, which can also be seen in Figure 4.10a. The solid lines represent the F-T model with the addition of the slip term (4.33) combined with the Lipscomb stress equation (4.24) [67]. It has also been shown that fitting the material parameters to the transient rheology allows for the qualitative prediction of the major features in the evolution of fiber orientation when the fiber orientation is quantitatively known at some point; i.e., initial condition or steady state [26, 67]. In the future fitting could potentially be used to determine unambiguous material parameters for a given fluid that could then be used in complex flow simulations to predict fiber orientation such as in injection molding.

***4.3.5.4 Extensional Viscosity*** The extensional behavior of fiber suspensions as determined experimentally remains elusive and rather poorly characterized. This is a direct consequence of the difficulty in measuring this material function. Tensile elongation samples tend to elongate in a nonuniform manner or "neck" which results in sample failure. Available data suggest the presence of fibers in a non-Newtonian fluid can lead to extensional thinning [100]. It is speculated that the thinning behavior is a result of a locally induced shear flow between the fibers. However, it is difficult to delineate material functions from the flow kinematics. Still, researchers have reported dramatic large extensional viscosities on the order of 20 to 40 times the shear viscosity [87, 101]. Research in this field would benefit from measurements performed using a lubricated squeeze flow apparatus.

## 4.4 FLEXIBLE FIBERS

A distinct class of fiber filled composites is one constructed from long fibers. The term *long fiber* is being used in this context to describe fibers that are able to bend or flex during processing and thus may affect both the flow behavior and material properties. The flexibility of a fiber can be easily characterized using (4.1). Because of the physical length of the long fibers, complications in measuring their rheology may exist. In this section, we discuss theory and experimental efforts of relating long fiber rheology.

### 4.4.1 Dilute Suspensions

***4.4.1.1 Semiflexible Fiber Kinematics*** The relatively limited published literature relating to flexible fibers is an indication of the difficulty associated with modeling such systems. One approach for semiflexible fibers, proposed by Strautins and Latz [102], is an extension of Jeffery's equation for rigid fibers. In this case the evolution equations are solved for two rigid rods (each rod of length $l_B$) connected by a joint. The joint and spring allows the two rigid rods to bend and exhibit flexibility as shown in Figure 4.11. In Figure 4.11, $\mathbf{p}$ and $\mathbf{q}$ are unit vectors associated with each of the rods of the flexible fiber. This is analogous to the unit vector $\mathbf{u}$ defined for rigid fiber theory, see, for example, Figure 4.3. However, for the case of a bead-rod fiber, there exists two rigid rods, each with an orientation dictated by either $\mathbf{p}$ or $\mathbf{q}$. The spring allows for flexibility and resistance to bending (spring stiffness coefficient $k$). The beads, on the other hand, provide surface area for hydrodynamic drag. This model is used

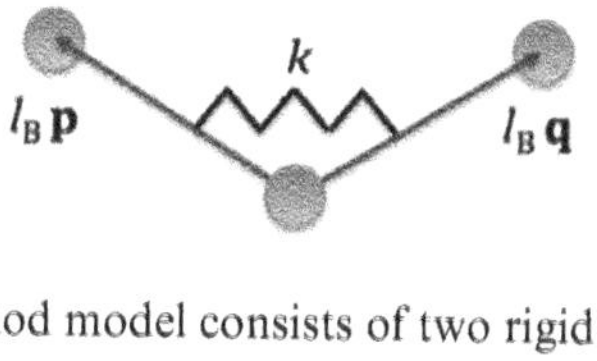


**FIGURE 4.11** The Bead-Rod model consists of two rigid rods connected to a pivoting bead.

in the formulation of a continuum model that describes the motion of fibers that can bend about its center of mass. In the treatment provided by Strautins and Latz, the fiber bending is limited to small angles, hence the theory is applicable for semiflexible systems. The theory consists of two orientation tensors and one orientation vector. One orientation tensor, similar to what exists in rigid-rod theory, describes the second moment of any one of the rods with respect to the orientation distribution function and is defined in (4.39). The other orientation tensor describes the mixed product of both rigid-rod vectors with the orientation distribution function (4.40). Last, the orientation vector is defined as the first moment of the distribution function, using either orientation vector, defined in (4.41).

$$\mathbf{A} = \iint \mathbf{pp}\psi(\mathbf{p}, \mathbf{q}, t)d\mathbf{p}d\mathbf{q} \tag{4.38}$$

$$\mathbf{B} = \iint \mathbf{pq}\psi(\mathbf{p}, \mathbf{q}, t)d\mathbf{p}d\mathbf{q} \tag{4.39}$$

$$\mathbf{C} = \int \mathbf{p}\psi(\mathbf{p}, \mathbf{q}, t)d\mathbf{p} \tag{4.40}$$

It is important to note that **C** does not vanish in the case of the Bead-Rod model as it does for the purely Rigid-Rod model. The equations that describe how a given flow field affects the orientation functions are then defined below:

$$\begin{aligned}\frac{D\mathbf{A}}{Dt} &= (\mathbf{W}\cdot\mathbf{A} - \mathbf{A}\cdot\mathbf{W}) + (\mathbf{D}\cdot\mathbf{A} + \mathbf{A}\cdot\mathbf{D} - (2\mathbf{D}:\mathbf{A})\mathbf{A}) + \\ &\quad \ldots \frac{l_B}{2}[\mathbf{Cm} + \mathbf{mC} - 2(\mathbf{m}\cdot\mathbf{C})\mathbf{A}] - 2k[\mathbf{B} - \mathbf{A}tr(\mathbf{B})]\end{aligned} \tag{4.41}$$

$$\begin{aligned}\frac{D\mathbf{B}}{Dt} &= (\mathbf{W}\cdot\mathbf{B} - \mathbf{B}\cdot\mathbf{W}) + (\mathbf{D}\cdot\mathbf{B} + \mathbf{B}\cdot\mathbf{D} - (2\mathbf{D}:\mathbf{A})\mathbf{B}) + \\ &\quad \ldots \frac{l_B}{2}[\mathbf{Cm} + \mathbf{mC} - 2(\mathbf{m}\cdot\mathbf{C})\mathbf{B}] - 2k[\mathbf{A} - \mathbf{B}tr(\mathbf{B})]\end{aligned} \tag{4.42}$$

$$\frac{D\mathbf{C}}{Dt} = \nabla\mathbf{v}^t\cdot C - (\mathbf{A}:\nabla\mathbf{v}^t)\mathbf{C} + \frac{l_B}{2}[\mathbf{m} - \mathbf{C}(\mathbf{m}\cdot\mathbf{C})] - k\mathbf{C}[1 - tr(\mathbf{B})] \tag{4.43}$$

$$\mathbf{m} = \frac{\partial^2 v_i}{\partial x_j \partial x_k} A_{jk}\mathbf{e}_i \tag{4.44}$$

As defined earlier, the velocity gradient $\nabla\boldsymbol{v} = \partial v_j/\partial x_i$, and tr() represents the trace of a specified tensor. Because this model accounts only for viscous flow effects, flexibility caused by flow will occur only when second-order spatial derivatives of the velocity term exist in (4.44). As a result, in simple flows, the model does not predict flow induced bending. It has been shown that this theory exhibits larger flow induced orientation when compared to that of the F-T model [102]. This was said to be a direct cause of the flexibility of the fiber because the flexibility forced the fiber to orient more in the center of the channel flow. Other than this example, the Bead-Rod model has gone unexamined to our knowledge.

#### 4.4.1.2 Flexible Fiber Kinematics

To simulate perfect flexibility a fiber must be discretized into multiple nodes of junction points, as described above, complicating the use of orientation tensors. Here the common approach is to perform a dynamic simulation of an individual fiber, or multiple fibers within a simulation box. Hinch [91] pioneered the work on flexible fibers by describing the motion of an inextensible but perfectly flexible fiber. To determine an equation of motion for the fiber, a force balance was performed on an infinitely thin thread where the tension forces, $T$, in the thread were set equal to the viscous forces. This resulted in two equations, one governing the tension within the thread, and the other governing the time rate of change of the position vector $\mathbf{x}(s, t)$ with respect to the arc length $s$

Evolution equation
$$\dot{\mathbf{x}} = \frac{\partial \mathbf{x}}{\partial t} = \nabla \mathbf{v} \cdot \mathbf{x} + \frac{\partial T}{\partial s}\frac{\partial \mathbf{x}}{\partial s} + \frac{1}{2} T \frac{\partial^2 \mathbf{x}}{\partial s^2} \tag{4.45}$$

Equation for tension
$$\frac{\partial^2 T}{\partial s^2} - \frac{1}{2}\left(\frac{\partial^2 \mathbf{x}}{\partial s^2}\right)^2 T = -\frac{\partial \mathbf{x}}{\partial t} \cdot \mathbf{D} \cdot \frac{\partial \mathbf{x}}{\partial t} \tag{4.46}$$

Equations (4.45) and (4.46) can be solved given the initial fiber orientation and with boundary conditions that the tension is zero at the ends of the thread: $T = 0$ at $s = +/- L$. The theory predicts that the fiber straightens to align itself in the flow direction for simple shear or stretch direction for simple elongation.

### 4.4.2 Nondilute Suspensions

#### 4.4.2.1 Theory and Simulations

There is limited literature that attempts to couple a stress to orientation in flexible fiber systems. In what does exist, researchers attempt to extend the use of stress equations based on short fiber theory to flexible systems, more out of convenience than from a theoretical basis. To apply the stress a decision must be made as to how to describe the orientation of a long fiber. For example, some authors have used the end-to-end vector to describe the orientation of long fiber [97]. Other authors have used the tangential orientation of the fiber in combination with direct simulation methods [103].

The GENERIC framework of Grmela [78, 104], which is based on a free energy approach, does allow for the addition of flexibility. Specifically, the GENERIC model uses entropy expressions gained from kinetic theory of polymers and dissipation to phenomenologically describe effects of fiber flexibility, excluded volume effects, and fiber and polymer mobility. Qualitatively, the model predicts an enhancement in the transient stress overshoot and viscosity with increased flexibility and concentration. In addition, increased flexibility reduces the steady state alignment in the flow direction. Rajabian [78] used this model to simulate the steady state properties of Kevlar fibers in polyethylene and reported it to have good agreement with the viscosity measurements but poor agreement with the first normal stress difference. Further, the model looses accuracy at higher shear rates for both properties. It is also important to note that the accuracy of the orientation model predictions provided by the GENERIC model has not been reported in the literature [78, 104].

Perhaps the first work on nondilute flexible fiber systems was performed by Goddard and Huang [105], who extended the work of Hinch [91] for a noninteracting perfectly flexible fiber. The authors chose to express the viscous drag by a mobility tensor, which can be seen in the following equations

$$\dot{\mathbf{x}} = K_L \left( \frac{\partial T}{\partial s} \right) \frac{\partial \mathbf{x}}{\partial s} + T\mathbf{K}_T \cdot \frac{\partial^2 \mathbf{x}}{\partial s^2} + \nabla \mathbf{v} \cdot \mathbf{x} \tag{4.47}$$

$$K_L \frac{\partial^2 T}{\partial s^2} - \left| \frac{\partial^2 \mathbf{x}}{\partial s^2} \right|^2 K_N T = - \frac{\partial \mathbf{x}}{\partial s} \cdot \nabla \mathbf{v} \cdot \frac{\partial \mathbf{x}}{\partial s} \tag{4.48}$$

where $\mathbf{K}_\mathrm{T}$ is the transverse mobility tensor while $K_\mathrm{L}$ and $K_\mathrm{N}$ are the lateral and normal components of the mobility tensor, respectively, defined by

$$K_L = \frac{\partial \mathbf{x}}{\partial s} \cdot \mathbf{K}_T \cdot \frac{\partial \mathbf{x}}{\partial s} \text{ and, } K_N = \left( \frac{\partial^2 \mathbf{x}}{\partial s^2} \cdot \mathbf{K}_T \cdot \frac{\partial^2 \mathbf{x}}{\partial s^2} \right) \left| \frac{\partial^2 \mathbf{x}}{\partial s^2} \right|^{-2} \tag{4.49}$$

The mobility tensor represents the inverse of the hydrodynamic resistance per unit length. This model has been used to help understand long fiber configuration in a variety of flow fields [106]. However, no attempt was made to associate a stress with the fiber orientation.

Direct simulation methods can be used to track the dynamic behavior of a large number of interacting fibers, and in contrast to continuum simulations, they have the potential to inherently describe very detailed kinematics [72]. Simulations, however, carry a large computational cost. For this reason, direct simulation methods are usually employed to study a specific behavior of a fiber suspension such as lubrication and hydrodynamic forces, and effects of concentration and flexibility [92–94, 107–109]. For example, Skjetne et al. [93] explored a variety of flexible fiber behaviors by modeling the fibers as a connection of prolate spheroids. Similarly, Joung [97] used direct simulations to predict how fiber curvature affected the relative viscosity of a long fiber suspension. They found that even small changes in fiber curvature resulted in large changes in the relative viscosity. Tang and Advani [94] simulated long fibers in simple shear flow by modeling the fiber as spheres connected by massless rigid rods with ball and socket joints. Their simulations resulted in viscosity predictions between those calculated from transversely isotropic fluid theory and slender body theory [94]. Nevertheless, direct simulations are of great importance in the quest of understanding specific phenomena; however, due to their computational cost they are presently of little use to most commercial processes.

***4.4.2.2 Rheological Characterization*** Characterizing the rheology of long glass fibers systems poses the same challenges realized by short glass fiber suspensions. However, in many cases the relative ratio of the rheometer geometry to the characteristic length of the fiber is greatly increased. Typically, this complicates measurements to a greater degree than short glass fiber systems. Published experimental literature for long fiber systems is limited but researchers have used rotational equipment, such as parallel-disk ignoring the inhomogeneous shear field, or cone and plate rheometers ignoring the existence of wall boundary effects. In addition to rotational equipment, sliding plate rheometers have shown some effectiveness when dealing with long concentrated fiber systems [110].

Thomasset [111] performed steady-state rheological experiments of glass fiber reinforced polypropylene samples using a parallel-disk rheometer. The authors found a more pronounced yield-like behavior for the long fiber systems than in similar short glass fiber suspensions. Keshtkar et al. [98] and Goto et al. [112] found the steady state viscosity of a fiber suspension to increase with both fiber concentration and flexibility. It was suggested that this was a result of increased likelihood of fiber–fiber interactions. Another interesting characteristic of long fibers reinforced within polymeric melts remains in their ability to flocculate. Keshtkar et al. [98] report that at low shear rates the viscosity of flexible fiber suspensions is quite high, but at high shear rates the viscosity does not depend on the aspect ratio. They attributed this behavior to the notion that flexible fibers may network or flocculate and result in more fiber–fiber contacts at low shear rates, thus enhancing properties [92]. Also significant shear thinning behavior of long fiber suspensions, particularly in the semiconcentrated regime and at low shear rates, is reported in literature [98]. This behavior has also been attributed to floc formation, and fiber–fiber interlocking. Similar flexibility effects are found on the magnitude of the primary normal force, which is found to increase with both fiber concentration and flexibility [13, 88, 98].

The work by Keshtkar et al. [98] is also among one of the very few experiments concerning the stress growth of long fibers using a parallel-disk plate geometry. In their experiments, they showed that in stress growth experiments fiber flexibility enhances both the magnitude and width of the stress overshoot, when plotted with strain. Similar characteristics were reported by Agarwal et al. [113], who used a sliding plate rheometer to analyze the startup rheology of long fiber reinforced polypropylene. Keshtkar et al. [98] ascertain that increased fiber flexibility creates more fiber–fiber interaction than an equivalent sample made of rigid fibers. Further, in flow reversal experiments Keshtkar et al. [98] report similar effects of flexibility. Specifically, they found that the delayed undershoot started quicker and the magnitude was more pronounced as flexibility increased. In response to this, they suggest that fiber reorientation begins at lower strain values for higher flexible systems, thus causing the longer transient behavior in the stress undershoot measurements. This may be due to lower fiber orientation of the flexible fibers on flow reversal, versus their rigid fiber counterparts. Again, the increase in magnitude of the overshoot is due to higher fiber–fiber interaction.

## 4.5 CONCLUSIONS

In this chapter we have presented an introduction to the theory for describing the flow behavior and experimental rheology of fiber suspensions. It is without doubt that the flow behavior and associated stresses of these complex fluids are intimately linked to the fiber structure. As such, the numerical treatment of rigid and flexible fiber suspensions is much different in addition to the expected experimental rheology, as currently, little is known. This is a direct result of the stored and released energy by the flexible fibers subject to deformation, and complex fiber configuration.

The theory for rigid fiber suspensions consists of two parts: (1) an equation governing the motion of the fibers and, (2) an equation to calculate the stress. Theoretical treatment for the evolution of fiber orientation has relied on an extension of rigid body theory to a statistical population of like rods. Modification of this theory to extend predictions to nondilute systems has relied on phenomenological expressions, which are able to qualitatively display similar trends to what has been shown experimentally. The equations used to describe the extra stress are less evolved, for concentrated suspensions and are essentially identical to that of dilute suspensions. It is expected that future efforts will be made to include fiber contact to the stress and elucidate the effect of elasticity on non-Newtonian suspending mediums.

Rheological measurements of glass fiber suspensions are complex but have proven to be insightful in understanding the connection between flow and fiber orientation. Specifically, transient shear measurements will likely continue to be a measurement used in the progression of model development as predictions are compared to experiments. In this regard, it cannot be stressed enough that reported measurements should be performed on well-defined suspensions. Furthermore, steady shear measurements of nondilute suspensions should be performed only in rheometric geometries that induce a homogeneous shear field. On a similar topic, the extensional rheology of fiber suspensions is ill defined. This area of research would largely benefit from a well designed series of experiments that accurately characterize the elongational behavior of fiber suspensions. Potentially, this could be accomplished with a lubricated squeeze flow, or semihyperbolic die rheometer [114].

## 4.6 NOMENCLATURE

| | |
|---|---|
| $\mathbf{A}$ | Second-order orientation tensor |
| $\mathbf{A}_4$ | Fourth-order orientation tensor |
| $a_r$ | Aspect ratio: defined as the particles (major axis)/(minor axis) |
| $a_{re}$ | Equivalent aspect ratio |
| $C$ | Constant for the analytical solution of Jeffery's equation |
| $C''$ | Shaqfeh and Fredrickson model constant |
| $C_I$ | Folgar-Tucker constant |
| $c_1$ | Fiber stress constant in Lipscomb model |
| $\mathbf{D}$ | Rate of strain tensor |
| $d$ | Fiber diameter |
| $D/Dt$ | Material derivative |
| $D_r$ | Rotational diffusion constant |
| $E_Y$ | Young's modulus |
| $F_z$ | Normal force |
| $G_v$ | Phan-Thien and Graham model function |
| $g$ | Acceleration due to gravity |
| $h$ | Spacing between fibers |
| $k$ | Constant for the analytical solution of Jeffery's equation |
| $k_B$ | Boltzmann's constant |
| $\mathbf{K}_T$ | Transverse mobility tensor |
| $K_L$ | Lateral component of the mobility tensor |
| $K_N$ | Normal component of the mobility tensor |
| $L$ | Fiber length |
| $L_n$ | Number average fiber length |
| $L_w$ | Weight average fiber length |
| $M$ | Ellipsoid major axis or torque |
| $m$ | Ellipsoid minor axis or extensional flow viscosity constant |
| $m_1$, $m_2$ | Fit parameters |
| $N$ | Hydrodynamic stress constant |
| $N_1$ | First normal stress difference |
| $N_1^+$ | First normal stress difference growth function |
| $n$ | Number of fibers per unit volume |
| $S_m$ | Average spacing between fibers |
| $T$ | Period of rotation or tension in Hinch equations |
| $t$ | Time |
| $t_s$ | Sedimentation time scale |
| $\mathbf{u}$ | Unit vector used to represent fiber orientation |
| $\dot{u}$ | Material derivative of $\mathbf{u}$ |
| $v$ | Velocity |
| $\mathbf{W}$ | Vorticity tensor |
| $\mathbf{x}$ | Position vector |

**Greek Symbols**

| | |
|---|---|
| $\alpha$ | Folgar-Tucker constant |
| $\gamma$ | Strain |
| $\dot{\gamma}$ | Shear rate |
| $\xi_r$ | Rotational friction factor |
| $\rho$ | Density |
| $\eta$ | Non-Newtonian steady-state shear viscosity |
| $\eta_s$ | Suspending medium viscosity |
| $\eta^*$ | Complex viscosity |
| $\eta^+$ | Shear stress growth coefficient |
| $\eta_0$ | Zero shear viscosity |
| $\eta_r$ | Reduced viscosity |
| $\Delta\eta_r{}^*$ | Reduced complex viscosity |
| $\eta_{sp}$ | Specific viscosity |
| $\theta$ | Zenith angle |
| $\theta_0$ | Zenith angle initial condition |
| $\boldsymbol{\sigma}$ | Total stress tensor |
| $\boldsymbol{\sigma}^+$ | Shear stress growth function |
| $\boldsymbol{\sigma}^-$ | Shear stress decay function |
| $\boldsymbol{\tau}$ | Extra stress tensor |
| $\tau_{critical}$ | Shear stress at which a fiber buckles |
| $\varphi$ | Azimuthal angle |
| $\varphi_0$ | Azimuthal angle initial condition |
| $\phi$ | Fiber volume fraction |
| $\psi$ | Probability distribution function |
| $\omega$ | Frequency |
| $\Omega$ | Constant for $D_{ro}$ equation |

## DISCLAIMER

Certain commercial materials are identified in this paper to foster understanding. Such identification does not imply recommendation or endorsement by the National Institute of Standards and Technology, nor does it imply that the materials or equipment identified are necessarily the best available for the purpose.

## REFERENCES

1. A.P.R. Eberle et al., *Ind. Eng. Chem. Res*, 47, 3470–3488 (2008).
2. M.A. Zirnsak et al., *J. Non-Newt. Fluid Mech.*, 54, 153–193 (1994).
3. R.G. Larson, *The Strucuture and Rheology of Complex Fluids*, Oxford University Press, New York, 1999.

4. D.H. Chung and T.H. Kwon, *Korean-Australian Rheol. J.*, 14, 175–188 (2002).
5. S.G. Advani and C.L. Tucker, *J. Rheol.*, 31, 751–784 (1987).
6. C.J.S. Petrie, *J. Non-Newt. Fluid Mech.*, 87, 369–402 (1999).
7. J.M. Dealy, *J. Rheol.*, 39, 253–265 (1995).
8. M. Chaouche and D.L. Koch, *J. Rheol.*, 45, 369–382 (2001).
9. M. Doi and S.F. Edwards, *The Theory of Polymer Dynamics*, Oxford University Press, New York, 1988.
10. J.M. Burgers, *Kon. Ned. Akad. Wet. Verhand (Erste Sectie)*, 16, 113–184 (1938).
11. G.K. Batchelor, *J. Fluid Mech.*, 46, 813–829 (1971).
12. M. Doi and N.Y. Kuzuu, *J. Poly. Sci. Part B, Polym. Phys.*, 18, 409–419 (1980).
13. L.H. Switzer and D.J. Klingenberg, *J. Rheol.*, 47, 759–778 (2003).
14. M.W. Darlington and G.R. Smith, *Polymer*, 16, 495–462 (1975).
15. M.W. Darlington, *J. Mat. Sci.*, 10, 906–909 (1975).
16. F. Lisy et al., *J. Appl. Polym. Sci.*, 52, 329–352 (1994).
17. K. Thomas and D.E. Meyer, *Plast. Rubber Compos. Process. Appl.*, 1, 35–36 (1976).
18. A. Mavrich et al., *J. Adhesion*, 46, 91–102 (1994).
19. A. Kriete, *Visualization in Biomedical Microscopies—3D Imaging and Computer Applications*, VCH Publishers, New York, 1992.
20. P.J. Hine et al., *Comp. Sci. Tech.*, 53, 125–131 (1993).
21. A.R. Clarke et al., *Comp. Sci. Tech.*, 55, 75–91 (1995).
22. P.J. Hine and R.A. Duckett, *Polym. Compos.*, 25, 237–254 (2004).
23. A.R. Clarke et al., *J. Microscopy—Oxford*, 69–79 (1993).
24. G. Fischer and P. Eyerer, *Polym. Compos.*, 9, 297–304 (1988).
25. Y.H. Lee *et al.*, *Mater. Res. Innov.*, 6, 65–72 (2002).
26. A.P.R. Eberle et al., *J. Non-Newt. Fluid Mech.*, 165, 110–119 (2010).
27. R.S. Bay and C.L. Tucker, *Polym. Eng. Sci.*, 32, 240–253 (1992).
28. B. Moginger and P. Eyerer, *Composites*, 22, 394–398 (1991).
29. E. Lefranche et al., *Adv. Polym. Technol.*, 24, 114–131 (2005).
30. C.W. Macosko, *Rheology Principles, Measurements, and Applications*, Wiley-VCH, New York, 1994.
31. M.A. Nawab and S.G. Mason, *J. Phys. Chem.*, 62, 1248–1253 (1958).
32. W.R. Blankeney, *J. Colloid Interface Sci.*, 22, 324 (1966).
33. A. Attanasio et al., *Trans. Soc. Rheol.*, 16, 147–154 (1972).
34. L.A. Utracki, *Two-Phase Polymer Systems*, Oxford University Press, New York, 1991.
35. A.P.R. Eberle et al., *J. Rheol.*, 53, 1049–1068 (2009).
36. A.J. Giacomin et al., *Polym. Eng. Sci.*, 29, 499–504 (1989).
37. G. Segre and A. Silberberg, *J. Fluid Mech.*, 14, 115–135 (1962).
38. G. Segre and A. Silberberg, *J. Fluid Mech.*, 14, 136–157 (1962).
39. G.B. Jeffery, *Proc. R. Soc. Lond. A*, 102, 161–179 (1922).
40. A. Einstein, *Ann. Phys.*, 19, 289–306 (1906).
41. S.E. Barbosa and M.A. Bibbo, *J. Poly. Sci., Part B, Polym. Phys.*, 38, 1788–1799 (2000).

42. K.B. Moses et al., *Rheol. Acta*, 40, 296–306 (2001).
43. F.P. Bretherton, *J. Fluid Mech.*, 14, 284–304 (1962).
44. R.G. Cox, *J. Fluid Mech.*, 45, 625–657 (1971).
45. H. Brenner, *Int. J. Muliphase Flow*, 1, 195–341 (1974).
46. B.J. Trevelyan and S.G. Mason, *J. Colloid Sci.*, 6, 354–367 (1951).
47. R.L. Powell, *J. Stat. Phys.*, 62, 1073–1091 (1990).
48. H. Giesekus, *Rheol. Acta*, 2, 101–112 (1962).
49. H. Giesekus, *Rheol. Acta*, 2, 112–122 (1962).
50. G.L. Hand, *Arch. Rational Mech. Anal.*, 7, 81–86 (1961).
51. S.M. Dinh and R.C. Armstrong, *J. Rheol.*, 28, 207–227 (1984).
52. M. Doi and S.F. Edwards, *The Theory of Polymer Dynamics*, Clarendon Press, Oxford, 1987.
53. E.J. Hinch and L.G. Leal, *J. Fluid Mech.*, 52, 683–712 (1972).
54. G.L. Hand, *J. Fluid Mech.*, 13, 33–62 (1962).
55. E.J. Hinch and L.G. Leal, *J. Fluid Mech.*, 76, 187–208 (1976).
56. D. Barthes-Biesel and A. Acrivos, *Int. J. Muliphase Flow*, 1, 1–24 (1973).
57. S.G. Advani and C.L. Tucker, *J. Rheol.*, 34, 367 (1990).
58. D.H. Chung and T.H. Kwon, *Polym. Compos.*, 22, 636–649 (1999).
59. D.A. Jack and D.E. Smith, *J. Comp. Mat.*, 38, 1851–1871 (2004).
60. G.G. Lipscomb et al., *J. Non-Newt. Fluid Mech.*, 26, 297–325 (1988).
61. M. Sepehr et al., *J. Polym. Eng.*, 24, 579–610 (2004).
62. Y. Ivanov et al., *J. Rheol.*, 26, 213–230 (1982).
63. A. Okagawa et al., *J. Colloid Interface Sci.*, 45, 303 (1973).
64. E. Ganani and R.L. Powell, *J. Compos. Mater*, 19, 194 (1985).
65. Y. Iso et al., *J. Non-Newt. Fluid Mech.*, 62, 115–134 (1996).
66. Y. Iso et al., *J. Non-Newt. Fluid Mech.*, 62, 135–153 (1996).
67. A.P.R. Eberle et al., *J. Rheol.*, 53, 685–705 (2009).
68. E. Ganani and R.L. Powell, *J. Rheol.*, 30, 995–1013 (1986).
69. M.C. Altan, *J. Thermoplast. Compos. Mater.*, 3, 275–313 (1990).
70. F.P. Folgar and C.L. Tucker, *J. Reinf. Plast. Compos.*, 3, 98–119 (1984).
71. R.S. Bay, "Fiber Orientation in Injection Molded Composites: A Comparison of Theory and Experiment," thesis, University of Illinois, Urbana-Champaign, 1991.
72. R.S. Bay and C.L.T. III, *Polym. Compos.*, 13, 332–341 (1992).
73. N. Phan-Thien et al., *J. Non-Newt. Fluid Mech.*, 103, 251–260 (2002).
74. H.M. Huynh, "Improved Fiber Orientation Predictions for Injection-Molded Composites," Master's thesis, University of Illinois, Urbana-Champaign, 2001.
75. M. Sepehr et al., *J. Non-Newt. Fluid Mech.*, 123, 19–32 (2004).
76. J. Wang et al., *J. Rheol.*, 52, 1179–1200 (2008).
77. A. Ramazani et al., *J. Rheol.*, 45, 945–962 (2001).
78. M. Rajabian et al., *Rheol. Acta*, 44, 521–535 (2005).
79. E.S.G. Shaqfeh and G.H. Fredrickson, *Phys. Fluids A*, 2, 7–25 (1990).
80. N. Phan-Thien and A.L. Graham, *Rheol. Acta*, 30, 44–57 (1991).

81. T. Kitano and M. Funabashi, *Rheol. Acta*, 25, 606–617 (1986).
82. A. Ramazani et al., *J. Non-Newt. Fluid Mech.*, 73, 241–260 (1997).
83. J. Azaiez, *J. Non-Newt. Fluid Mech.*, 66, 35–54 (1996).
84. H.M. Laun, *Colloid & Polym. Sci.*, 262, 257–269 (1984).
85. R. Guo et al., *Polym. Eng. Sci.*, 45, 385–399 (2005).
86. A.T. Mutel, "Rheological Behavior and Fiber Orientation in Simple Flow of Glass Fiber Filled Polypropylene Melts," Ph.D. thesis, McGill University, Montreal, Quebec, Canada, 1989.
87. C. Mobuchon et al., *Polym. Comp.*, 26, 247–264 (2005).
88. J. Thomasset et al., *J. Non-Newt. Fluid Mech.*, 125, 25–34 (2005).
89. J.P. Greene and J.O. Wilkes, *Polym. Eng. Sci.*, 35, 1670–1681 (1995).
90. J.K. Kim and J.H. Song, *J. Rheol.*, 41, 1061–1085 (1997).
91. E.J. Hinch, *J. Fluid Mech.*, 74, 317–333 (1976).
92. M. Djalili-Mogaddam and S. Toll, *J. Non-Newt. Fluid Mech*, 132, 73–83 (2005).
93. P. Skjetne et al., *J. Chem. Phys*, 107, 2108–2121 (1997).
94. W.Z. Tang and S.G. Advani, *CMES*, 8, 165–176 (2005).
95. W.P. Cox and E.H. Merz, *J. Polym. Sci.*, 28, 619–622 (1958).
96. A.T. Mutel and M.R. Kamal, *Polym. Compos.*, 5, 29–35 (1984).
97. C.G. Joung et al., *J. Non-Newt. Fluid Mech.*, 102, 1–17 (2002).
98. M. Keshtkar et al., *J. Rheol.*, 53, 631–650 (2009).
99. M. Sepehr et al., *J. Rheol.*, 48, 1023–1048 (2004).
100. M.R. Kamal et al., *Polym. Compos.*, 5, 289–298 (1984).
101. G.B. Weinberger and J.D. Goddard, *Int. J. Multiphase Flow*, 1, 465–486 (1974).
102. U. Strautins and A. Latz, *Rheol. Acta*, 46, 1057–1064 (2007).
103. R.F. Ross and D.J. Klingenberg, *J. Chem. Phys.*, 106, 2949–2960 (1997).
104. M. Grmela et al., *J. Chem. Phys.*, 109, 6973–6981 (1998).
105. J.D. Goddard and Y.-H. Huang, *J. Non-Newt. Fluid Mech.*, 13, 47–62 (1983).
106. R. Balasubramanyam, *Composites*, 20, 14–20 (1989).
107. G. Ausias et al., *J. Non-Newt. Fluid Mech.*, 135, 46–57 (2006).
108. J. Ferec et al., *J. Rheol.*, 53, 49–72 (2009).
109. X.J. Fan et al., *J. Non-Newt. Fluid Mech.*, 74, 113–135 (1998).
110. K.A. Ericsson et al., *Rheol. Acta*, 36, 397–405 (1997).
111. J. Thomasset et al., *J. Non-Newt. Fluid Mech.*, 125, 25–34 (2005).
112. S. Goto et al., *Rheol. Acta*, 25, 246–256 (1986).
113. N. Agarwal, *Transient Rheology of Polymer Melts Containing Long Glass Fibers*, MS thesis, Virginia Tech, Blacksburg, Virginia, USA, 2009.
114. D.G. Baird and I. Huang, *Appl. Rheol.*, 16, 312–320 (2006).

**CHAPTER 5**

# RHEOLOGY AND PROCESSING OF POLYMER NANOCOMPOSITES

RAMANAN KRISHNAMOORTI[1] and TIRTHA CHATTERJEE[2]
[1] Department of Chemical and Biomolecular Engineering, University of Houston, Houston, TX, USA
[2] Materials Research Laboratory, University of California, Santa Barbara, CA

## CONTENTS



## 5.1 INTRODUCTION

The development of advanced lightweight polymeric materials that possess significantly improved properties due to dispersed nanoparticles has attracted significant interest over the last two decades and is now starting to reach

*Applied Polymer Rheology: Polymeric Fluids with Industrial Applications*, First Edition.
Edited by Marianna Kontopoulou.

commercial applications [1–3]. Understanding the rheological properties and processing characteristics of such nanocomposites is crucial to gain a fundamental understanding of the processability and structure–property relations for these materials. While the highly mature fields of filled polymers, containing large amounts of added micron-size inorganic fillers, and colloidal dispersions that have been studied and understood for the last 50 years provide a good foundation to understand nanocomposites based on nanoparticles, several unique physical and chemical aspects of polymer nanocomposites set them apart. The nanoscopic dimensions and large aspect ratios often encountered with engineering nanoparticles employed in nanocomposites result in six interrelated characteristics [4]: (1) low percolation threshold ($\sim$0.05 to 2 vol %); (2) orientational and positional correlation between particles occurring at low volume fractions; (3) high number density of particles per unit volume ($10^6$–$10^8$ particles/$\mu m^3$); (4) large interfacial area to volume ratio for the nanoparticles ($10^3$–$10^4$ $m^2$/mL); (5) short distances between particles; and (6) comparable size scales of the nanoparticle, distance between nanoparticles, and the dimensions of the polymer. These have profound influences on the rheological properties and processing of polymer nanocomposites and therefore set such nanocomposites apart from other multiphase and heterogeneous materials.

In this review, we describe the linear and nonlinear viscoelastic properties and processing characteristics for such nanocomposites and correlate these properties to the nanoscopic and mesoscopic structures of the hybrid materials. We try to provide general principles that guide the flow properties of a broad class of nanoparticle based composites, including those containing carbon black; silica; layered silicates; grapheme; single-and multi-walled carbon nanotubes; and a variety of inorganic-, organic-, metallic- and metal oxide–based spherical nanoparticles. We begin the review with a brief overview of the dispersion of nanoparticles in polymers, and subsequently describe overarching principles that describe the linear and nonlinear rheological properties in such nanocomposites. Last we relate these rheological properties to the processing of such nanocomposites and the ability to produce materials with controlled three-dimensional structure.

## 5.2 DISPERSION OF NANOPARTICLES AND THEIR CHARACTERIZATION

The primary challenge for achieving successful dispersions of aggregated or crystallized (into predominantly low-dimensional structures) nanoparticles without compromising the inherent advantages of the nanoparticles, such as their unique and sometimes outstanding mechanical, thermal, and optical properties remains and dominates the development of such materials. Moreover, methods to *quantitatively* characterize the dispersions of the nanoparticles, challenged by structures that are hierarchical, continue to be outstanding issues that need to be addressed.

The dispersion of nanoparticles in polymer matrices is largely controlled by the enthalpy of mixing and is seldom driven exclusively by entropic driving forces. The interparticle interactions between nanoparticles, which are important in the determination of the enthalpy of mixing, are a function of chemical nature and the size, shape, and functionalization of the nanoparticles. Even for uncharged nanoparticles, the van der Waals interaction energy per pair of particles scales with the size of the particles and inversely with the distance between the two particles [5]. Moreover, the specific dependence of the interactions on the size and distance vary with particle geometry. For anisotropic rod and plate-like nanoparticles following the arguments of Onsager, it is clear that beyond a critical volume fraction that scales $\sim 1/\alpha$ ($\alpha$ being the aspect ratio), in the absence of strong matrix-nanoparticle attractive interactions, the nanoparticles form nematic, smectic, and columnar phases [6]. Such ideas have been refined using self-consistent and field-theoretical methods and indicate that to obtain well-dispersed anisotropic nanoparticles at modest to high concentrations requires significant efforts to reduce particle–particle attraction and increase particle–matrix attractive interactions. Thus for anisotropic nanoparticles, at relatively modest concentrations, entropic effects will not result in a dispersion of the nanoparticles [7].

Typical experimental strategies to create nanoparticle dispersions in polymers using functionalization of the nanoparticle have included weak interaction based steric stabilization of nanoparticles aided by van der Waals or $\pi$ stacking interactions between an organic ligand and the nanoparticle, ionic interactions between ligand and nanoparticle, and covalent functionalization of the nanoparticle [4]. In addition, polymers grafted to and grafted from nanoparticles provide an effective mechanism to disperse nanoparticles in polymers. A significant limitation is placed on this strategy due to entropy driven segregation, because of the poor penetrability of grafted polymers, especially at high grafting densities, by matrix (or free) polymers that are significantly higher in molecular weight than the grafted chains. Mechanical methods, such as melt-state shear, ultrasonication in solution and solid-state pulverization, have proven to be effective methods to achieve nanoparticle dispersions in polymers [4] but might also result in some degradation of the polymers and, therefore, their properties.

The characterization of the dispersion state of nanocomposites has included electron and force microscopy, light, X-ray and neutron scattering, chemical spectroscopy, electric and dielectric spectroscopy, and a range of mechanical and physical methods. While the hierarchical nature of many of the nanoparticles and the wide range of length scales associated with the dispersion state of many of these nanocomposites are very complex, some overall quantitative mechanisms for structural characterization are being developed. For instance, to describe the mesoscale structure in such nanocomposites, in the dilute regime where particle–particle overlap is not expected, the presence of well-dispersed nanoparticles is established quantitatively through scattering methods and microscopy measurements that reveal the effective aspect ratio and interparticle distance. In the semidilute regime (i.e., concentrations above the geometric overlap concentration and below the formation of ordered or liquid-crystalline phases), even well-dispersed nanoparticle systems are characterized by the formation of flocs or clusters. Again such fractal structures are well-characterized by ultra-small and small-angle scattering methods that lead to the determination of the floc or cluster size, the fractal dimension associated with such clusters, and the mesh size characterizing the intra-floc structure [8]. On the other hand, to characterize the structure or dispersion at the nanometer length scale—especially in the context of anisotropic nanoparticles that order onto a lattice structure, wherein partial and complete dispersion are termed as *intercalated* and *exfoliated* structures, respectively—descriptors such as the average number of nanoparticles that organize themselves in a packet or tactoid and the number of individualized nanoparticles are often used [3].

## 5.3 LINEAR VISCOELASTIC PROPERTIES

The rheological properties of dispersions of nanoparticles in organic solvents and water are intimately linked to the mesoscopic structure of the inorganic filler in the dispersions. For instance, in the case of naturally occurring 2:1 smectite based layered-silicates (either pristine or organically modified) dispersed in low molecular weight solvents, negative thixotropy has been observed and attributed to the breakdown and slow reformation of the "house of cards" structure due to the application and subsequent removal of shear [3, 9, 10]. Further, due to the mesoscopic structure present under quiescent conditions, many of these systems are also known to exhibit finite yield stresses. While the house of cards structure has been attributed to the electrostatic attraction between edges of positively charged and negatively charged faces, others have suggested that the layers are associated with each other in tactoids and the face–face mediated interactions result in a granular mesoscopic structure [10].

As noted previously, for polymer-based nanocomposites the melt-state rheological properties are dictated by a combination of the mesoscopic structure and the strength of the interaction between the polymer and the nanoparticle [11, 12]. The mesoscopic structure would be crucially dependent not only on the strength of the polymer–nanoparticle interaction but also on the inherent viscoelastic properties of the matrix in which the nanoparticles are dispersed.

The melt-state linear shear viscoelastic properties for a wide range of polymer nanocomposites, covering a broad range of polymer matrices and nanoparticles, including isotropic (spherical) to rod-like to plate-like nanoparticles, have been examined and reported over the last 20 years [2, 13–22]. For systems with homogenous dispersion of nanoparticles, the following generalized observations have been noted for these materials:

1. A transition from liquid-like to solid-like rheological behavior for nanocomposites at relatively low nanoparticle loading is observed, as shown in Figure 5.1 for the case of organically modified layered silicates

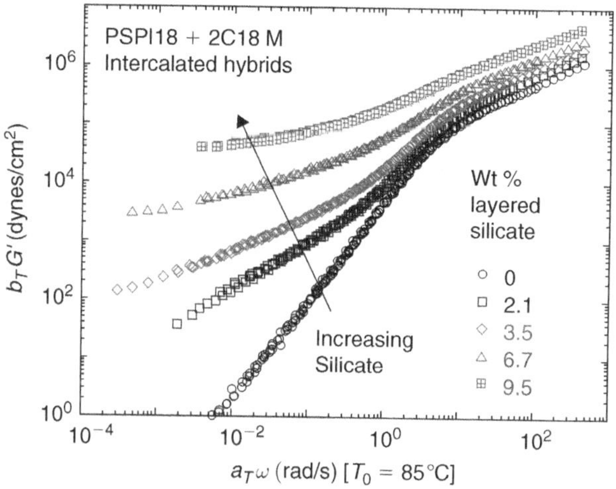


**FIGURE 5.1** Time-temperature superposed linear dynamic storage modulus ($G'$) for a series of nanocomposites containing organically modified layered silicate (montmorillonite) dispersed in a disordered styrene-isoprene block copolymer [18]. At high frequencies ($\omega > 1/\tau_{relax}$), the qualitative viscoelastic behavior of the polymer is unaffected, but at low frequencies, the frequency dependence of the moduli gradually change from liquid-like to solid-like. Reprinted with permission from Ref. 18.

dispersed in a styrene-based polymer [18]. These are manifested through the presence of a long-time elastic plateau in the relaxation behavior and simultaneously a diverging viscosity behavior at long times or equivalently low frequencies.

2. The applicability of the principle of time-temperature superposition for preparing viscoelastic master curves, with frequency shift factors comparable to that of the unfilled polymer, is often observed.
3. Quantitative agreement between linear dynamic oscillatory shear measurements and linear stress relaxation measurements using an empirical two-point collocation based on Ferry's [23] formulation is observed, as shown in Figure 5.2.

The change in viscoelastic behavior from liquid-like for the unfilled polymers (i.e., the storage modulus $G' \propto \omega^2$ and the loss modulus $G'' \propto \omega^1$, at frequencies well below the terminal relaxation time) to solid-like for nanocomposites (i.e., $G'$ & $G'' \propto \omega^0$) with low concentrations of nanoparticles is reminiscent of other soft concentrated dispersions [24] and has been attributed to the formation, in the quiescent state, of a geometrically percolated network structure. In general, a broad class of materials, termed *soft glassy materials* [25], exhibit similar behavior and demonstrate the slow progression from liquid-like to solid-like character with addition of a solid dispersed component. It has been theoretically predicted that for the case of weak interactions between polymers and

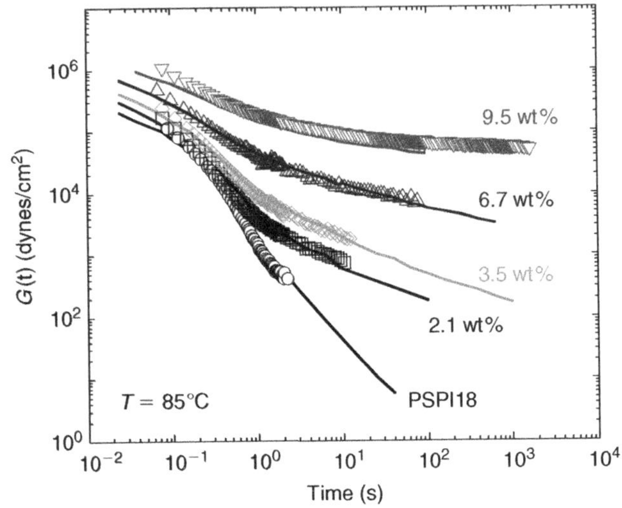


**FIGURE 5.2** Linear stress relaxation modulus $G(t)$ for the series of nanocomposites shown in Figure 5.1. The symbols represent the data from the stress relaxation measurements, and the lines represent predictions using the two-point collocation method [23] captured as $G(t) = G'(\omega) - 0.4G''(\omega) + 0.014G''(10\omega)\,|_{\omega=1/t}$, and the dynamic viscoelastic data reported in Figure 5.1. Reprinted with permission from Ref. 18.

the nanoparticles, liquid-like behavior should dominate as the polymer bridges are not long-lived to help sustain the stress [11]. In fact, experiments of PEO based water solutions with dispersed unmodified layered silicates [26, 27] and PDMS–organically modified layered silicate based nanocomposites [9] demonstrate this phenomenon, but such experimental realizations are rare as most nanoparticle–polymer hybrids are formed and stabilized by attractive polymer–nanoparticle interactions rather than relying on entropy driven mixing [4].

That the application of the Boltzmann time-temperature superposition and the shift factors (and consequently the flow activation energies) are unaffected by the amount of added layered silicate has been understood in terms of the lack of a temperature-dependent relaxation associated with the nanoparticle due to the strong attractive interaction between the nanoparticles and the polymer that are responsible for the effective dispersion of the nanoparticles [15, 18, 28]. This does not preclude the presence of two distinct dynamic populations of polymer chains with a fraction of solid-like chains (with either no perceptible relaxation time or with relaxation times beyond the experimental range) and the remaining fraction of matrix-like chains [18]. However, should the interactions between the nanoparticles and the polymer change with temperature, as would be observed for the case of systems for which the blends are stabilized by the formation of hydrogen bonds between the nanoparticles and the polymer at low temperatures and broken at high temperatures, then time-temperature superpositioning is not expected to be followed as observed experimentally. For instance in the case polycarbonate-layered silicate nanocomposites [29] and poly(hydroxyl-styrene)-based layered silicate nanocomposite [30], the strength of the hydrogen bonding changes with temperature and is reflected in the absence of time-temperature superposability of the melt-state rheological data.

Thus the development of robust time-temperature superposed mastercurves that show a gradual progression from liquid-like to solid-like behavior is observed for a large class of polymer nanocomposites and is due to percolation of the nanoparticles or their aggregates with or without the aid of polymer bridges. In fact, on the basis of these changes in linear viscoelastic properties with concentration, it has been suggested that melt-state rheology is an effective metrology tool for quantifying the dispersion of such nanocomposites [31]. The development of a network superstructure is manifested in the low-frequency complex viscosity, $\eta^*$. The values of $\eta^*$ diverge at a finite value of the complex modulus ($G^*$) and are consistent with the behavior of a material that exhibits a yield stress. Further, the transition from liquid to solid-like behavior is better visualized—in this case for the dispersion of single-walled carbon nanotubes in a low molecular weight PEO matrix [2, 32]—through an examination of the composition dependence of the reduced viscosity ($\eta_r^*$), defined as $\eta_r^* = \frac{\eta^*}{\eta_0^*}$, where $\eta_0^*$ is the zero shear viscosity of the polymer and $\eta^*$ is the viscosity of the nanocomposite at a chosen frequency (Fig. 5.3). The data demonstrate a

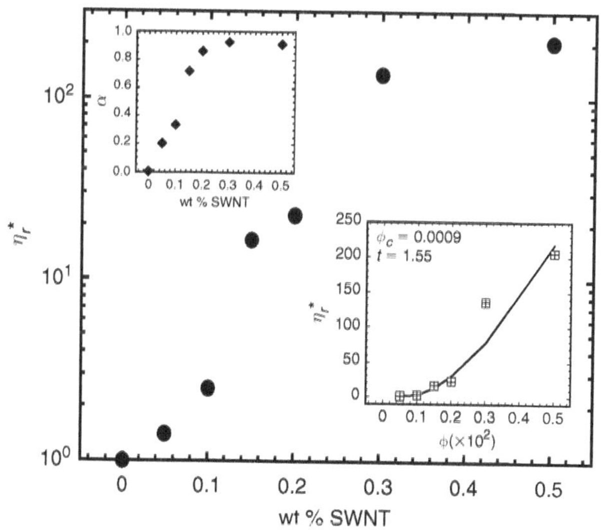


**FIGURE 5.3** The composition dependence of the reduced viscosity ($\eta_r^*$) for a series of nanocomposites based on nanotubes dispersed in a PEO matrix [32]. The classical sigmoidal nature of the curve indicates the changing nature of reinforcement with increasing SWNT loading and the presence of a geometrical percolation of the SWNT. *Top Inset*: Power law exponent for low frequency dependence of $\eta^*$ ($\eta^* \propto \omega^{-\alpha}$), *Bottom Inset*: geometrical percolation threshold volume fraction ($p_c$) is obtained from plot of ($\eta_r^*$) vs. ($p - p_c$). Reprinted with permission from Ref. 32.

sigmoidal dependence that is attributed to a changing reinforcement mechanism with increasing SWNT concentration. In the case of the PEO with dispersed nanotubes, at low nanotube loadings, the nanoparticles act as isolated objects and the viscosity (or modulus) is dominated by the matrix contribution and can be modeled along the lines adopted by Guth [33]. Beyond the percolation of the nanotubes or their aggregates, the network superstructure dominates the viscoelastic response and follows typical power law like behavior associated with systems near the percolation threshold [34]. On the basis of the above arguments, the composition dependence of the reduced viscosity at low and intermediate nanoparticle concentration is modeled as

$$\eta_r^* = 1 + 0.67(\alpha p) + 1.62(\alpha p)^2 + m(p - p_c)^t \tag{5.1}$$

The linear and quadratic terms result from Guth's modification [33] of Einstein's relationship for anisotropic fillers in dilute solution and the power term is the scaling law of structural properties near the percolation threshold, with $p$ being the volume fraction and $p_c$ being the geometrical percolation threshold. The aspect ratio ($\alpha$) is related to the geometrical percolation threshold (for instance using or extrapolating the calculation of percolating ellipsoids in the absence of excluded volume by Garboczi et al. [35]) and requires an iterative solution of the above equation. A model fit of experimental data using (5.1) yields a value for $p_c$ of $9 \times 10^{-4}$ ($\sim$ 0.09 wt %, considering SWNTs density 1 gm cm$^{-3}$) [36], a scaling exponent ($t$) of 1.55, and an effective aspect ratio ($\alpha$) of 650 [2, 32].

While the dispersion state, the formation of a percolated network, and the growth of the network dominate the properties of polymer nanocomposites in the dilute regime (i.e., below overlap) and near the percolation concentration, the behavior in the semidilute regime and in the concentrated regime are dominated by the mesoscale superstructure of the materials. For anisotropic nanoparticles, this semidilute regime dominates the range of compositions (typically between 0.1 and 10 vol % for many anisotropic materials) where many materials formulations are prepared [37]. Further, for anisotropic nanoparticles, at high concentrations liquid-crystalline structures (smectic or nematic) are formed, and those defect-ridden liquid-crystalline domains (due to heterogeneous and polydisperse nature of the nanoparticles) dominate the properties.

In the semidilute regime, the nanoparticles can form diffusion limited fractal aggregates much like in colloidal systems [8, 37–39]. For colloidal suspensions, it has been shown previously that the linear elastic modulus of the gelled system is closely related to the interactions between the particles and the fractal structure of the gel-like material [40, 41]. For nanoparticle dispersions with particle concentrations in excess of percolation ($p \gg p_c$, where $p$ is the volume fraction of the nanoparticles and $p_c$ is the value at percolation) a hierarchical fractal-like structure is observed through a combination of ultra-small and small angle scattering methods with mass fractal scaling ranging from 1.8 to 3.

For the case of nanotubes dispersed in polymeric matrices a two-level structural hierarchy is observed where inside the flocs, individual or small nanotube bundles overlap each other to form a relatively dense mesh [8]. With increasing nanotube loading, the network becomes denser and the mesh size ($\zeta$) decreases following a scaling relation ($\zeta \sim (p - p_c)^{\sim(-0.2 \text{ to } -0.4)}$) but the average floc size (R) is found to be largely independent of the particle concentration, of the order of $\sim$ 3–5 $\mu$m. On the other hand, the total number of flocs ($N$) increases ($N \sim (p)^{\sim 1.0}$) indicating that with addition of nanotubes, the network primarily grows through addition of new flocs.

Such semidilute nanocomposites demonstrate time-temperature-composition superposability in their linear viscoelastic response that is dominated by the gel-like character arising from the fractal superstructure of the nanoparticles (Fig. 5.4) [2, 38, 40, 42]. Similar composition superposable mastercurves were also observed in colloidal dispersions in low molecular weight dispersions and are thought to result from the self-similar hierarchical structure of such dispersions. The modulus of the network of nanoparticles scales as $(p - p_c)^{\delta}$, with $\delta$ ranging between 2.5 and 4.5 for most cases and is argued to be caused by the formation of fractal superstructures between weakly attractive contacts without significant dependence on the chemical and topological identity of the primary nanoparticles underlying the superstructure [2, 38, 40, 42].

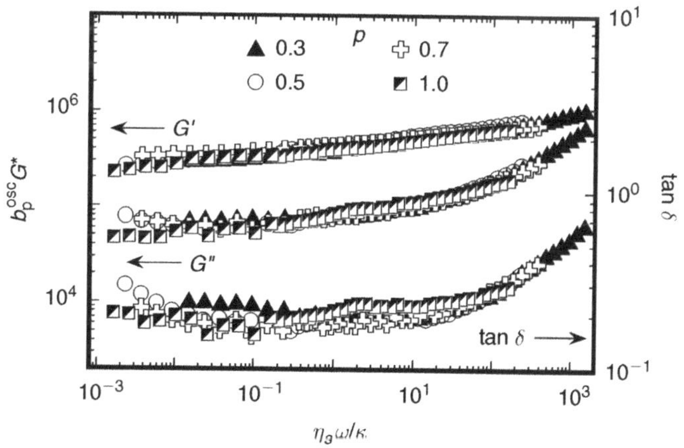


**FIGURE 5.4** Time-temperature-composition superposed mastercurves for linear dynamic rheological properties for the nanocomposites of nanotubes dispersed in low molecular weight PEO matrix in the semidilute regime (in terms of nanoparticle concentration) described in Figure 5.3. $\eta_s$ is the matrix viscosity, $\kappa$ is a frequency shift factor that captures the composition dependence of the relaxation spectrum, and $b$ is a modulus shift factor providing the scaling of the modulus with nanoparticle addition. Reprinted with permission from Ref. 2.

## 5.4 NONLINEAR VISCOELASTIC PROPERTIES

### 5.4.1 Alignment by Large Amplitude Oscillatory Shear and Disorientation Kinetics

For many anisotropic nanoparticle-based nanocomposites, prolonged application of large amplitude oscillatory shear results in dramatic changes in the linear viscoelastic properties and the alignment of the nanoparticles or their aggregates as a result of this shear. Such changes in viscoelastic properties and the underlying alignment of the nanoparticles appear to occur only in semidilute and concentrated dispersions and not in dilute dispersions of the nanoparticles. Moreover, for the case of well-dispersed single walled carbon nanotubes large-amplitude, even for semidilute dispersions, leads only to rather modest alignment, which is short lived after cessation of the large amplitude oscillatory shear [19, 21].

For the case of layered silicate and graphene based nanocomposites, application of large amplitude oscillatory shear leads to a decline in the moduli during shear that continues to gradually decrease with time before reaching a plateau [19, 43]. Previously, using small-angle neutron and X-ray scattering on layered silicate based nanocomposites it has been shown that the predominant alignment of the silicate layers following such a flow history was that of parallel layers, normally along the velocity gradient direction [17, 44, 45]. The small amplitude oscillatory viscoelastic response following such large amplitude oscillatory shear is considerably more liquid-like than that before flow alignment, as shown in Figure 5.5. In addition, both $\eta_r^*$ and $G'$ are significantly smaller (at all frequencies) than those of the unaligned sample. Further, the frequency dependence of both $G'$ and $G''$ for the aligned samples is much stronger than that of the unaligned samples and starts to resemble that of the free homopolymers.

This increased low-frequency dependence of the viscoelastic moduli suggests a breakdown of the percolated silicate network in the shear-aligned sample. In fact, for the case where the nanoparticles are aligned parallel to the shear direction, percolation would be expected when the effective objects form a percolated network structure in two-dimensions. For the case of overlapping disks and on the basis of continuum model calculations, it is expected that percolation would occur at a volume fraction of disks of $\sim$ 0.67 [46]. For the case of oblate ellipsoids with an effective aspect ratio of 30, this calculation would suggest a critical percolation threshold to occur at $\sim$ 44 wt % of the ellipsoids, far above the development of smectic structures in three dimensions and typically not studied. Thus, in close analogy to their small molecule counterparts, it appears that the polymer nanocomposites based on anisotropic nanoparticles possess a property defining mesoscopic structure that can be significantly altered by the application of large shear flow fields.

The preservation or destruction of these aligned nanocomposites is of significant interest in the estimation of property improvements in nanocomposites.

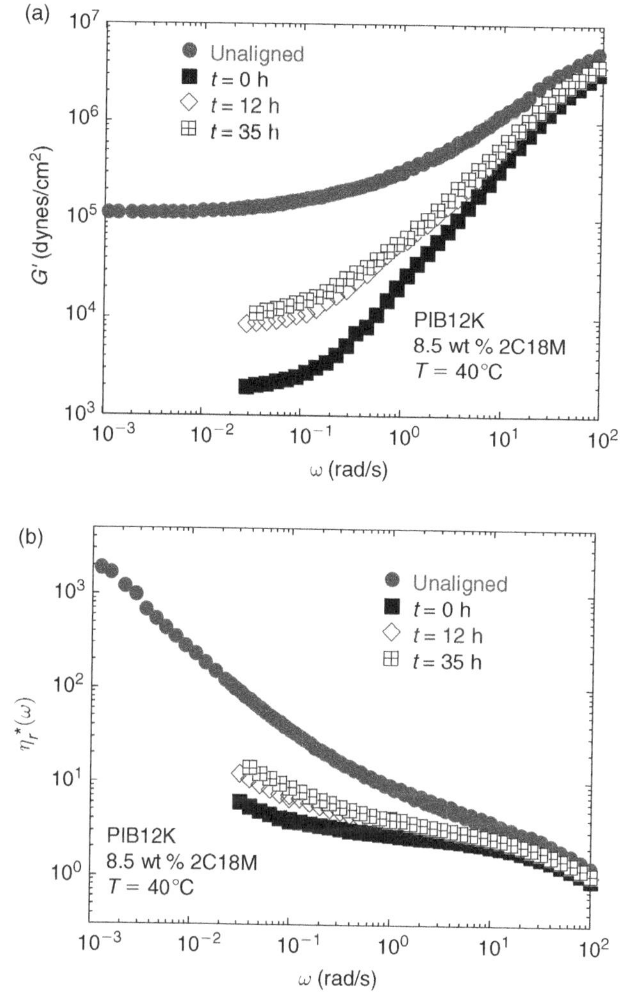


**FIGURE 5.5** Frequency dependence of the small amplitude oscillatory viscoelastic response for the unaligned, aligned after large amplitude oscillatory shear ($t = 0$) and at intermediate times of quiescent disorientation for a polyisobutylene-based organically modified montmorillonite (8.5 wt %) nanocomposite at 40°C [19]; (a) Storage modulus, (b) Reduced viscosity. The data at $t = 0$, were collected immediately after prolonged large amplitude oscillatory shear (LAOS) at $T = 40°C$, $\omega = 0.1$ rad/s, $\gamma_o = 1.5$ for 16 h. Reprinted with permission from Ref. 19.

Solomon and co-workers [28] studied the disorientation of polypropylene-based montmorillonite exfoliated nanocomposites after alignment and showed that the process is non-Brownian, with disorientation kinetics that are faster than those predicted by Brownian motion. In that case, it was argued that the nanocomposites prepared were intrinsically unstable (thermodynamically) and

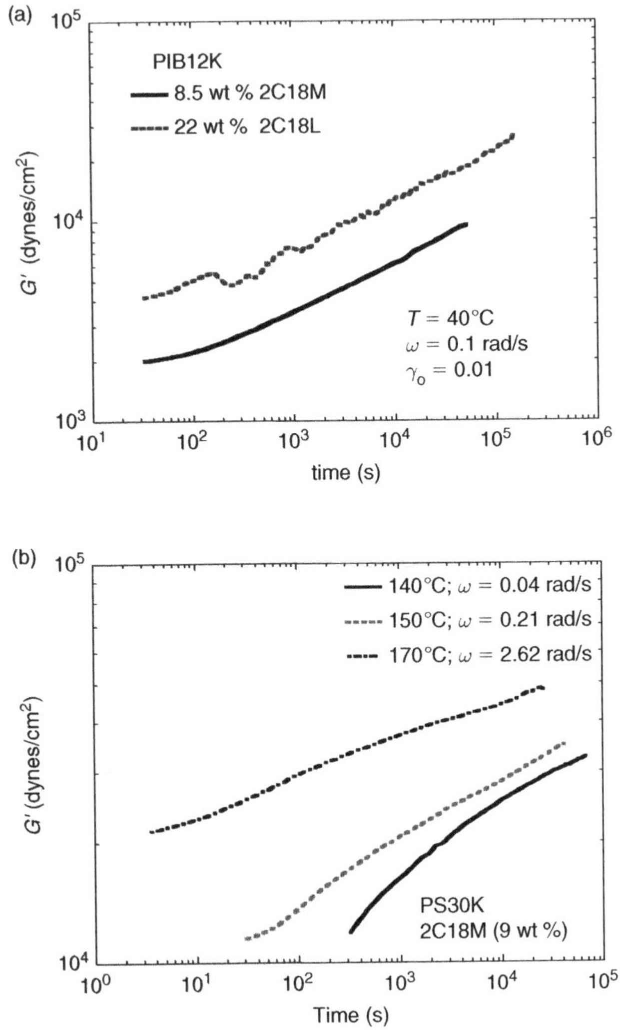


**FIGURE 5.6** (a) Small amplitude oscillatory elastic moduli ($G'$) monitoring of the recovery after shear alignment for the polyisobutylene–based nanocomposites (described in Figure 5.5) containing two different organically modified layered silicates. The nominal particle size (diameter) for the montmorillonite (2C18M) sample is $\sim$ 400 nm and for the fluorohectorite is $\sim$ 5 $\mu$m. (b) Effect of temperature and matrix viscosity on the disorientation of 9 wt % organically modified montmorillonite–based nanocomposites of polystyrene (monodisperse and molecular weight of 30 K; PS30K). The matrix viscosity at 140°, 150°, and 170°C are measured to be $2.7 \times 10^4$ Pa · s, $5.0 \times 10^3$ Pa · s and $3.7 \times 10^2$ Pa · s, respectively. The value of $G'$ ($\gamma_o = 1.2$ and $\omega = 0.1$ rad/s at 180°C) at the end of the LAOS alignment for the data in (a) are 1030, 940, and 930 dynes/cm$^2$ for $T = 140°$, 150°, and 170°C, respectively. (c) Effect of the changing matrix molecular weight on the disorientation kinetics of 9 wt % organically modified montmorillonite dispersed in polystyrene. Temperatures were chosen to maintain a constant matrix viscosity. The values of $G'$ ($\gamma_o = 1.0$ and $\omega = 0.1$ rad/s at 180°C) at the end of the LAOS alignment for the data are 390, 4,160, and 20,500 dynes/cm$^2$ for PS30K, PS152K, and PS290K (LAOS at $T = 185°$C instead of 180°C), respectively. Reprinted with permission from Ref. 19.

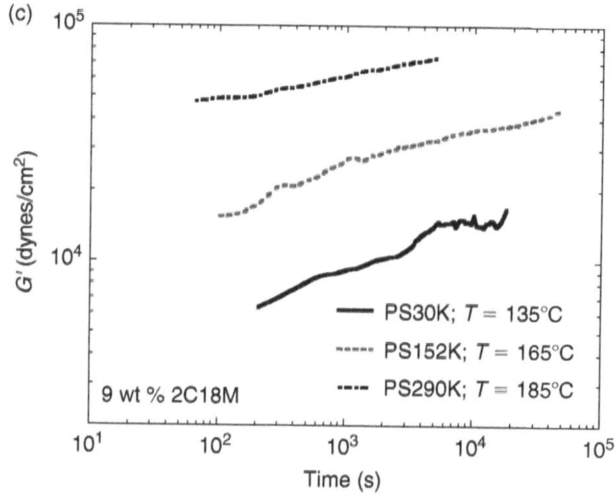


**FIGURE 5.6** Continued

the strong attraction between the silicates led to the rapid disorientation. On the other hand, Lele and co-workers [44] using in-situ X-ray scattering have suggested that compatibilized syndiotactic polypropylene nanocomposites with dispersed layered silicates initially disorient rapidly (much faster than Brownian motion) and then remain at a constant orientation state for a time period of $\sim$ 1500 s. It was argued that the rapid initial decrease in the orientation was a result of the coupling of the polymer chains to the silicates and not a result of silicate layer attraction.

Bonn and co-workers [47, 48] have recently studied the behavior of aqueous laponite dispersions using a combination of rheology and dynamic light-scattering measurements and have indicated that the recovery after large amplitude oscillatory preshear demonstrated signatures of cooperative relaxation and exhibit logarithmic scaling with time. On the other hand, Han and co-workers [26] have shown that the relaxation after flow-induced alignment for a similar aqueous dispersion of laponite in high molecular weight poly (ethylene oxide) is much faster than that expected from Brownian motion. They hypothesized that the poly(ethylene oxide) chains act as dynamic bridges between the laponite sheets and that the relaxation results from the elastic restoring force of the dynamically bound poly(ethylene oxide) chains. Ren et al. [19] in an extensive study demonstrated that for the case of structurally well-defined polymer nanocomposites based on different layered nanoparticles, the disorientation process exhibits signatures of aging observed in soft-colloidal glasses and is independent of temperature, nanoparticle size, and chemical details, viscoelasticity, and molecular weight of the polymeric matrix (Fig. 5.6). This independence of the disorientation kinetics with nanoparticle size, temperature, and polymer matrix indicates that the disorientation process is not governed by Brownian motion.

Because of the close analogy with the recovery observed for soft glassy materials, we compare the recovery process during disorientation to some of the energy landscape arguments that have been successfully used to describe soft glassy materials [25]. Application of shear is considered to change the energy landscape and allows for the system to access new metastable states [49]. For this reason, the application of large stresses (larger than the yield stress) is considered as a rejuvenating condition and in the context of vitrification can be considered to be analogous to heating above the glass transition temperature [24]. Upon removal of the stress (equivalent to quenching below the glass transition), it is anticipated that the energy landscape is again changed, requiring the system to find a new metastable state consistent with the changed landscape. This recovery from the liquid state shows many similarities to physical aging including the observed logarithmic dependence of the recovery modulus and the independence of the disorientation process from the detailed physical characteristics of the layered silicate and the polymer. While such an energy landscape–based argument appears to be valid for isotropic particles dispersed in a Newtonian small molecule matrix, the extension of these ideas to highly anisotropic particles is experimentally robust as demonstrated by the results of Ren et al. [19], Oberhauser and co-workers [50] on polymer nanocomposites and those of Bonn et al. on aqueous laponite dispersions [48].

### 5.4.2 Strain Dependent Nonlinear Behavior

The dependence on strain amplitude of the viscoelastic properties of polymer nanocomposites has been studied by an examination of the stress-relaxation in response to step shear as a function of applied strain. The studies have focused on the semidilute dispersions of anisotropic nanoparticles, where the fractal-like mesoscale structure dominates the linear viscoelastic response.

For these nanocomposites, the strain amplitude dependence of the stress-relaxation is shown in Figure 5.7 and briefly summarized as follows: (1) there is a linear regime that occurs at relatively low strain amplitude ($\gamma_o$) values where the $G(t)$ data are independent of $\gamma_o$; (2) increasing the strain amplitude beyond the linear region ($\gamma_{critical}$) leads to a strain-softening behavior with conservation of the relaxation spectrum, suggesting the possibility of applying time-strain separability for these data (i.e., $G(t,\gamma) = h(\gamma)G(t)$, where $h(\gamma)$ is the damping function); (3) the critical strain amplitude for the onset of softening behavior ($\gamma_{critical}$) decreases with increasing nanoparticle concentration; and (4) application of strain amplitudes much larger than $\gamma_{critical}$, leads to a stress relaxation spectrum that is no longer conserved and time-strain superposability is no longer valid.

The damping function $h(\gamma)$ scales according the empirical relationship $\sim 1/(1 + m\gamma^2)$ for many of these polymer nanocomposites, where $m$ is a constant. Moreover the data for the damping function superpose somewhat reasonably by normalizing the bulk strain $\gamma$ by $\gamma_{critical}$. Further, $\gamma_{critical}$ scales with the volume fraction as $(p)^{\sim(-2)}$ and suggests that the breaking of nanoparticle

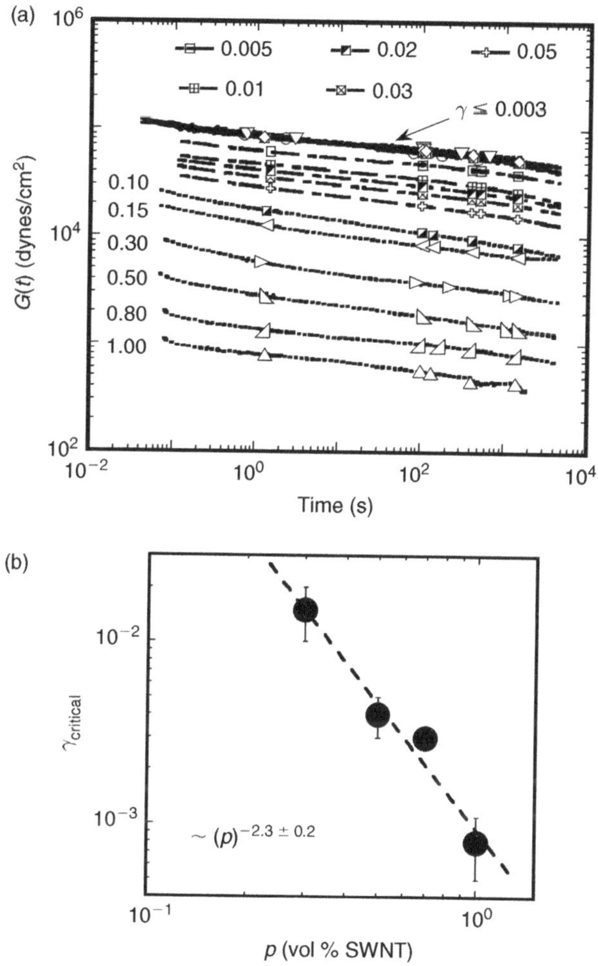


**FIGURE 5.7** Representative stress relaxation behavior for nanotubes dispersed in a PEO matrix, with a particle concentration $p$ of 0.7 vol % as a function of the applied (bulk) strain amplitude [2]. For low-amplitude strain, linear behavior is observed followed by a time-strain superposable zone. At higher strain amplitude time-strain superposability is violated. (b) The composition dependence of the critical strain for the onset of shear thinning is shown; the scaling behavior is consistent with the short-range interactions dominating these nanocomposites. Reprinted with permission from Ref. 2.

interactions (either direct or mediated through the polymer) and the breakdown of multiple connections between percolating network elements dominate the onset of shear thinning. This scaling of $\gamma_{\mathrm{critical}}$ and the previously demonstrated strong scaling of the elastic modulus are typical of fractal networks, such as those of colloidal gels [41], layered silicate [20], flocculated silica spheres [51], and multiwalled carbon nanotube dispersions [38, 39], which with increasing density of nanoparticles become stiffer and more fragile. Similar scaling is anticipated from theoretical efforts [52] examining the three-dimensional percolation of random percolating elements and from computer simulations [53] considering individual bonds resist both bending and stretching (i.e., enthalpic networks [12]). Using the development of Shih and co-workers [41] for fractal networks with $p \gg p_c$, where the interactions between flocs dominate over those within a floc (the strong link regime), it is anticipated that the scaling for the elasticity ($G_p$) and critical strain ($\gamma_c$) follow $p^{\mu}$ and $p^{-\delta}$, respectively, with $\mu = (D + d_b)/(D - d_f)$ and $\delta = (1 + d_b)/(D - d_f)$, where $d_b$ and $d_f$ are the backbone and fractal dimensions of the network and $D$ is the Euclidian dimension. Similar relations are also available for the case where interactions within a floc dominate; however, for the case of many polymer nanocomposites, the strong link regime appears to be more appropriate. Thus the linear and onset to nonlinear viscoelasticity can be used to quantitatively determine the fractal scaling of the network structure in such nanocomposites.

For polymer nanocomposites, the strain experienced is nonhomogenous and in particular the strain experienced by a single nanoparticle inside the network is a function of network size (or nanoparticle loading) and different from the bulk strain applied. To accommodate the network size effect, a local strain is calculated where the bulk strain is modified for the dispersed filler effect along the lines adopted by Watanabe et al. [54]

$$\gamma_{\text{local}} = [1 + 0.67(\alpha p/100) + 1.62(\alpha p/100)^2]\gamma_{\text{bulk}} \tag{5.2}$$

where $\alpha$ is the effective anisotropy of the nanoparticle [32]. Clearly, the local strain used here is reflective of the local stress on the network objects. Under these conditions, the stresses that arise from the collective network (i.e., stress contributions from the percolated structures) are neglected to calculate the local strain as demonstrated in (5.2). The damping function, $h(\gamma)$, for a nanocomposite of PEO with dispersed nanotubes (discussed above) as a function of local strain shows excellent superpositioning (Fig. 5.8). Similarly, a layered silicate–based nanocomposite with a disordered polystyrene–polyisoprene diblock [20], dominated by short-range interactions, exhibits a superpositioning of $h(\gamma)$ with local strain and the onset of shear thinning at a similar value as the local strain ($\gamma_{\text{local}} \sim 0.1$). On the other hand, for the cases of long-range interacting systems—such as those observed for a brominated paramethylstyrene–isobutylene polymer containing dispersed carbon black [22], aqueous solutions of unmodified clays dispersed in PEO [27], and silica nanoparticles dispersed in PEO [13]—such a simple superpositioning fails and perhaps reflects the long-range interactions (due to ionic, H-bonding, and bridging interactions caused by long-chain polymers bridging between nanoparticles) that dominate those systems. This nearly universal behavior for the damping function for weakly interacting polymer nanocomposites suggests that the linear and nonlinear viscoelasticity are dominated by a combination of the quiescent state network structure and the local strain experienced by the

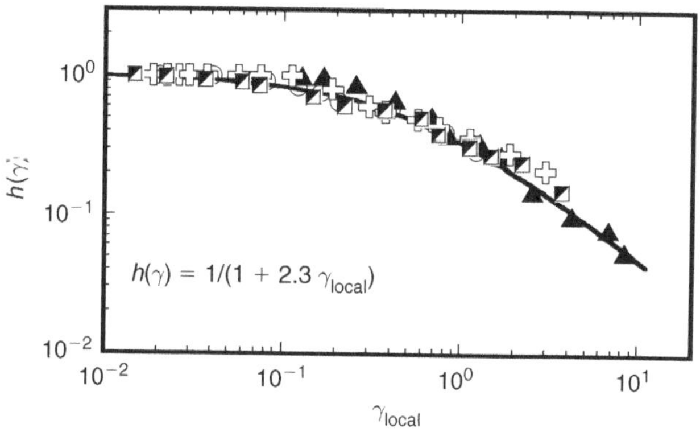


**FIGURE 5.8** Damping function $h(\gamma)$ required for time-strain superposition for the systems of nanotubes dispersed in PEO and shown in Figure 5.7 [2]. Plotting the damping function $h(\gamma)$ with respect to the local strain leads to a collapse of all the data onto a single mastercurve. The onset of the shear thinning is observed at $\gamma_{\mathrm{local}} \sim 0.1$ and is similar to other nanocomposite systems with short-range interactions. Reprinted with permission from Ref. 2.

network elements. The onset of shear thinning is accompanied by the removal of these additional pathways of connecting the backbone of the network and not by the breakdown of the backbone of the network. On the other hand, the onset of the failure of the time-strain superposability is perhaps related to the irreversible deformation of the backbone of the fractal network.

### 5.4.3 Steady Shear Response

The steady shear response of layered silicate–based polymer nanocomposites has important consequences on the potential processability of the materials. In this context, the melt state steady shear viscoelastic behavior has been systematically studied for various polymer nanocomposites, with the most interest being focused on ones based on anisotropic nanoparticles with concentrations in the dilute and semidilute regime.

The response to constant-rate steady shear is studied as a transient response from rest using startup of steady shear measurements. For the case of polymer nanocomposites with dispersed anisotropic nanoparticles with particle concentration in the semidilute range, under continuous shear the time dependence of the shear stress shows an initial stress overshoot that equilibrates to a well-defined steady state value ($\sigma_\infty$) without any oscillations, as seen in Figure 5.9 [55]. The polymer matrix is not responsible for the stress overshoot, as such overshoots are observed for nanocomposites prepared with low molecular weight (nonentangled) polymer matrices. Since the stress overshoot appears only in the case of semidilute dispersions, wherein the percolated network structure of the nanoparticles is well established, it is conceivable that the stress

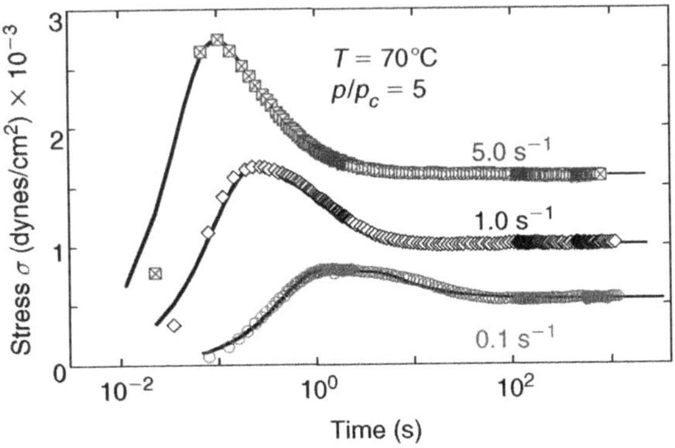


**FIGURE 5.9** Transient shear stress response obtained during startup of steady shear measurements for a SWNT-PEO dispersion with $p/p_c = 5.0$ [55]. For all shear rates, the stress data exhibit an initial overshoot arising from the shear-induced rearrangement of the cluster structure and at long times the network reaches an equilibrium state to exhibit a steady viscoelastic response. Solid lines are model fit to the experimental data, as described in (5.3). Reprinted with permission from Ref. 55.

overshoot originates from a structural alteration of the percolated network structure due to the application of the steady shear. This conjecture is strengthened by a calculation of the nondimensional Peclet number that typically is significantly $> 1$. This indicates that the convective transport regime dominates the processes and where it is expected that the shear rate controls the structures and the relaxation process of the network. In the case of dilute dispersions, at low shear rates (where the Peclet number is $< 1$), Brownian motion can dominate the relaxation process, and the resulting rotation, and tumbling of anisotropic particles can lead to the observation of time-dependent oscillating shear and normal stresses in startup measurements.

Immediately after startup of steady shear, the material acts as a solid and generates a stress that is proportional to the total strain ($= \dot{\gamma} \times t$). The strain also induces structural rearrangement of the nanoparticle network, similar to that noted in shear rejuvenation processes of soft-glassy materials, and these structural rearrangements and altered network structure would act to dissipate the stress generated or stored in the system. Whittle and Dickinson using Brownian dynamics have observed stress overshoot in model particle based gels and suggested a semiempirical model [56]:

$$\sigma(\mathrm{t}) = \sigma_0 \frac{t/\tau_1}{1 + (t/\tau_1)^{1+\delta}} + \sigma_\infty \left\{ 1 - \exp\left( \frac{t}{\tau_2} \right) \right\} \tag{5.3}$$

where $\tau_1$ and $\tau_2$ are the characteristic time scales associated with the rearrangement of the network structure under shear and the network breakup (or flow), respectively, and $\delta$ is a phenomenological parameter. $\sigma_0$ is related to the shear rate dependent elastic modulus of the material, and $\sigma_\infty$ is the steady-state shear stress at the shear rate. As noted by Whittle and Dickinson, in the intermediate zone, i.e., $t > \tau_1$, the first term reduces to a power law decay process, whose rate depends on the numerical value of $\delta$. For the case of nanotubes dispersed in a PEO matrix, $\delta$ gradually increases with increasing shear rate from a value of 0.5 to a value of 1.4. This implies that with increasing shear rate the decay process becomes increasingly more important compared to the elastic response. On the other hand, the second term (exponential term) in (5.3), describes the long time behavior and is related to the network relaxation process that leads to the steady flow (with an equilibrium number of stress-bearing members or contacts) with an equilibrium stress value ($\sigma_\infty$) at long time. Fitting of the steady shear startup data for the nanotubes dispersed in PEO indicates that the values of $\tau_1$ and $\tau_2$ are independent of nanotube concentration and strongly dependent on shear rate. Further, the characteristic time $\tau_2$, corresponding to the stress dissipation, always exceeds the value of $\tau_1$, corresponding to the structural rearrangement. This observation suggests that under continuous shear, the fractal nanotube network first rearranges the structure to bear the evolved stress and beyond a threshold total strain of $\sim \dot{\gamma} \times t_{max}$, the nanocomposite flows.

For dense colloidal suspensions under continuous strain, Silbert et al. [57] found an inhomogeneous distribution of the stress with few high-stress bearing clusters and the population of such clusters essentially controls the stress response. On the other hand, recent nonequilibrium molecular dynamics simulation studies of polymer nanocomposites have shown that the stress overshoot results from the elastic stretching of the particle-polymer network [58]. More important, they also found an inhomogeneity in stress distribution and only a small number of bonds actually carry the excess stress developed. We conjecture that under steady shear, the flocs locally rearrange in response to the applied deformation and results in cluster–cluster collisions and jamming of the network elements that gives rise to the observed stress overshoot. With continuous shearing and when the local stress exceeds the yielding stress, the network bonds break and the network flows until a final steady state is reached where an equilibrium between bond formation and bond breaking occurs [56]. On the other hand, the process to a steady flowing behavior is governed by an establishment of equilibrium between bond-breaking and bond-formation processes.

For most polymer nanocomposites the steady shear viscosity $\eta(\dot{\gamma}) = \sigma_\infty/\dot{\gamma}$ is significantly different from the complex viscosity ($\eta^*(\omega)$) obtained from linear viscoelastic measurements using small amplitude oscillatory shear when compared at $\omega = \dot{\gamma}$ (Fig. 5.10). At comparable values of frequency ($\omega$) and shear rate $\dot{\gamma}$, $\eta(\dot{\gamma})$ is always less than $\eta^*(\omega)$. Further, $\eta(\dot{\gamma})$ typically shows a weaker scaling with $\dot{\gamma}$ as compared to $\eta^*(\omega)$ with $\omega$. Moreover, the nanoparticle concentration scaling of $\eta(\dot{\gamma})$ is weaker than that of $\eta^*(\omega)$ and indicates that the application of continuous steady shear results in significant changes in the dissipation mechanism as compared to the quiescent or near-quiescent state structure that is dominated by the mass fractal based network

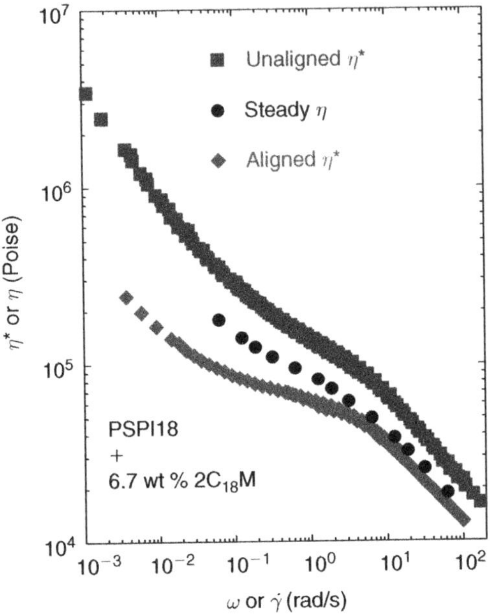


**FIGURE 5.10** Test of the empirical Cox–Merz rule for a nanocomposite belonging to the series of samples described in Figures 5.1 and 5.2, with 6.7 wt % organically modified layered silicate [20]. Mastercurves of the dynamic complex viscosity obtained from linear dynamic oscillatory measurement of the as-loaded randomly oriented nanocomposite ($\eta^*_{\text{linear}}$) are compared to the steady shear viscosity $\eta(\dot{\gamma})$, with $\eta^*_{\text{linear}}$ exceeding the$\eta(\dot{\gamma})$for comparable shear rates. On the other hand, the linear dynamic complex viscosity after prolonged large amplitude shear–induced alignment $\eta^*_{\text{aligned}}$ is consistently lower than the $\eta(\dot{\gamma})$ at comparable shear rates. We thus conclude that steady shear, even at the lowest shear rates examined, leads to some alignment of the anisotropic silicate layers, resulting in the failure of the Cox–Merz rule for the nanocomposites. Reprinted with permission from Ref. 20.

of flocs or aggregates. On the basis of the weaker network (i.e., lower viscosity) and weaker scaling with nanotube concentration of the steady shear viscosity, it is clear that some of the network elements (or junctions) that bear stress are eliminated during the application of continuous shear. Due to the fact that these nanocomposites in fact do flow when subjected to continuous shear, it is possible that the network structures that resist deformation in the flow direction are eliminated. Assuming an initial isotropic state under quiescent conditions, the network elements resisting the displacement along the flow direction would nominally constitute a third of all elements, and their elimination (from the stress bearing function) would decrease the reinforcement by that factor. The nanoparticle concentration dependence for $\eta$ and the same for $\eta^*$ are certainly consistent with the elimination of stress-bearing network elements as is the frequency or shear rate dependence of the viscosity [55].

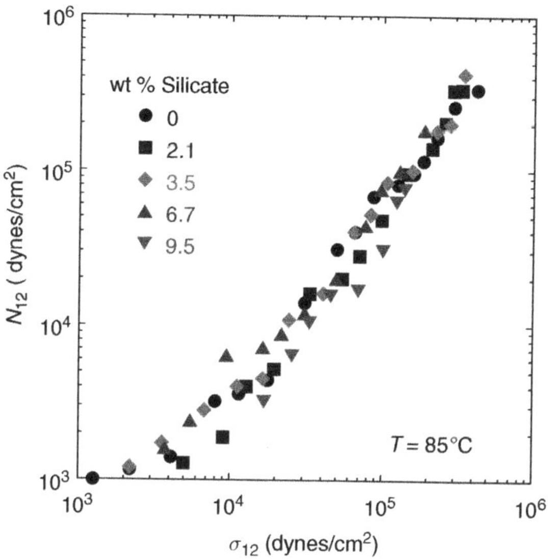


**FIGURE 5.11** The first normal stress difference ($N_{12}$) plotted as a function of the shear stress ($\sigma_{12}$) for the polymer melt and the nanocomposites described in Figures 5.1, 5.2, and 5.10 [16]. The data indicate that at the same values of $\sigma_{12}$, the first normal stress difference ($N_{12}$) is independent of the silicate loading and similar to that of the unfilled polymer, suggesting that the silicate layers do not enhance or diminish the elasticity of the hybrids. Reprinted with permission from Ref. 16.

For the nanocomposites with a nanoparticle network structure under quiescent conditions, form the steady state shear stress $\sigma(\dot{\gamma})$ measured as a function of shear rate ($\dot{\gamma}$), a yield stress ($\sigma_y$) can be estimated for each of the nanocomposites by fitting to

$$\sigma(\dot{\gamma}) = \sigma_y + k(\dot{\gamma})^n \tag{5.4}$$

where $k$ and $n$ are positive numbers. The estimated yield stresses are for the semidilute dispersions are significantly smaller than the long-time modulus observed in stress-relaxation measurements. On the other hand, for the nanocomposites below percolation the yield stress, within experimental error, is zero.

The first normal stress difference ($N_{12} = \sigma_{11} - \sigma_{22}$), a measure of the elasticity, is independent of nanoparticle loading for a layered-silicate based nanocomposite when compared at the same values of shear stress ($\sigma_{12}$), and identical to that of the unfilled polymer (Fig. 5.11) [16]. Those measurements were typically restricted to high shear rates (to obtain reliable values of $N_{12}$), where the silicate layers are believed to be oriented and only relatively small changes are observed in the viscosity data. It has been suggested that the near independence of the recoverable strain ($= N_{12}/\sigma_{12}$) on the silicate loading and the near equivalence to that of the unfilled polymer melt is a result of the ability of the two-dimensional silicate layers to be preferentially oriented by shear flow [16]. Similarly, Kharchenko et al. [59] have observed a decrease in die swell for nanotube dispersed polypropylene nanocomposites, indicating a decrease in elastic normal stresses in those nanocomposites.

Thus the steady shear viscoelasticity provides complementary information to the linear and nonlinear dynamic oscillatory shear measurements regarding the quiescent structure and the ability of shear to deform and orient these nanocomposites. For low shear rates, the mesostructure remains unaffected by the imposed flow, and the viscoelastic response reflects the quiescent structure, particularly for the high silicate content where a yield stress is observed and indicates a solid-like response. At high shear rates, on the other hand, the mesoscopic structure is oriented by the imposed shear and suggests that the layers, in the absence of tethering and hence mechanical coupling, orient in the flow direction and contribute negligibly to the viscoelastic properties of the nanocomposite.

## 5.5 SUMMARY

The viscoelastic properties of polymer nanocomposites provide critical information regarding the state of dispersion of the nanoparticles and the mesoscale structure that the nanoparticles form in the polymer matrix. For instance, for nanocomposites with particle concentration in the semidilute regime, the linear viscoelastic response obeys the principle of time-temperature-composition superpositioning. Furthermore, the viscoelastic properties also provide critical insight into the processing characteristics of these materials and could help predict the effect of nanoparticles on the final properties. While significant insight has grown over the last 20 years toward understanding polymer nanocomposites, several outstanding problems regarding their rheology remain that require critical advances in experimental and theoretical efforts. For instance, in the dilute limit of nanoparticles, the effect of chemical interactions and particle topology on fundamental alterations to the dynamics of the polymer matrix remains an outstanding issue. In spite of some recent elegant measurements [60], the extensional flow properties of polymer nanocomposites remain to be fully sorted out. Last, the effect of nanoparticles on the local dynamics, especially in the glassy state and therefore their effect on aging in polymer nanocomposites remains to be fully understood.

## 5.6 ACKNOWLEDGMENTS

RK acknowledges the partial financial support from the Air Force Office for Sponsored Research (FA9550-06-1-0422) and the National Science Foundation (CMMI-0708096). TC thanks Mitsubishi Chemicals Center for Advanced Materials at UCSB for partial support.

## REFERENCES

1. P.M. Ajayan, J.-C. Charlier, and A.G. Rinzler, *Proc Nat Acad Sci*, 96, 14199 (1999); P.M. Ajayan, L.S. Schadler, C. Giannaris, and A. Rubio, *Adv. Mat.*, 12, 750 (2000); R. Krishnamoorti, R.A. Vaia, *J. Polym. Sci. Part B: Polym. Phys.*, 45, 3252 (2007); M. Alexandre, P. Dubois, *Mater. Sci. Eng., R.*, 28, 1 (2000).
2. Chatterjee, T.; Krishnamoorti, R. *Phys. Rev. E*, 75, 050403 (2007).
3. E.P. Giannelis, R. Krishnamoorti, and E. Manias, *Adv. Polym. Sci.*, 138, 107 (1999).
4. R. Krishnamoorti, *MRS Bulletin*, 32, 341 (2007).
5. J.N. Israelachvili, *Intermolecular and Surface Forces*, Academic Press, San Diego, (2006).
6. L. Onsager, *Ann. N. Y. Acad. Sci.*, 51, 627 (1949).
7. S.K. Kumar and R. Krishnamoorti, *Ann. Rev. Chem. Biomolec. Eng*. 1, 37 (2010).
8. T. Chatterjee, A. Jackson, and R. Krishnamoorti, *J. Am. Chem. Soc.*, 130, 6934 (2008).
9. R. Krishnamoorti, R.A. Vaia, and E.P. Giannelis, *Chem. Mater.*, 8, 1728 (1996).
10. B.K.G. Theng, *The Chemistry of Clay-Organic Reactions*, John Wiley & Sons, New York, 1947; B.K.G. Theng, *Formation and Properties of Clay Polymer Complexes*, Elsevier, New York, 1979.
11. S. Salaniwal, S.K. Kumar, and J.F. Douglas, *Phys. Rev. Lett.*, 89, 258301 (2002).
12. M. Surve, V. Pryamitsyn, and V. Ganesan, *Phys. Rev. Lett.*, 96, 177805 (2006).
13. Q. Zhang and L.A. Archer, *Langmuir*, 18, 10435 (2002).
14. V. Goel, T. Chatterjee, L. Bombalski, et al., *J. Polym. Sci. Part B: Polym. Phys.*, 44, 2014 (2006); R. Krishnamoorti and K. Yurekli, *Curr. Opin. Colloid Interface Sci.*, 6, 464 (2001); C.A. Mitchell, J.L. Bahr, S. Arepalli, et al., *Macromolecules*, 35, 8825 (2002); C.A. Mitchell and R. Krishnamoorti, *Macromolecules*, 40, 1538 (2007); D.J. Carastan and N.R. Demarquette, *Int. Mater. Rev.*, 52, 345 (2007); S.A. Madbouly and J.U. Otaigbe, *Prog. Polym. Sci.*, 34, 1283 (2009); A.Y. Malkin, *Polym. Sci. Ser. A*, 51, 80 (2009); M. Moniruzzaman and K.I. Winey, *Macromolecules*, 39, 5194 (2006); M. Okamoto, *J. Ind. Eng. Chem.*, 10, 1156 (2004); S.S. Ray and M. Okamoto, *Prog Polym Sci*, 28, 1539 (2003); G. Schmidt and M.M. Malwitz, *Curr. Opin. Colloid Interface Sci.*, 8, 103 (2003); R. Sengupta, S. Chakraborty, S. Bandyopadhyay, et al., *Polym. Eng. Sci.*, 47, 1956 (2007); M.J. Solomon and Q. Lu, *Curr. Opin. Colloid Interface Sci.*, 6, 430 (2001).
15. R. Krishnamoorti and E.P. Giannelis, *Macromolecules*, 30, 4097 (1997).
16. R. Krishnamoorti, J. Ren, and A.S. Silva, *J. Chem. Phys.*, 114, 4968 (2001).
17. R. Krishnamoorti and A.S. Silva, in *Polymer-Clay Nanocomposites*, John Wiley & Sons, New York, (2000).
18. J. Ren, A.S. Silva, and R. Krishnamoorti, *Macromolecules*, 33, 3739 (2000).
19. J.X. Ren, B.F. Casanueva, C.A. Mitchell, and R. Krishnamoorti, *Macromolecules*, 36, 4188 (2003).
20. J.X. Ren and R. Krishnamoorti, *Macromolecules*, 36, 4443 (2003).
21. L. Xu, S. Reeder, M. Thopasridharan, et al., *Nanotechnology*, 16, S514 (2005).
22. K. Yurekli, R. Krishnamoorti, M.F. Tse, et al., *J. Polym. Sci. Part. B: Polym. Phys.*, 39, 256 (2001).

23. J.D. Ferry, *Viscoelastic Properties of Polymer*, John Wiley & Sons., New York, (1980).
24. M. Cloitre, R. Borrega, and L. Leibler, *Phys. Rev. Lett.*, 85, 4819 (2000).
25. P. Sollich, *Phys. Rev. E*, 58, 738 (1998).
26. G. Schmidt, A.I. Nakatani, P.D. Butler, et al., *Macromolecules*, 33, 7219 (2000).
27. G. Schmidt, A.I. Nakatani, and C.C. Han, *Rheol. Acta*, 41, 45 (2002).
28. M.J. Solomon, A.S. Almusallam, K.F. Seefeldt, et al., *Macromolecules*, 34, 1864 (2001).
29. K.M. Lee and C.D. Han, *Polymer*, 44, 4573 (2003).
30. S. Choi, K.M. Lee, and C.D. Han, *Macromolecules*, 37, 7649 (2004); K.M. Lee, and C.D. Han, *Macromolecules*, 36, 7165 (2003).
31. S.K. Basu, A. Tewari, P.D. Fasulo, and W.R. Rodgers, *Appl. Phys. Lett.*, 91, 053105 (2007); D.F. Eckel, M.P. Balogh, P.D. Fasulo, and W.R. Rodgers, *J. Appl. Polym. Sci.*, 93, 1110 (2004); T.D. Fornes and D.R. Paul, *Polymer*, 44, 4993 (2003); D.L. VanderHart, A. Asano, and J.W. Gilman, *Macromolecules*, 34, 3819 (2001); A. Vermogen, K. Masenelli-Varlot, R. Seguela, et al., *Macromolecules*, 38, 9661 (2005); J. Zhao, A.B. Morgan, and J.D. Harris, *Polymer*, 46, 8641 (2005).
32. T. Chatterjee, K. Yurekli, V.G. Hadjiev, and R. Krishnamoorti, *Adv. Funct. Mater.*, 15, 1832 (2005).
33. E.J. Guth, *J. Appl. Phys.*, 16, 20 (1945).
34. D.W. Schaefer and C.Y. Chen, *Rubber Chem. Technol.*, 75, 773 (2002).
35. E.J. Garboczi, K.A. Snyder, J.F. Douglas, and M.F. Thorpe, *Phys. Rev. E*, 52, 819 (1995).
36. M.J. O'Connell, S.M. Bachilo, C.B. Huffman, et al., *Science*, 297, 593 (2002).
37. L. Moreira, R. Fulchiron, G. Seytre, et al., *Macromolecules*, 43, 1467 (2006).
38. E.K. Hobbie and D.J. Fry, *Phys. Rev. Lett.*, 97, 036101 (2006).
39. E.K. Hobbie and D.J. Fry, *J. Chem. Phys.*, 126, 124907 (2007).
40. V. Prasad, V. Trappe, A.D. Dinsmore, et al., *Faraday Discussion*, 123, 1 (2003); V. Trappe and D.A. Weitz, *Phys. Rev. Lett.*, 85, 449 (2000).
41. W.H. Shih, W.Y. Shih, S.I. Kim, et al., *Phys. Rev. A*, 42, 4772 (1990).
42. C.J. Rueb and C.F. Zukoski, *J. Rheol.*, 41, 197 (1997).
43. R. Krishnamoorti and E.P. Giannelis, *Langmuir*, 17, 1448 (2001).
44. A. Lele, M. Mackley, G. Galgali, and C. Ramesh, *J. Rheol.*, 46, 1091 (2002).
45. H.Y. Chen, S.P. Chum, A. Hiltner, and E. Baer, *Macromolecules*, 34, 4033 (2001).
46. M.B. Isichenko, *Rev. Mod. Phys.*, 64, 961 (1992).
47. D. Bonn, H. Kellay, H. Tanaka, et al., *Langmuir*, 15, 7534 (1999).
48. D. Bonn, P. Coussot, H.T. Huynh, et al., *Europhys. Lett.*, 59, 786 (2002).
49. G. Gagnon, J. Patton, and D.J. Lacks, *Physical Review E*, 64, 051508 (2001); D.J. Lacks and J.R. Wienhoff, *J. Chem. Phys.*, 111, 398 (1999).
50. M.A. Treece and J.P. Oberhauser, *Polymer*, 48, 1083 (2007); M.A. Treece and J.P. Oberhauser, *Macromolecules*, 40, 571 (2007).
51. M. Chen and W.B. Russel, *J. Colloid Interface Sci.*, 141, 564 (1991).

52. S. Feng, P.N. Sen, B.I. Halperin, and C.J. Lobb, *Phys. Rev. B*, 30, 5386 (1984); Y. Kantor and I. Webman, *Phys. Rev. Lett.*, 52, 1891 (1984).
53. S. Arbabi and M. Sahimi, *Phys. Rev. B*, 47, 695 (1993); M. Sahimi and S. Arbabi, *Phys. Rev. B*, 47, 703 (1993).
54. Y. Aoki, A. Hatano, T. Tanaka, and H. Watanabe, *Macromolecules*, 34, 3100 (2001).
55. T. Chatterjee and R. Krishnamoorti, *Macromolecules*, 41, 5333 (2008).
56. M. Whittle and E. Dickinson, *J. Chem. Phys.*, 107, 10191 (1997).
57. L.E. Silbert, R.S. Farr, J.R. Melrose, and R.C. Ball, *J. Chem. Phys.*, 111, 4780 (1999).
58. J.D. Thomin, P. Keblinski, and S.K. Kumar, *Macromolecules*, 41, 5988 (2007).
59. S.B. Kharchenko, J.F. Douglas, J. Obrzut, et al., *Nat. Mater.*, 3, 564 (2004).
60. A.J. Schmidt, M. Chiesa, D.H. Torchinsky, et al., *Appl. Phys. Lett.*, 92, 244107 (2008).

# CHAPTER 6

# RHEOLOGY OF WOOD–PLASTICS COMPOSITES

TIEQI LI
NOVA Chemicals Corporation, Calgary, AB, Canada

## CONTENTS



---

*Applied Polymer Rheology: Polymeric Fluids with Industrial Applications*, First Edition. Edited by Marianna Kontopoulou.

## 6.1 INTRODUCTION

Natural fiber composites, containing natural fibers and other lignocellulosic fillers have attracted a great deal of attention from both industry and academia in the past decades due to their potential ecological and economic advantages [1–5].[1] With relatively lower density and costs, natural fibers usually have specific stiffness and strength comparable to their synthetic counterparts, such as glass fibers [4, 5], and can be effective reinforcing agents for composites [6, 7]. It is expected that the rheology of discontinuous and continuous natural fiber reinforced thermoplastics will be a subject of growing interest, as their use becomes more widespread. While the consistency and processibility of the long or continued natural fiber composites are an on-going issue that needs to be resolved [4, 7], lignocellulosic fillers of low aspect ratio, have been explored extensively due to their low cost and their potential to impart higher stiffness when combined with a thermoplastic matrix.

Wood–plastics composites are of especial interest in the category of natural fiber composites. Originally started in North America, the wood–plastics industry experienced significant growth between the late 1990s and 2006 [8–10] and has become an important sector in the manufacture of products for construction and other applications. Major commercial products include extruded deck boards, fence boards, posts, and injection-molded smaller parts. Other products on their way of finding market acceptance are compression molded roof plates, shingles, and other type of panels. The extruded products are available with either solid or hollow cross-sections. Several unformed products have also been explored and found applications. The matrix is usually virgin and/or recycled post-consumer polyethylene because of its good toughness and low cost. Polypropylene and poly(vinylchloride) are also used when higher stiffness or dimensional/thermal stability are required. Persistent efforts have

[1]The opinions expressed in this chapter are solely the author's and do not represent NOVA Chemicals' point of view.

been made in both industry and academia to improve the durability and the moisture resistance of wood–plastics through proper formulation [11], process optimization, and/or using low cost alternative fillers.

As the wood–plastics composites are mainly manufactured by extrusion, injection, or compression molding, the rheology of these filled plastics is of essence for both industrial research and development as well as academic studies. Rheological characterization serves one or more of the following purposes:

- Measurement of viscosity for quality control in manufacturing
- Generation of data needed for die design and process control
- Use of viscoelastic properties as a probe of the composite structure/morphology to guide the development of formulations and the use of appropriate additives, such as coupling agents and lubricants.

The melt rheology of wood-filled plastics, however, is not readily predictable from the knowledge of the conventional thermoplastics. The major difference between these filled thermoplastics from other thermoplastic composites is the very high filler content. As an example, commercial products typically use wood filler contents of 30–40% by weight for molded solid products or foamed extrusion profiles, or 40–65 wt % for solid extrusion formulations. In terms of volume concentration, such wood contents are 20–50% by volume, much higher than the filler content of conventional thermoplastic composites containing mineral fillers, which typically have filler a volume content of <20%. On the other hand, the melt rheology of wood–plastics composites is different from that of conventional concentrated suspensions. In wood–plastics, linear polymers of higher molecular weight are required to impart stiffness and impact resistance to the composites, in contrast to slurries within a Newtonian medium, or suspensions having a lower molecular weight polymer as a carrier. Unlike other concentrated suspensions such as inorganic solids in linear polymers of lower molecular weight [12–15], elevated elasticity is an important feature in wood–plastics rheology. The rheology of wood plastics is not the same as that of the filled rubbers [16, 17] either because, unlike rubber compounds, the matrix polymers used for wood plastics are not cross-linked. Based on the above, it can be deduced that the high volume contents and the moderately high molecular weight matrices make wood–plastics composite a unique family of materials in terms of their rheology.

In this chapter, the melt rheology of wood–plastics composites is briefly summarized. In Section 6.2, published results of both shear and extensional rheometry obtained through capillary measurements are summarized, with the emphasis on appropriate use of capillary rheometry in characterizing the flow performance of wood plastics formulations. In Section 6.3, linear viscoelastic experiments are briefly discussed as a means of nondestructive characterization of the wood plastics composites. Section 6.4 analyzes the nonlinear viscoelastic characteristics, using an integral constitutive equation in combination with extensional flow data obtained in hyperbolic dies and shear flow data obtained in capillary dies. Examples of the nonlinear rheological analysis are then provided for a few extruded wood–plastics formulations to show the capability of this approach to reliably characterize formulations and its potential for flow modeling. It is my hope that the wood–plastics rheology outlined here would also provide the prerequisite knowledge and methodologies for further research on discontinuous long-fiber-reinforced natural fiber composites that are emerging as a more environment friendly replacement for conventional composite with synthetic fibers.

## 6.2 CAPILLARY RHEOMETRY OF WOOD–PLASTICS COMPOSITES

Capillary rheometry data of thermoplastics containing wood or other lignocellosic fillers have been present in the literature since the 1980s [18–21]. However, the majority of the earlier studies were performed on melts with relatively low filler contents, e.g., about 20 wt % or lower. At these concentrations, the presence of fillers resulted in an increase in shear viscosity and somewhat higher levels of shear thinning compared to the matrix, according to the values of the power law exponent [19], similar to melts containing mineral fillers.

With the rapid growth of the wood–plastics industry at the turn of the century, there was a demand for reliable wood–plastics rheology both to aid formulation development and to support process modeling and control. Systematic efforts were initiated in 2001 at the Wood Materials and Engineering Laboratory at Washington State University with government and industry support [9, 10], leading to industrially relevant research and many scientific publications [22–25]. Extensive and in-depth research has also been performed at McMaster University [26–29] with a focus on the role of matrix polymer architecture, origin of wall slip, and possible remedies.

The intention here is not to provide a systematic historical review but to describe the key rheological methods that are applicable in the characterization of the flow properties of wood–plastics compositions pertinent to the wood–plastics industry. This section summarizes the evaluation through capillary rheometry of polyethylene melts filled with 30–60% by weight of wood, based mainly on data available in the literature. Extensional viscosity measurements obtained using capillary rheometers equipped with hyperbolic dies are also included, as extensional flow is an important component of the actual flow and offers the capability of differentiating between wood–plastics formulations. A limited portion of previously unpublished data is also included for illustrative purposes when necessary. The main objective is to review the capability of capillary rheometry to provide reliable viscosity data that characterize the flow of wood–plastics as a function of the composition. A focus is placed on identifying wall-slip characteristics in shear flow, both for interpretation of the rheometry data and for qualitatively predicting the real flow in processing.

Section 6.2.1 summarizes the observations of significant wall slip and methods for its characterization. Section 6.2.2 introduces the hyperbolic die methods for collecting the uniaxial extensional viscosity data for wood plastics. Section 6.2.3 provides cases of using both shear and extensional viscosity data from capillary rheometry to characterize typical wood plastics formulations. A more quantitative analysis of both shear and extensional flow is presented in Section 6.3.

### 6.2.1 Capillary Rheometry and Wall-Slip Phenomena

To obtain reliable data on the rheology of wood plastics using capillary rheometry, the following questions must be addressed:

- Under what conditions can viscosity data be measured in a repeatable manner?
- Are the measured viscosity data representative of the strain rates and shear levels typical of the actual processing so that they can be used for modeling or process optimization purposes?

This section reviews the general features of shear flow in capillary rheometers and presents a suitable method obtaining reliable shear viscosity data based on the Mooney analysis.

#### *6.2.1.1 Discontinuous Flow Curves and Appropriate Test Protocol*

The shear flow of polyethylene–wood composites in piston-driven capillary rheometers is usually characterized by a discontinuous flow curve and a dependence of the results on test protocol [23, 24, 26]. This unique feature is exemplified in Figure 6.1 for a melt containing 50 wt % of fine maple wood flour in high-density polyethylene (HDPE). As shown in the figure, the flow curve—shown in terms of the apparent shear rate, $\gamma_a'$, vs, the shear stress, $\tau$, on double logarithmic scales—includes two branches and a transitional zone in between. At shear stress levels above the transitional zone, the flow rate appears to be much higher than that expected by extrapolating the ($\tau$, $\gamma_a'$) data of the lower branch. In the transitional zone, the melt pressure for $\gamma_a'$ oscillates instead of reaching a plateau level (Fig. 6.2). Consistent with the pressure curves presented in Figure 6.2, the extrudates also show a distinctively different surface appearance, depending on the value of $\gamma_a'$ (Fig. 6.3), as follows:

- Along the high branch of the flow curve, e.g., at $\gamma_a' = 638\ s^{-1}$ or higher, the extrudate shows alternate smooth and gross fractured surfaces.
- On the transitional zone, e.g., at $\gamma_a'$ between 150 and 251 $s^{-1}$, the extrudate also shows alternate smooth and fractured surfaces. However, the fractured portion is much smoother than the one observed at higher rates.
- Along the lower branch at moderately low shear rate, e.g., at $\gamma_a' = 63.8\ s^{-1}$, the extrudate shows only smooth surface.
- At a very low rate of 1 $s^{-1}$, uniform gross melt fracture occurs again.

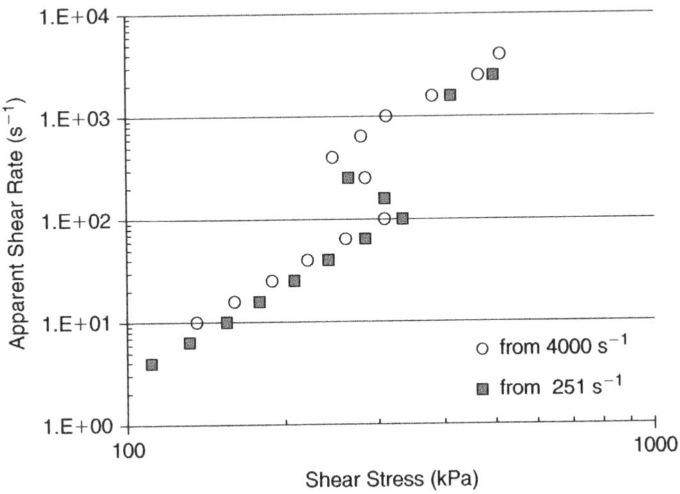


**FIGURE 6.1** Flow curve of HDPE melt containing 50 wt %, 140-mesh maple flour at 180°C starting from different apparent shear rate. (For sample details see Ref. 24. The tests beginning from 4000 $s^{-1}$ were measured stepwise from high to low values. Those beginning from 251 $s^{-1}$ were high to low runs followed by additional tests at higher rates.

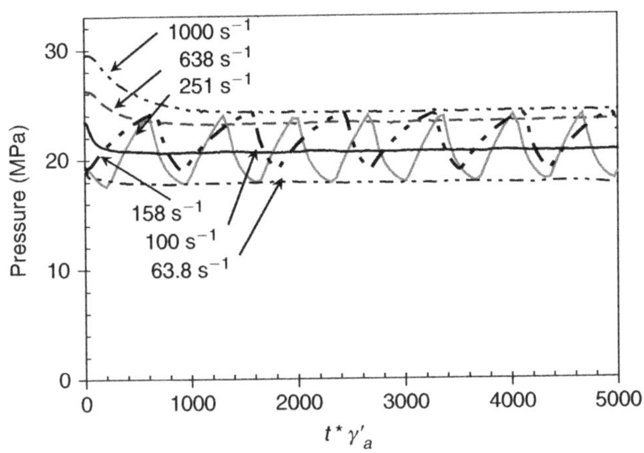


**FIGURE 6.2** Typical melt pressure evolution of the HDPE/wood melt shown in Figure 6.1 along the higher branch (1000 and 638 $s^{-1}$), transitional zone (158 $s^{-1}$), and lower branch (100 $s^{-1}$ and 63.8 $s^{-1}$) of the flow curve.

The appearance of the extrudate is similar to that of the neat HDPE for the majority of the flow curve with the exception of the melt fracture at the lowest shear rate. Similarly as for the neat HDPE [30–32], flow instabilities associated with the disentanglement of the high molecular linear chains on the capillary

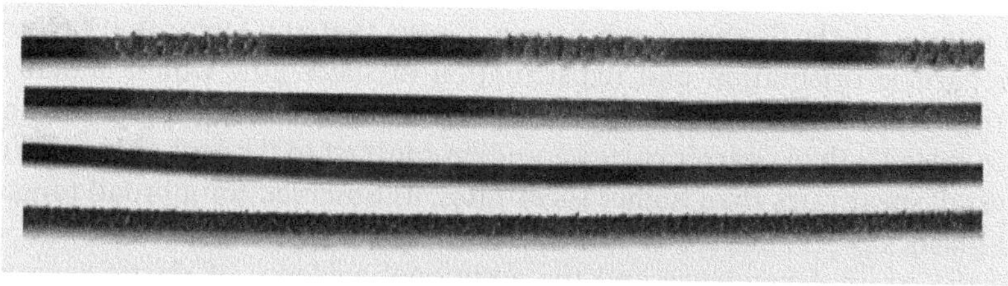


**FIGURE 6.3** Extrudates of the HDPE/wood melt after 180°C capillary shear experiments (tungsen carbide dies of 2 mm in diameter by 30 mm in length) at different apparent shear rates.

wall should be the origin of the discontinued flow curve and the melt fracture at shear stress levels in the transitional zone or above. The role of wood filler is to increase the shear stress, leading to the occurrence of the instability at a lower apparent shear rate [22, 23].

The difference of the flow curve from that of the neat HDPE is observed both at stress levels well above the transitional zone and at very low shear rates. In contrast to the neat resin, the pressure oscillation does not occur with the HDPE–wood composites at the very high shear stress, though similar melt fracture is still observed. Based on the Mooney analysis [22, 26, 28], the volumetric flow rate at such high stress level consists mainly of the contribution from the wall-slip component. That is, the apparent steady melt pressure originates from the plug flow nature. The stick-slip mechanism at the wall is still observable from the alternate fractured/smooth appearance (Fig. 6.3a); but the overall steady/spurt flow does not happen as a result of the plug flow nature.

At very low stress levels, where the apparent shear rate is also very low (1 $s^{-1}$ in Figure 6.1), abnormal melt fracture is observed for the composites. This melt fracture is unique to the highly filled melt as it results from the presence of filler particles. For the simple case shown in Figure 6.1 for the melt containing only fine particles, the fractured extrudate is characterized with protruding wood particles seen on surface by naked eye. Based on the color and on microscopic observation, a thin layer of polymer may be present on the top surface of the wood in this case. It would thus appear that the fracture occurs between the polymer matrix and the wood particles or within the polymer that exists near the surface of wood particles.

This interfacial or interphacial melt fracture can be explained from the viewpoint of the orientation and distribution of the wood particles in shear flow. In polymer suspensions, high shear rates are expected locally near the interface between polymer and filler, on a microscopic scale [33–36]. However, in contrast to most common polymer suspensions, the matrix used in wood plastics has a high molecular weight. The energy due to high interfacial shear stress cannot be dissipated during the same time frame over which the wood particle orientation or redistribution takes place. As a result, the stored elastic energy is released at the die exit and causes fracture if it exceeds the strain capacity of the material. Such an interfacial elastic energy mechanism means that any factor that favors the filler particle orientation and redistribution in shear flow would actually impose risk of melt fracture. This mechanism predicts more melt fracture with finer filler particles than coarser particles and, in contrast to the neat polymer, at lower overall shear rates than higher shear rates, as observed for a broad range of wood plastics [9, 10, 26, 28].

While the strain dependence and the additional evidence of elastic energy stored during macroscopic shear flow is further discussed in the following sections, for the practical purpose of obtaining reproducible viscosity data, it is recommended that:

- A screening test be performed to determine the transitional $\gamma_a'$ range for an unknown sample.
- The test be started at an initial $\gamma_a'$ value just above the transition zone (to make sure the material is well condensed and soaked) and followed by lower rates.
- Only the viscosity data along the lower branch be used for comparing different materials, unless a pressure-driven rheometer is used.

***6.2.1.2 Wall-Slip Correction with Mooney Analysis and Strain Dependence*** The Mooney analysis, which is a common method used to characterize the flow with significant wall slip, has been proven applicable for wood–plastics [22–29]. In certain cases the wall-slip velocity calculated with the Mooney analysis can even be useful as a screening tool to evaluate the presence of external lubrication [10]. To perform the Mooney analysis, capillary data are collected using three or four sets of dies of different diameter to generate the shear rate–shear stress data for each die diameter. Entrance pressure can be corrected with the zero-land length dies, while the Rabinowitsch correction is performed to count on the shear thinning effect. The Mooney plot, which is the plot of shear rate against reciprocal of the diameter of the die at a constant shear stress, is constructed. When the Mooney plot can be approximated as a linear relation, the slope and the intercept allows the wall-slip velocity and the wall-slip corrected shear rate to be calculated. The detailed test and analysis can be found in the literature [12–15, 22, 26, 28, 30] and will not be repeated here.

The actual mechanism of wall slip is actually not well understood. The slip layer thickness of HDPE containing 40 and 60 wt % of 40-mesh maple flour had been estimated to be on the order of 0.01 to 0.1 μm. This dimension is well above the characteristic chain dimension of polyethylene. The slip layer encountered with 40 wt % pine filler is much higher. It is likely that the chain disentanglement is not the only cause of the wall slip. A depletion of filler near the wall like the case for colloidal suspensions [37] or polymer melt with inorganic fillers [14], and a possible enrichment of the extractable [22] can all be among the potential causes of such phenomenological wall slip layer.

It is worth noting that the Mooney analysis can be applied, in spite of the pronounced strain dependence resulting in different extrudate appearance, as explained above. In other words, the Mooney analysis is applicable regardless of how the data on different dies are collected and is applicable to wood–plastics no matter whether a fixed die length [22, 23, 25] or a fixed length: diameter ratio [26, 28] is used. The wall-slip velocity–shear stress relations obtained from the two methods appear to be similar. It is likely that the formation of a phenomenological slip layer in capillary shear flow happens either during extensional flow at the entrance or at a very short distance within the capillary die. The shear stress growth analysis of Section 6.4.1 suggests the latter is true due to the significant nonlinear nature of the flow. If a physical slip layer does exist, whether consisting of materials with lower wood content or of higher concentration of small molecules than the bulk, phase segregation is more likely a result of the extensional flow at the entrance and not related to the extent of the shear flow in the die. More research would shed some light on this topic for reaching a simpler method for wall-slip correction.

In brief, the Mooney analysis is a useful tool for evaluating the phenomenological wall slip of wood–plastics in shear flow. The wall-slip velocity and the contribution of slip to the overall flow can allow qualitative estimations of the compositional changes during flow. The Mooney analysis and extrudate observation can together provide an appropriate empirical characterization of the shear flow for the wood–plastics.

### 6.2.2 Extensional Flow in Hyperbolic Dies

The importance of measuring extensional viscosity is twofold:

- Extensional flow is a common type of flow encountered during processing.
- If appropriately measured, it can be used to characterize the status of filler dispersion and the polymer-filler interface.

Higher extensional viscosity has been observed in melts containing long fibers, due to the orientation of fibers and the larger microscopic shear flow near the fiber surface [33–36]. For wood–plastics, the aspect ratio is usually low compared to the long fiber reinforced polymer composites, therefore the effect of orientation is not expected to be as significant. The extensional viscosity is hence mainly governed by the filler dispersion and the extent of interfacial stress transfer: the stress in the matrix is higher when the wood particles are dispersed better and when coupling between the polymer matrix and the wood filler exists. Therefore knowledge of the extensional viscosity can serve as a useful tool for formulating wood–plastics.

A comparison between macroscopic extensional and shear viscosities is especially useful in characterizing the state of the filler dispersion and their interface. The Trouton ratio, normally defined as the ratio between the steady state uniaxial extensional viscosity and the shear viscosity, can be used as the relevant parameter for this purpose [22, 36]. However, the shear viscosities measured by capillary rheometry may vary with both wall slip and total strain [24]. Following Stange et al. [38] with a minor modification, a strain hardening factor $S_H$ is defined as the ratio of the extensional viscosity, $\eta_E^+(\dot{\varepsilon}, t)$, at a given rate, $\dot{\varepsilon}$, and time, $t$, over the linear shear stress growth function, $\eta^0(t)$, at the same time, $t$, to better serve this purpose

$$S_H = \eta_E^+(\dot{\varepsilon}, \mathrm{t})/\eta^0(t) \tag{6.1}$$

It is worth noting that this parameter is a third of the one originally used by Stange et al. [38] to account for the possible softening of the highly filled melt. This parameter reveals whether the tested sample has reached the strain hardening stage at the time for the strain rate. The determination of this parameter and its use are provided in Section 6.4 and Appendix 6.A.2 for representative wood–plastics formulations.

To measure the uniaxial extensional viscosity, the hyperbolic die technique can be used. The hyperbolic dies are manufactured to have a flow channel with the cross-section area inversely proportional to the axial distance from the start of the flow, hence providing a constant extensional strain rate, $\dot{\varepsilon}$. Ideally a hyperbolic die of radius, $r_z$, at an axial location, $z$, would satisfy [22, 39]

$$(\pi r_z^2) = a/(b + z) \tag{6.2}$$

where $a$ and $b$ are geometry constants for the die, and the extensional strain rate, $\dot{\varepsilon}$, is constant and can be determined from the volumetric flow rate $\dot{Q}$ and the geometry of the die

$$\dot{\varepsilon} = \dot{Q}/a \tag{6.3}$$

As a practical approximation, the geometry is more conveniently specified with the radius $R_0$, the barrel radius, and the total extensional strain (Hencky strain) $\varepsilon_L$. The exit radius $R_L$ is usually measured to ensure the total extensional strain $\varepsilon_L$ is of desired value according to

$$\varepsilon_L = 2\ln(R_0/R_L) \tag{6.3a}$$

When the die is used with a piston-driven capillary rheometer and the flow rate, $\dot{Q}$, is a constant, the extensional viscosity is defined based on the flow rate, $\dot{Q}$, the pressure drop, $\Delta P$, and the die dimensions, $R_0$ and $R_L$, following Collier's original definition through a potential energy argument [39] as follows

$$\eta_E = \frac{\Delta P}{2\left[\frac{\dot{Q}}{\pi R_0^2}\frac{1}{L}\right]\left[(R_0/R_L)^2 - 1\right]\ln(R_0/R_L)} \tag{6.4}$$

This equation actually defines the extensional viscosity in a hyperbolic die as the ratio of the pressure gradient over the strain rate. For each die with a specific $\varepsilon_L$, the extensional viscosity evaluated by (6.4) is obviously an average over the time when the strain grows from 0 to $\varepsilon_L$.

The hyperbolic die technique described above is a useful tool for characterizing the extensional flow performance in close resemblance to the actual flow in wood–plastics processing. The experimental determination of the extensional viscosity with conventional free surface methods (melt drawing techniques) is difficult, if not impossible, for wood–plastics melts. The major obstacle is the very low strain limit beyond which wall slip or melt fracture occurs [24], similar to other highly filled polymer melts [40]. The low strain limit does not allow the melt to be drawn to the extent comparable to what can be achieved in a confined geometry. Hyperbolic dies in capillary rheometry provide suitable flow conditions to allow the measurement of the extensional viscosity. These measurements have been used as an effective tool in qualitatively comparing the effects of wood filler characteristics, volume content and additive formulations as reviewed in following section. When multiple tests with dies of different strain are used, the experiments also allow a quantitative flow analysis, as shown in Section 6.4.

### 6.2.3 Applications to Wood–Plastics Formulations

As discussed in the previous sections, capillary rheometry can characterize the shear flow of wood–plastics melts by considering a wall-slip parameter together with a shear viscosity at each shear rate. Hyperbolic dies also enable the evaluation of extensional viscosity at a given rate, $\dot{\varepsilon}$, and strain, $\varepsilon_L$. In this section, a summary of the common features of representative wood–plastics formulations is provided first, followed by a brief review of the effects of important formulating variables, such as wood filler characteristics and volume content, coupling agents, lubricants or dispersing agents, and matrix architecture.

It is worth noting that actual wood–plastics composites can consist of vastly different polymers, fillers, and additives. The characteristics of the flow as measured by capillary rheometry can, therefore, be distinctively different. This section outlines only observations on wood–plastics formulations containing the basic ingredients used in the industry—that is, linear polyethylene as the matrix, particulate wood fillers, coupling agents, and lubricants. Other reinforcing agents, additives, and minor polymer components are not discussed.

***6.2.3.1 Common Features of Shear and Extensional Flow of Wood–Plastics*** The flow of typical wood–plastics composites for extruded profiles generally has the following major characteristics:

- Significant wall slip at both the lowest and higher shear rates. As discussed in the previous section, wall slip is observed at all relevant shear rates. With 40 wt %, 40-mesh maple filler, the relative magnitude of the contribution of wall slip to the overall flow rate is comparable to that of the real shear induced flow. For HDPE containing 60 wt % of maple filler or 40 wt % of pine, the flow can be dominated by wall slip—that is, it resembles a plug flow [22].

- Presence of yield stress in shear flow [22, 23, 25]. When corrected for wall slip through the Mooney analysis, an unbounded shear viscosity is obtained at the lower shear rates corresponding to a limiting value of shear stress below which the flow cannot occur.
- Surface appearance is a function of both the shear rate and the total strain level and also depends on filler characteristics and the use of additives [26, 28]. The extrudates tend to be smoother at a moderately high range of shear rates, slightly below the oscillating transitional zone. The appearance can be greatly improved with the use of lubricants [10, 25] and, if affordable, processing aids [29].
- Extensional viscosity varies strongly with both wood content and additives. It is not sensitive though to the type of fillers in well mixed compounds [22–25].

***6.2.3.2 Effects of Composition*** The effects of the composition on the flow performance of typical wood composites can be summarized as below:

- The type of wood species has strong effects on shear and extensional viscosity [9, 22]. Soft wood fillers such as pine, having larger surface areas, can result in higher viscosity than the hard wood fillers at about 40 wt % filler loading. At very high filler loadings, e.g., 60 wt % or higher, the difference is minimal.
- Wood mesh size alone does not appear to be an important factor affecting shear or extensional viscosity [23]. However, presence of even a small fraction of larger particles affects the melt strength significantly and impacts the extrudate appearance and process stability [26].
- Wood content is a major factor affecting the flow performance of wood plastics. Increasing wood loading results in an increase of both shear and extensional viscosity. The wood content dependence qualitatively follows the rules known for suspensions. The increase with wood content, however, is not as drastic as observed in other suspensions containing inorganic fillers. The compressibility and deformability of wood fillers at the processing temperatures have been postulated as the possible cause for this milder rise of viscosity [23].
- Coupling agents act as internal lubricants. The effect of coupling agent, such as maleated polyethylene in HDPE with hardwood fillers, depends on filler content [23, 25]. To be effective as a coupling agent, a maleated or acrylic additive would have base molecular weight slightly larger than the limit for chain entanglements of the matrix. For an efficient coupling agent, the additive would cause higher shear and extensional viscosities at high wood loadings, for example 50 and 60 wt % for HDPE with maple

fillers. At lower filler loading levels, the additive can result in a decrease in the viscosities due to its role as an internal lubricant in facilitating the orientation and distribution of wood particles in flow. An example of the effects of the base resin molecular weight had been shown by Hristov and Vlachopoulos [27]. A quantitative view of the role of a typical coupling agent in HDPE–maple composites is provided in Section 6.4.4.

- Lubricants and processing aids help external lubrication and, in some cases dispersion [10, 25]. Lubricants and processing aids are used mainly for improving flow rate and surface appearance. An optimized additive can also act as a dispersion aid, reducing viscosity and improving extrudate appearance [10, 25]. The role of lubricants in shear flow is mainly to promote desired wall slip, as evident through the Mooney analysis of capillary shear flow data. The internal lubricating effect of the optimized dispersion aids is quantitative analyzed in Section 6.4.4.
- The matrix polymer architecture can affect flow performance. Different matrices can result in different flow performances, affecting especially the onset of flow instability and the extrudate appearance [27]. For practical formulation of the wood composites, it is essential to choose resins of broad molecular weight distribution for processibility, while trying to maintain higher molecular weight for end-use properties.

## 6.3 LINEAR VISCOELASTICITY MEASUREMENTS AND STEP RATE TESTS IN ROTATIONAL RHEOMETERS

Capillary rheometry has been widely used to characterizing flow performance of wood plastics because of both the widespread availability of suitable instruments and the similarity between the flow encountered in tests and in actual processing operations. However, the analysis of capillary rheometry tests is complicated due to the presence of wall slip. Both the wall slip–corrected shear flow curves and the extensional viscosity data can be generated only using multiple sets of dies. This process is time-consuming. Alternative tests using parallel-plate geometry, such as small strain frequency sweep, step strain shear experiments, and step rate stress growth tests are useful if the purpose of the rheological characterization is to qualitatively compare the materials instead of obtaining precise viscosity data for flow modeling. These tests also provide data for more rigorous flow analysis exemplified in next section.

The results obtained through use of the conventional frequency sweep experiments, have been discouraging in general. Rheological evaluation in the linear viscoelastic (LVE) regime is based on oscillatory measurements in the regime where the rheological properties are strain independent. Unfortunately, routine strain sweep tests and carefully evaluated step strain experiments have shown that the strain limit up to which the strain independency is valid is very low [24, 25]. Moreover the validity of the Cox-Merz rule is ambiguous, even

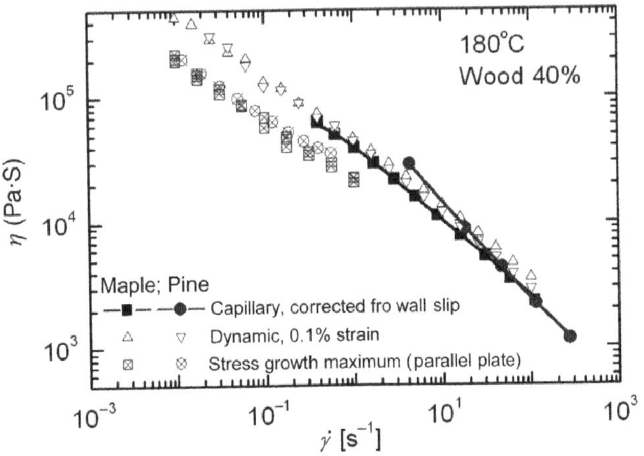


**FIGURE 6.4** Comparison of complex viscosity to slip-corrected shear viscosity measured by capillary rheometry. The stress growth maximum obtained by a parallel plate rheometer is also shown. The material is HDPE with 40 wt % wood [10].

though it was found to approximately apply (Fig. 6.4) in HDPE composites containing 40-mesh wood fillers. Despite the limitations in linearity, frequency sweeps can still serve as a powerful non destructive test that can differentiate the subtle differences in structure, if measured carefully. Some good examples are documented in the literature [24, 25] and will not repeated here. To practically apply oscillatory measurements to compare formulations or additives, one needs to use the right size of fixtures, while considering the rheometer torque capacity, depending on the materials to be tested.

As a prerequisite for in-depth understanding of the flow of highly filled wood–plastics melts, frequency sweep and the step strain tests have been successfully used to confirm that the nonlinearity starting from very low strain is a general feature. In addition to the frequency sweep data, step shear strain rate experiments become an effective means to probe the mechanism underlying the phenomena observed with capillary rheometry. As shown in Figure 6.4, the maximum transient viscosity measured at the maximum torque is much lower than the slip-corrected shear viscosity obtained by capillary measurements and the complex viscosity measured by frequency sweeps. It is worth noting that the stress maxima are observed well before any fracture is visible macroscopically. To summarize, the nonlinear characteristics are dominant in the rheology of wood plastics and can be observed with the following experiments:

- Low strain limit for strain-independent dynamic modulus.
- High damping factors observed with step strain experiments in parallel plates.
- Low stress maximum and early decay of stress in step rate tests in parallel plates.

The nonlinear rheology is hence crucial for characterizing the flow of wood plastics no matter whether the purpose is for materials comparison or for predicting the processibility and will be discussed indepth below.

## 6.4 NONLINEAR RHEOLOGY OF THE WOOD–PLASTICS MELTS

As reviewed in the previous sections, the rheology of wood plastics is characterized by a strong dependence of the shear or extensional stress on both strain rate and strain. The rheological characterization taking into consideration the nonlinear nature of the flow provides more insight into the flow mechanism of these highly filled composites having a linear, high molecular polymer matrix. This section explores the nonlinear rheology of wood–plastics melts based on an analysis using an integral constitutive equation.

As explained in detail in Appendix 6A, using an integral constitutive equation—namely, the Wagner model—the stress growth over time in a simple shear or uniaxial extensional flow can be quantitatively characterized with:

- The memory function based on the relaxation spectrum determined from the LVE characterization.
- The evolution of strain as described by the Finger tensor, based on a finite strain description of the flow.
- The damping function that accounts for a decay of the resistance to flow at increasing strains.

This constitutive equation allows the prediction of the flow for any melt of wood–plastics with a known relaxation spectrum and damping function. Alternatively, it can also be used to determine the damping function and other nonlinear characteristics to compare formulations, if the relaxation spectrum can be determined. Examples of simple extensional and shear flow for HDPE melt with 30 to 60 wt % wood fillers are provided below. The materials preparation and test details, including capillary shear, frequency sweep, step-rate shear stress growth, and the hyperbolic die extensional viscosity data, are have been published [24].

### 6.4.1 Extensional Flow in Hyperbolic Dies

As reviewed in Section 6.2, the extensional viscosity measured using hyperbolic dies is a function of both strain rate, $\dot{\varepsilon}$, and total strain, $\varepsilon_L$. The extensional viscosity data on HDPE wood–plastics melt containing 140-mesh maple flour [24] at different $\dot{\varepsilon}$ and $\varepsilon_L$ have been analyzed using the K-BKZ constitutive equation (Fig. 6.5). The damping exponent, $\beta$, and the strain hardening factor, $S_H$, have been calculated by minimizing the difference between the calculated

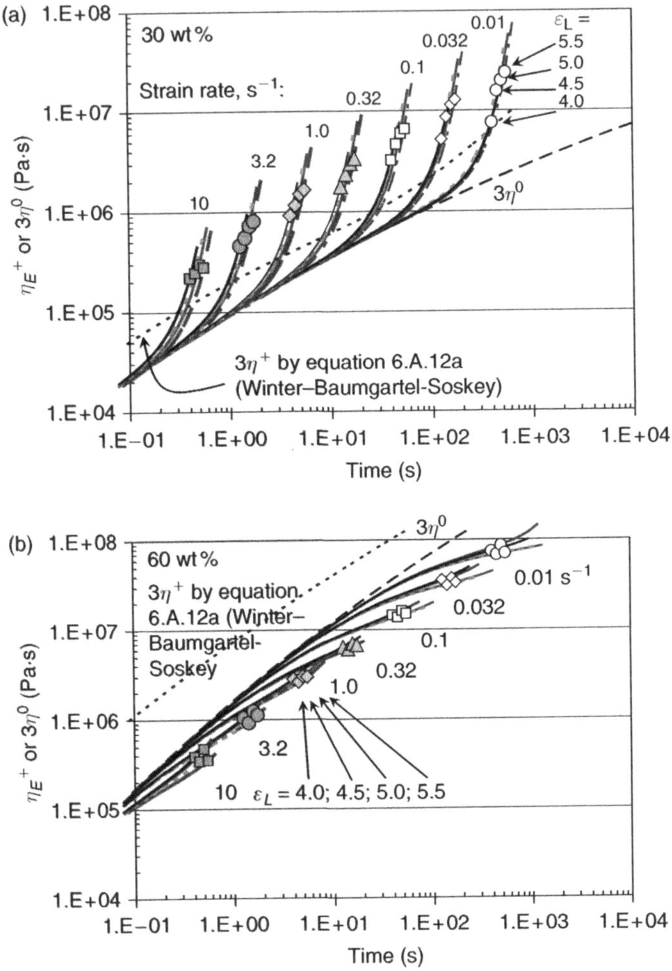


**FIGURE 6.5** Uniaxial extensional viscosity at 180°C of the HDPE melt containing (a) 30 wt % or (b) 60 wt % of 140-mesh maple flour, compared to the K-BKZ fits. The fitted parameter $\beta$ is shown in Figure 6.6. *Dash-dotted* lines are the LVE baselines used for calculating $S_H$; *dotted line*, based on Winter–Baumgartel-Soskey equations (6.A.12a) for comparison only.

stress growth function and the measured extensional viscosity at each ($\varepsilon_L$, $\dot{\varepsilon}$). The extensional stress growth functions are presented in Figure 6.5, along with the LVE shear stress growth function baselines, to compare the strain hardening behavior. The calculated parameters $\beta$ and $S_H$ are plotted in Figure 6.6 to compare the strain hardening as a function of wood loading. According to the figures:

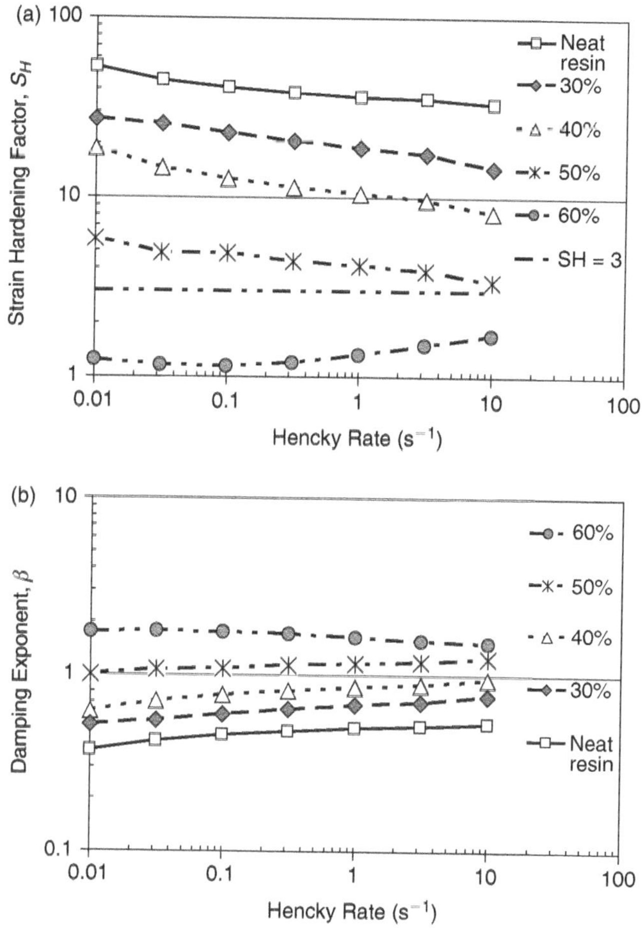


**FIGURE 6.6** (a) Strain hardening factor and (b) damping exponent at $\varepsilon_L = 5.0$ for data shown in Figure 6.5.

- At each given rate, $\dot{\varepsilon}$, the calculated extensional stress growth function, $\eta_E^+(t, \dot{\varepsilon})$, agrees with the extensional viscosity data measured with various dies generating different $\varepsilon_L$. That is, within the $\varepsilon_L$ range used here (4.0 to ~5.5), the actual extensional flow depends more on strain rate, $\varepsilon'$, than on total strain: $\eta_E^+(t, \dot{\varepsilon}, \varepsilon_L) \sim \eta_E^+(t, \dot{\varepsilon})$.
- The direct measurement of extensional viscosity, $\eta_E(\dot{\varepsilon})$, does depend on the $\varepsilon_L$ generated by each die at each individual rate. However, when compared for the same die, the extensional viscosity characterizes well the strain rate dependence and the effects of composition (wood content), validating previous qualitative analysis [22–24].

The strain hardening effect weakens when wood loading increases from 30 to 60 wt %, as evident by the extensional stress growth function (Fig. 6.5) and the strain hardening factor, $S_H = \eta_E(t, \dot{\varepsilon})/\eta^0(t)$ (Fig. 6.6a). At the highest wood weight content of 60% it is observed that $\eta_E^+(t, \dot{\varepsilon}) < \eta^0(t)$ (Fig. 6.5d) and $S_H < 3$ (Fig. 6.6a), meaning that strain hardening is not observed within the entire $\varepsilon_L$ range.

It is worth noting that the damping exponent $\beta$ is larger at higher wood loading and is in good correlation with the decreasing $S_H$. This increase in nonlinearity in extensional flow is similar to that observed for unfilled polymers with different contents of long-chain branching [41]. Both $\beta$ and $S_H$ can be used to quantitatively characterize the strain hardening with increasing wood loading.

### 6.4.2 Simple Shear Flow

Step-rate shear experiments have been performed using a parallel plate rheometers to obtain the transient shear viscosity, $\eta^+(\dot{\gamma}, t)$, for the shear rate, $\dot{\gamma}$ [24]. The step-rate shear stress growth data, $\eta^+(\dot{\gamma}, t)$, were compared with the K-BKZ analysis fits for the same HDPE melt containing 140-mesh wood filler at different loading. The point of maximum torque at each shear rate is used to determine the damping function, $h(\gamma)$, and the damping exponent, $\beta$. As shown in Figure 6.7, the K-BKZ prediction based on the point of maximum torque characterizes the step-rate shear stress growth data well for the melt with 50 and 60 wt % wood filler content, which is the common range used for commercial extruded profiles without foaming. For the melt containing 40 wt % wood, the K-BKZ fit does not result in good fits at the shortest times.

For the melt containing 30 wt % wood, the method is not useful because the measured transient shear viscosity, $\eta^+(\dot{\gamma}, t)$, is always higher than the linear stress growth function, $\eta^0$, throughout the tested time frame, beyond which melt fracture in terms of cracks or void formation on surface is evident. The excessive stress can be related to the presence of a yield stress as evident from the capillary shear results. The yield stress is not accounted for in the memory function due to the assumption that the wood plastics melt is in general a viscoelastic fluid. The excessive stress at 30 or 40 wt % wood content remains a subject of future study either through experimental determination of the relaxation spectrum on high frequencies or including the unaccounted component in the memory function.

Despite the uncertainty at very short time or high rates with the low wood content, the analysis in general provides limited insight in the strain dependence observed with capillary experiments as mentioned also in Section 6.2. As evidenced in Figure 6.7c, the transient shear viscosity depends on the strain for each strain rate in these step-rate tests. More important, the time required to reach the maximum torque is very short. One can postulate that in capillary

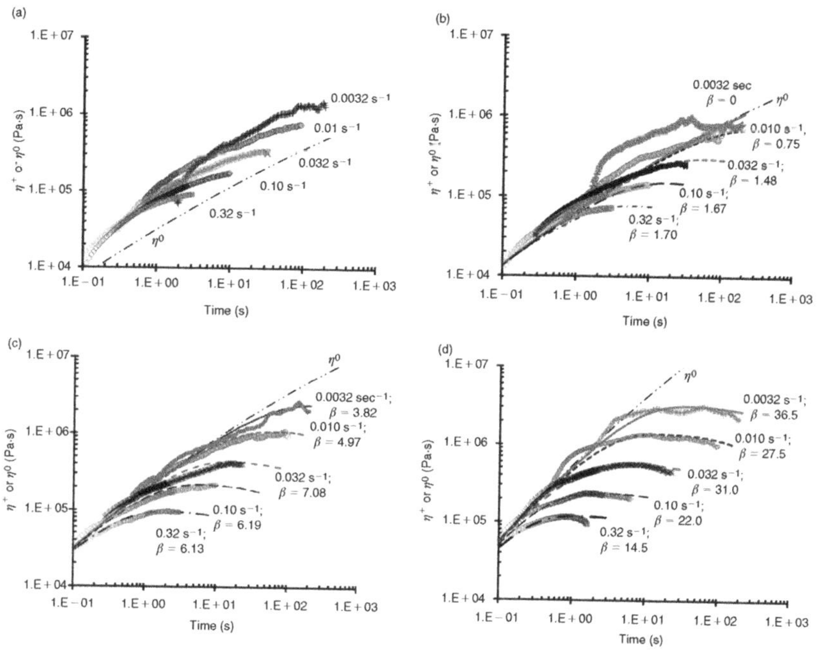


**FIGURE 6.7** Shear stress growth function obtained from step-rate shear experiments at 180°C, compared to K-BKZ fits and LVE baseline for HDPE melt with 140-mesh wood at (a) 30 wt %, (b) 40 wt %, (c) 50 wt %, and (d) 60 wt %. (Materials are the same as in Figures 6.5 and 6.6.)

dies, the damping effect may be still present but weaker; the maximum available shear is always reached well before the full length of the dies is reached. This is in good agreement with the fact that the Mooney analysis applies well, no matter if the dies of the same length and different diameter [22, 23, 25] or of the same aspect ratio [26, 28, 29] are used. It is worth noting that the shear growth experiments were performed in parallel plate rheometers, where the hydrostatic pressure is atmosphere, while the capillary tests would involve much higher hydrostatic pressure. Future analysis by taking the hydrostatic pressure into consideration is needed to help explain better the shear flow of these filled melts containing moderate amounts of 30 or 40 wt % wood filler within a high molecular weight matrix.

### 6.4.3 Practical Applications—Role of Coupling Agents and Lubricants

While the shear stress growth for the melt with moderate wood contents remains the subject of further investigation, the K-BKZ analysis of the hyperbolic die data is very useful for quantitatively evaluating the uniaxial extensional flow, both for the purpose of materials characterization and for the modeling of the flow of the wood plastics in contracting channels. Examples are given here to demonstrate the role of coupling agent, lubricants and their combinations.

#### 6.4.3.1 *Maleated Polyethylene (Coupling Agent)*

Maleated polyethylene (MAPE) is used as a coupling agent and is a common ingredient in wood plastics. Its role depends on the architecture of both MAPE and matrix polymer and can also vary at different wood contents. The strain hardening factor and the damping exponent of formulations containing various amounts of MAPE are compared to the original data for the melt of an HDPE with a 40-mesh maple wood filler shown in Figure 6.5 [25]. The results shown in Figure 6.8 reveal clearly that at all wood contents of interest, the MAPE alleviates the loss of strain hardening due to the wood filler. Because the filler size and aspect ratio in these samples are small, the major origin of strain hardening is from the matrix polymer. The result in Figure 6.8 hence suggests that the MAPE component at the wood–polymer interface helps prevent the wood particles from hindering the growth of extensional strain of matrix polymer. In other words, a coupling agent in the melt state actually plays an internal lubricating role [22–25].

#### 6.4.3.2 *Different Lubricants and Combination with the Coupling Agent*

The K-BKZ analysis of the hyperbolic die data can also provide quantitative comparison between the different lubricating systems. As shown in Figure 6.9, the analysis reveals the following different trends for two different lubricating systems:

- An ester type additive (OP) is a effective dispersion aid, given that $\eta_E^+(\dot{\varepsilon}, t)$ is higher than that of the control sample and closer to the linear limit, $3\eta^0(t)$ (comparing Figures 6.9a and 6.9c).
- A zinc stearate/*N*,*N*-ethylene *bis*-stearamide lubricant (ZnSt-EBS) appears to show less effect on $\eta_E^+(\dot{\varepsilon}, t)$ than the OP additive (comparing Figures 6.9c and 6.9e).
- OP and MAPE have synergistic effects of internal lubrication, seen by the more significant strain hardening in the melt with both OP (Fig. 6.9d) and MAPE than the one with MAPE alone (Fig. 6.9b).
- ZnSt-EBS and the MAPE do not show synergistic effects (Fig. 6.9b, f).

The difference between the additives is further illustrated in Figure 6.10. These findings can be used as a practical quantitative measure of the effectiveness of internal lubrication using different lubricants and combination with coupling agents, such as MAPE.

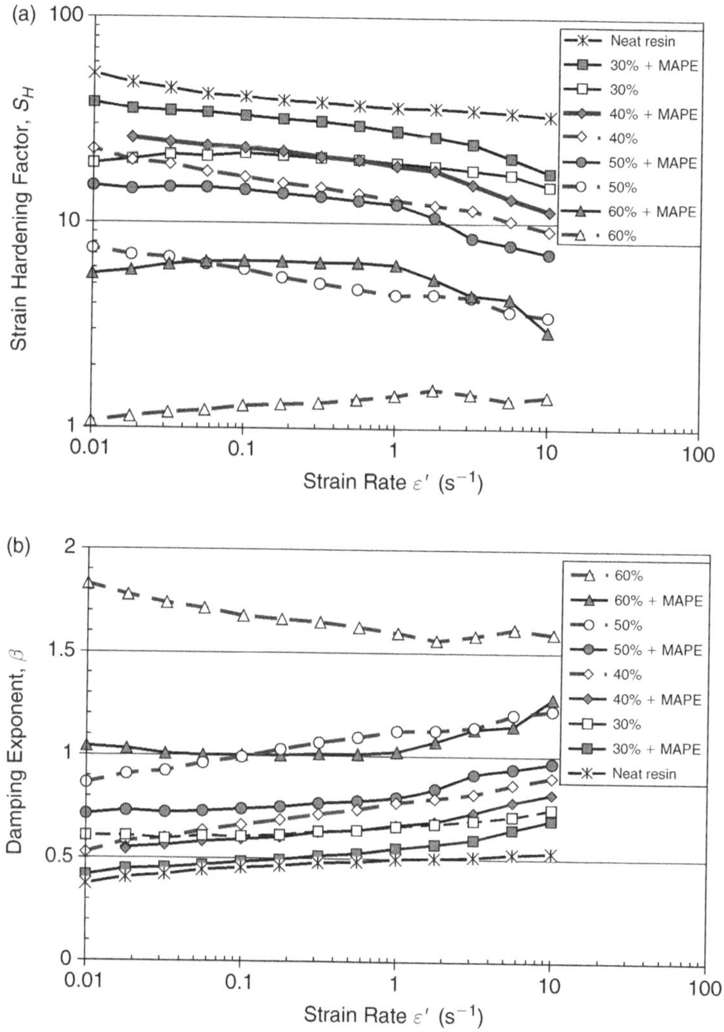


**FIGURE 6.8** (a) Stain hardening factor and (b) damping exponent (b) of extruded HDPE formulations with or without MAPE at various contents of 40-mesh maple filler. $\varepsilon_L = 5$ at 180 °C. For materials and extrusion see Ref. 25.

## 6.5 SUMMARY AND OUTLOOK

This chapter reviewed the rheology of wood–plastics composites melts has been reviewed. In general, the shear flow in capillary dies shows strong wall slip and can result in melt fracture at both high and very low shear rates. The wall slip in capillary shear flow results from the disentanglement of the linear matrix polymer at the wall at high shear stress and may also have contribution from the

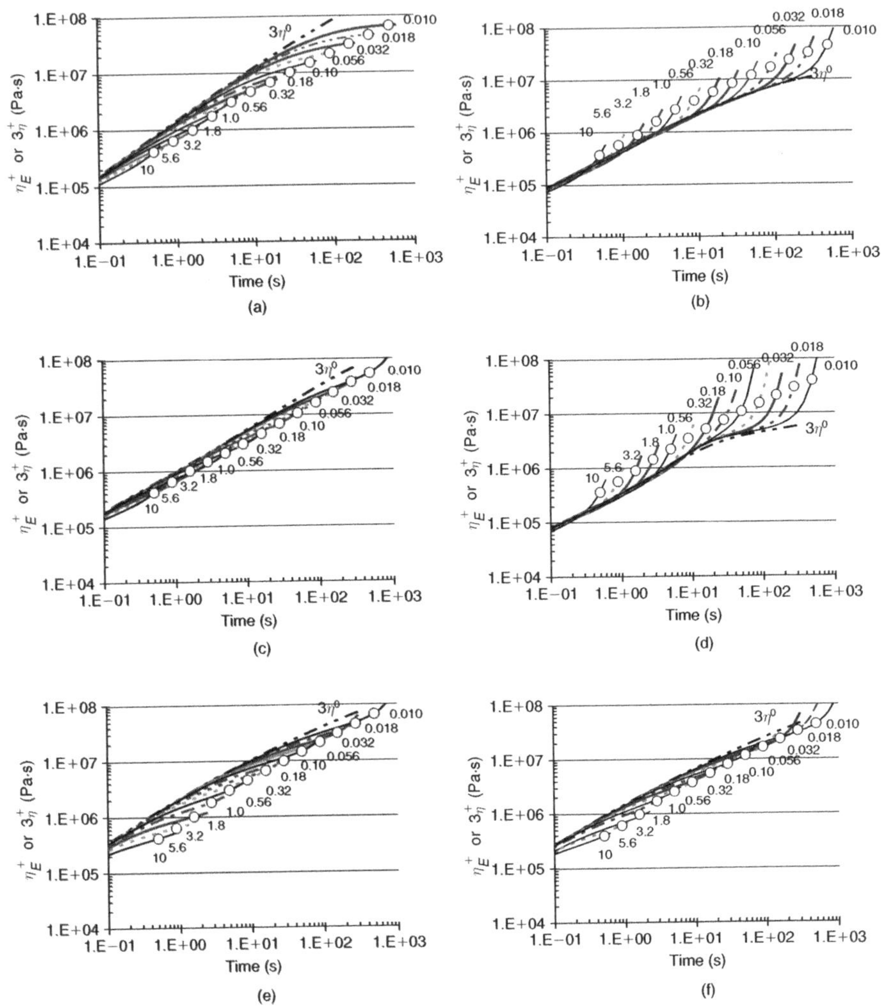

**FIGURE 6.9** Extensional viscosity by hyperbolic die of Hencky strain of $\varepsilon_L=5$ at 180°C (symbols) and K-BKZ fits (lines) for extruded HDPE–wood formulations: (a) HDPE–Wood, (b) HDPE–MAPE–Wood, (c) HDPE–OP–Wood, (d) HDPE–MAPE–OP–Wood, (e) HDPE–ZnSt-EBS–Wood, and (f) HDPE–MAPE–ZnSt-EBS–Wood. Numbers indicate strain rate. *MAPE.*, maleated polyethylene; *OP.*, OP100 lubricant package of formulated ester technology; *ZnSt-EBS*, zinc stearate/*N*,*N*-ethylene bis-stearamide lubricants. For details, see Ref. 25.

depletion of wood filler from the surface and possible wood extractive from wood. The melt fracture occurs either from the instability due to the matrix polymer or due to the excessive energy stored near the wood-polymer interface.

The Mooney analysis can be used to satisfactorily characterize the shear flow of wood plastics in capillary dies. The shear flow data combined with the

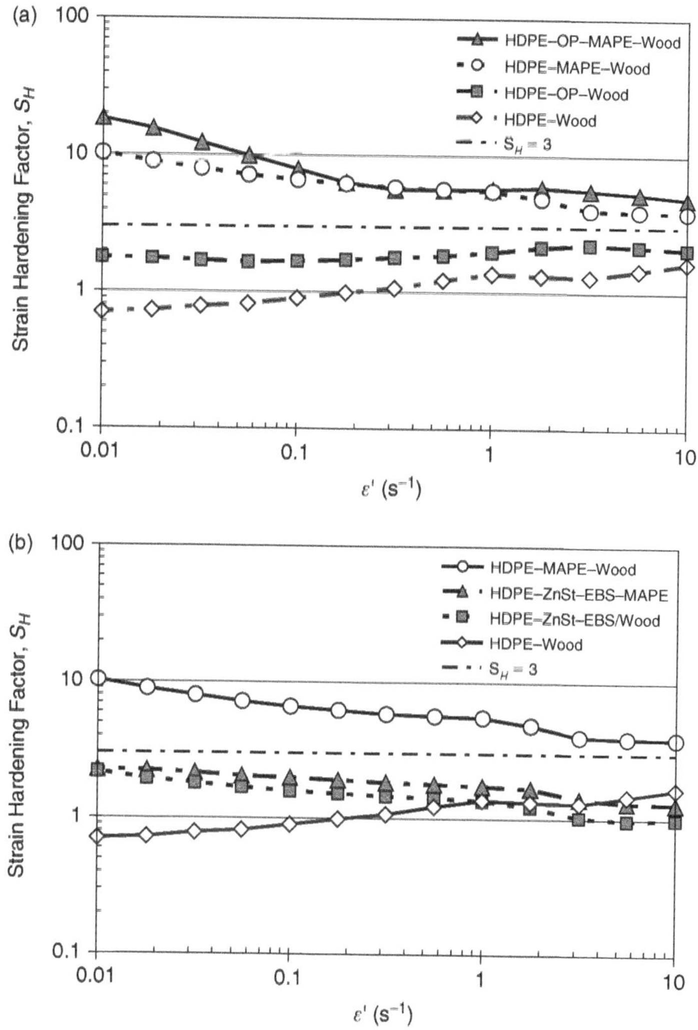


**FIGURE 6.10** Strain hardening factor of formulations shown in Figure 6.9.

Mooney analysis and the extensional viscosity data obtained using hyperbolic dies have been established as an effective empirical means to evaluate the role of various ingredients in the flow performance of wood plastics formulations. The application of small strain frequency sweeps, strain sweeps, step strain experiments, and the step-rate shear tests is also briefly reviewed as a probe to compare wood–plastics with different compositions.

The nature of the wood–plastics flow was further evaluated by analyzing the stress growth functions in either simple uniaxial or simple shear flow with constant strain rates using a K-BKZ type integral constitutive. The analysis validates the hyperbolic die method as a reliable means to characterize the uniaxial extensional viscosity of the wood plastics. A strain hardening factor is used to compare the extensional stress growth data from hyperbolic die to the linear viscoelastic stress growth function. The strain hardening factor quantitatively characterizes the extensional flow, as a function of wood content and additives. The analysis appears to also apply to shear flow when the wood loading is very high, e.g., 50 or 60 wt %.

A few unresolved problems arise through the rheological studies of wood plastics both in modeling the actual wood plastic flow and in applying these techniques to polymer matrices containing long natural fibers as reinforcements. These are as follows:

- An excessive transient shear stress is observed at short times at moderately high wood loadings (30 or 40 wt %) in comparison to the stress growth function obtained from the the K-BKZ analysis, which is based on a memory function assuming the subject as viscoelastic liquid. It would be worth exploring if other memory functions or constitutive equations would help overcome this problem and provide improved predictions for the flow of wood plastics, both with respect to the yield stresses and with the viscoelastic response.
- The study of shear flow in confined channels for actual processing. Because the fracture or strain limit of these highly filled linear polymer melts can be expected to rely on hydrostatic pressure, it would be worth to explore the pressure dependence and to extend the K-BKZ analysis to the shear flow in capillary dies and slit dies.
- The extrudate morphology after capillary shear flow is found to depend on the total shear that the material experiences in the flow. This strain dependency would require the viscous and elastic responses of the matrix to be considered separately. The progress in this regard, possibly through analysis based on a microscopic model, will be especially useful for composites with long natural fibers as this excessive stress may directly affect the consolidation of the matrix polymer and hence affect the end properties of the composites.

## REFERENCES

1. K. Joseph, B. Kuriakose, C.K. Premalatha, and S. Thomas, *Plast. Rubber Compos. Process. Appl.*, 21, 237–245 (1994).
2. H. Chtourou, B. Riedl, and A. Ait-Kadi, *J. Reinf. Plast. Compos.*, 11, 372–394 (1992).
3. S.L. Bai, R.K.Y. Li, H.M. Zeng, and Y.W. Mai., *J. Mater. Sci. Lett.*, 17, 1805–1907 (1998).
4. A.K. Bledzki and G. Gassan, *Prog. Polym. Sci.*, 24, 221–274 (1999).
5. D. Nabi Saheb and J.P. Jog., *Adv. in Polym. Techn.*, 18, 351–363 (1999).

6. P. Wambua, J. Ivens, and I. Verpoest, *Compos. Sci. Techn.*, 63, 1259–1264 (2003).
7. B. Madsen, *Properties of Plant Fibre Yarn Polymer Composites—An Experimental Study*, Technical University of Denmark, Copenhagen, 2004.
8. P. Smith, in *Proceedings of the Sixth International Conference on Woodfiber-Plastic Composites,* Forest Products Society, Madison, WI, 2001.
9. M.P. Wolcott, M. Chowdhury, D. Harper, et al., in *Proceedings of the Sixth International Conference on Woodfiber-Plastic Composites,* Forest Products Society, Madison, WI, 2001.
10. T.Q. Li. and M.P. Wolcott, in *Proceedings of the seventh International Conference on Woodfiber-Plastic Composites,* Forest Products Society, Madison, WI, 2003.
11. T.Q. Li and N. Yan, *Compos. A.*, 38, 1–12 (2007).
12. U. Yilmazer, C. Gogos, and D.M. Kalyon, *Polym. Compos.*, 10, 242–248 (1989).
13. D.M. Kalyon, P. Yaras, B.K. Aral, and U. Yilmazer, *J. Rheol.*, 37, 35–53 (1993).
14. M. Allende and D. Kalyon, *SPE ANTEC*, 45, 1130–1137 (1999).
15. D.M. Kalyon, *J. Rheol.*, 49, 621–640 (2005).
16. J. L Leblanc, *Plast., Rubber Compos.*, 30, 282–293 (2001).
17. A. Mongruel, M. Cartault. *J. Rheol.*, 50, 115–135 (2006).
18. S.N. Maiti and M.R. Hassan, *J. Appl. Polym. Sci.*, 37, 2019–2032 (1989).
19. M. Sain, V. Khunova, J. Hurst, and J. Balatinecz, *Plasty Kauc.*, 35, 199–202 (1998).
20. Y.L. Joo and M.H. Cho, *Int. Polym. Process.*, 14, 10–20 (1999).
21. S. Dong, S. Sapieha, and H.P. Schreiber, *Polym. Eng. Sci.*, 32, 1734–1739 (1992).
22. T.Q. Li and M.P. Wolcott, *Compos. A.*, 35, 303–311 (2004).
23. T.Q. Li and M.P. Wolcott, *Polym. Eng. Sci.*, 45, 549–559(2005).
24. T.Q. Li and M.P. Wolcott, *Polym. Eng. Sci.*, 46, 114–121 (2006).
25. T.Q. Li and M.P. Wolcott, *Polym. Eng. Sci.*, 46, 464–473 (2006).
26. V. Hristov, E. Takács, and J. Vlachopoulos, *Polym. Eng. Sci.*, 46, 1204–1214 (2006).
27. V. Hristov and J. Vlachopoulos, *Macromol. Mater. Eng.*, 292(5), 608–619 (2007).
28. V. Hristov and J. Vlachopoulos, *Polym. Compos.*, 29, 831–839 (2008).
29. V. Hristov and J. Vlachopoulos, *Adv. Polym. Techn.*, 26, 100–108 (2007).
30. S.A. Hatzikiriakos and J.M. Dealy., *J. Rheol.*, 36, 703–741 (1992).
31. S.-Q. Wang and P.A. Drda, *Macromolecules*, 29, 2627–2632 (1996).
32. L.A. Utracki and R. Gendron, *J. Rheol.*, 28, 601–623 (1984).
33. Y. Chan, J.L. White, and Y. Oyanagi, *J. Rheol.*, 22, 507–524 (1978).
34. G.K. Batchelor, *J. Fluid. Mech*, 46, 813–829 (1971).
35. J.D. Goddard, *J. Non-Newt. Fluid Mech.*, 1, 1–17 (1976).
36. C. Lacroix C, M. Grmela M, and P.J. Carreau, *J. Non-Newt. Fluid Mech.*, 86, 37–59 (1999).
37. P.J.A.H. Kok, S.G. Kazarian, et al., *J. Rheol.*, 46, 481–493 (2002).
38. J. Stange, C. Uhl, and H. Munstedt, *J. Rheol.*, 49, 1059–1079 (2005).
39. J.R. Collier, S. Petrovan, N. Hudson, and X. Wei, *J. Appl. Polym. Sci.*, 105, 3551–3561 (2007).
40. H. Gevgilili and D.M. Kalyon, *J. Rheol.*, 45, 467–475 (2001).

41. A.D. Gotsis, B.L.F. Zeevenhoven, and C. Tsenoglou, *J. Rheol.*, 48, 895–914 (2004).
42. H.H. Winter, M. Baumgärtel, P. Soskey. in: Techniques in Rheological Measurement. Chapman & Hall London. Collyer AA (ed), 1993. p. 123–160.
43. J.M. Dealy and R.G. Larson, *Structure and Rheology of Molten Polymers,* Hanser Publishers, Munich, 2006.
44. K. Osaki, A. Murai, N. Bessho, and B. Kim, *J. Soc. Rheol. Jpn.*, 4, 166 (1976).
45. M. Kobayashi, T. Takahashi, J. Takimoto, and K. Koyama, *Polymer*, 37, 3745 (1996).
46. J.K. Jackson, C. Garcia-Franco, and H.H. Winter, *Proceedings of the Annual Technology Meeting of the Society of Plastic Engineering*, 2438–2442 (1992).
47. C. Friedrich, H. Braun, and J. Weese, *Polym. Eng. Sci.*, 35, 1661–1669 (1995).
48. D.W. Mead, *J. Rheol.*, 38, 1797–1827 (1994).

## APPENDIX 6. A STRESS GROWTH PREDICTIONS USING K-BKZ CONSTITUTIVE EQUATION AND LINEAR VISCOELASTICITY THEORY

### 6.A.1 Stress Growth Functions by K-BKZ Analysis

The stress growth functions in general can be analyzed with a factorized K-BKZ model, specifically the Wagner model in this work. Following Winter et al. [42], the factorized stress growth is written in terms of the memory term, $\mu(t, t')$, based on the relaxation spectrum, $g_i(\tau_i)$; the damping function, $h(I_1, I_2)$; and the Finger tensor, $\underline{\underline{C}}^{-1}$, as

$$\underline{\underline{\sigma}}(t) = \int_{-\infty}^{t} \sum_{i=1}^{N} \frac{g_i}{\tau_i} e^{-(t-t')/\tau_i} h(t', t) \underline{\underline{C}}^{-1}(t', t) dt \tag{6.A.1}$$

The damping function, $h(t, t')$, accounts for the fading of the memory with the growing strain and can be written as the general form $h(I_1, I_2)$ in terms of the first and second invariants, $I_1$ and $I_2$, of the Finger tensor. The damping function has been used mainly as a probe of the molecular mechanism of the flow. In this work, $h(t, t')$ is used in the simplest exponential decaying form for its simplicity, because the purpose here is to obtain a quantitative measure of the nonlinear behavior in either simple shear or simple extensional flow.

For the simple uniaxial extension [43]

$$\underline{\underline{C}}^{-1}(t', t) = \begin{pmatrix} e^{2[\varepsilon(t')-\varepsilon(t)]} & 0 & 0 \\ 0 & e^{-[\varepsilon(t')-\varepsilon(t)]} & 0 \\ 0 & 0 & e^{-[\varepsilon(t')-\varepsilon(t)]} \end{pmatrix} \tag{6.A.2}$$

In the uniaxial extensional flow with a constant Hencky strain rate, $\dot{\varepsilon}$, such as in hyperbolic dies

$$\varepsilon(t, t') = \dot{\varepsilon}(t - t') \tag{6.A.3}$$

Using the simplest exponential decaying form of the damping function ([41] and references therein), $h(t, t')$ is written as:

$$h(t, t') = e^{-\beta\dot{\varepsilon}(t-t')} \tag{6.A.4}$$

This leads to the extensional stress growth function as

$$\begin{aligned} \eta_E^+(t, \dot{\varepsilon}) &= \frac{1}{\dot{\varepsilon}}(\sigma_{11} - \sigma_{22}) \\ &= \frac{1}{\dot{\varepsilon}} \int_{-\infty}^{t} \sum_{i=1}^{N} \frac{g_i}{\tau_i} e^{-(t-t')/\tau_i} e^{-\beta\dot{\varepsilon}(t-t')/\tau_i} [e^{2\dot{\varepsilon}(t-t')} - e^{-\dot{\varepsilon}(t-t')}] dt' \end{aligned} \tag{6.A.5}$$

For simple shear, the Finger tensor is [43]

$$\underline{\underline{C}}^{-1}(t',t) = \begin{pmatrix} 1+[\gamma(t')-\gamma(t)]^2 & [\gamma(t')-\gamma(t)] & 0 \\ [\gamma(t')-\gamma(t)] & 1 & 0 \\ 0 & 0 & 1 \end{pmatrix} \tag{6.A.6}$$

In step-rate shear experiments, the strain is determined by the shear rate, $\dot{\gamma}$, as

$$\gamma(t,t') = \dot{\gamma}(t-t') \tag{6.A.7}$$

Assuming again the exponential decaying damping function for its simplicity

$$h(t,t') = e^{-\beta\dot{\gamma}(t-t')} \tag{6.A.8}$$

The shear stress growth function can be evaluated as:

$$\begin{aligned} \eta^+(t,\dot{\gamma}) &= \frac{\sigma_{12}}{\dot{\gamma}} \\ &= \frac{1}{\dot{\gamma}}\int_{-\infty}^{t}\sum_{i=1}^{N}\frac{g_i}{\tau_i}e^{-(t-t')/\tau_i}e^{-\beta\dot{\gamma}(t-t')/\tau_i}[\dot{\gamma}(t-t')]dt' \end{aligned} \tag{6.A.9}$$

Equations (6.A.5) and (6.A.9) have been used to determine the stress growth functions in the analysis of both the hyperbolic die extensional viscosity data and the step rate test results. It is worth noting that the damping exponent, $\beta$, in (6.A.4) and (6.A.9) is not the same.

Equations (6.A.4) (also for 6.A.9) can be used to determine the stress growth functions in either simple shear or simple extensional flow as follows:

- Obtain relaxation spectrum, $g_i(\tau_i)$, through LVE characterization, e.g., using small strain frequency sweep, step strain or creep experiments. Due to the large damping at higher rates [24], frequency sweep is used in this work.
- Select representative viscosity data, assume the stress growth function equals the measured viscosity at the given strain rate and time, and solve numerically for the damping exponent, $\beta$. This parameter is calculated for the viscosity, $\eta(\dot{\varepsilon}, \varepsilon_L)$, measured with the die of Hencky strain, $\varepsilon_L$, at the strain rate, $\varepsilon$. For step-rate shear experiments, the stress and time at the maximum torque is used for calculate, $\beta$.
- Calculate the stress growth function on the desired time frame under the strain rate.

The methodology for the determination of the relaxation spectrum is reported later in this appendix.

### 6.A.2 Strain Hardening Factor

The importance of the K-BZK analysis is that it provides the means to quantitatively compare the measured viscosity data to the predictions based on the linear viscoelastic properties. The comparison allows the determination of the damping characteristics (damping exponent, $\beta$, and, for shear experiments, the damping function, $h(\gamma)$) for materials characterization and also for predicting the flow in real processing operations. The damping characteristics, such as the damping exponent for either the simple shear or the uniaxial extensional flow, can be readily solved based on (6.A.5) or (6.A.9) for the given flow of the material with a known relaxation spectrum.

For extensional flow, a simple strain hardening factor can be readily defined as

$$S_H = \eta_E^+(\dot{\varepsilon}, t)/\eta^0(t) \tag{6.A.10}$$

$S_H = 3$ for Newtonian viscoelastic fluid. For the wood–plastics melt in hyperbolic dies, the parameter defines the strain hardening (or softening) at the given rate, $\varepsilon$, and after a total strain, $\varepsilon_L$, (or over the residence time, $t = \varepsilon_L/\dot{\varepsilon}$). In contrast to the Trouton ratio defined for steady state capillary shear viscosities [22], this hardening parameter reveals whether the flow is in a transient hardening stage.

### 6.A.3 Linear Viscoelastic Stress Growth Function $\eta^0(t)$

The determination of the straining hardening behavior requires the linear viscoelastic (LVE) stress growth function, $\eta^0$, as the baseline, which can be obtained as a special form of (6.A.1.) When the damping function is $h \equiv 1$, the shear stress, $\sigma_{12}$, and the first normal stress difference, $N_1$, for the linear regime is obtained as

$$\sigma_{12}^+(t) = \dot{\gamma}_0 \sum_{i=1}^{N} g_i \tau_i \left(1 - \left(1 + \frac{t}{\tau_i}\right) e^{-\frac{t}{\tau_i}}\right) \tag{6.A.11a}$$

$$N_1^+(t) = \dot{\gamma}_0^2 2 \sum_{i=1}^{N} g_i \tau_i^2 \left[1 - \left(1 + \frac{t}{\tau_i} + \frac{1}{2}\left(\frac{t}{\tau_i}\right)^2\right)\right] e^{-\frac{t}{\tau_i}} \tag{6.A.11b}$$

Equation (6.A.11a) generates the linear viscoelastic shear stress growth function, $\eta^0$, used for evaluating the strain hardening in extensional flow.

In the case of $t \gg \tau_i$, the more widely used formulas of the LVE stress growth functions as per Winter–Baugartel-Soskey are retrieved

$$\sigma_{12}^+(t) = \dot{\gamma}_0 \left(G_0 t + \sum_{i=1}^{N} g_i \tau_i \left(1 - e^{-\frac{t}{\tau_i}}\right)\right) \tag{6.A.12a}$$

$$N_1^+(t) = \dot{\gamma}_0^2\left\{G_0t^2 + 2\sum_{i=1}^{N} g_i\tau_i^2\left[1-\left(1+\frac{t}{\tau_i}\right)\right]e^{-\frac{t}{\tau_i}}\right\} \tag{6.A.12b}$$

with the solid elastic constant, $G_0$, taken as 0 for the viscoelastic liquids of interest.

For melt of many different types of polyethylenes, the Winter–Baumgartel-Soskey form gives the linear stress growth function that is close to but slightly lower than the Kobayashi–Osaki approximation [44, 45]:

$$\eta^+(t)|_{\omega = 1/t} = t[G''(\omega) + 1.12G'(0.5\omega)-0.200G'(\omega)] \tag{6.A.13}$$

where $G''(\omega)$ and $G'(\omega)$ are the loss and storage modulus, respectively. For melts with short reptation times, e.g., <0.02 s, both the Winter–Baumgartel-Soskey equations and the Kobayashi–Osaki equation give good consistency between step-rate data and the prediction from frequency sweep.

The difference between (6.A.11) and (6.A.12) depends on the viscoelastic nature of the material under consideration. For unfilled resins of narrow molecular weight distribution and short reptation times, the difference is marginal. For resins with broad molecular weight distribution, high molecular weight and especially with high filler loadings, the Winter–Baumgartel-Soskey equations are found to overestimate the stress on the time frame common to the step-rate tests, which is typically from a fraction of a second to a few hundreds of seconds. Equations (6.A.11a) and (6.A.11b) are hence adopted for calculating the LVE baseline for determining the strain hardening factor, $S_H$.

### 6.A.4 Determination of Relaxation Spectrum $g_i(\tau_i)$

The discrete relaxation spectrum, $g_i(\tau_i)$, of 33 modes has been obtained for each material in this work. The $\tau_i$ range used starts from the minimum value of $10^{-10}$ s and increases at an interval of 0.6 mode per decade. The spectrum is converted from the frequency sweep results by assuming that the $\log \tau_i \sim \log g_i$ relation follows one second-order polynomial function or two such functions added together. The extrapolation by assuming a smooth overall shape of the kernel for relaxation spectrum is based on the original proposal by Winter et al. [46] and Friedrich et al. [47]. The relaxation time interval was adopted following the suggestion of Mead [48] and was validated on representative spectra. It allows the reliable determination of the relaxation spectrum for polyethylenes from frequency sweep data in a very limited frequency range.

# CHAPTER 7

# BLOCK COPOLYMERS IN EXTERNAL FIELDS: RHEOLOGY, FLOW-INDUCED PHENOMENA, AND APPLICATIONS

MICHEL CLOITRE[1] and DIMITRIS VLASSOPOULOS[2]
[1] ESPCI ParisTech, Matière Molle et Chimie, Paris, France
[2] FORTH, Institute of Electronic Structure and Laser, Heraklion, Crete, Greece and University of Crete, Department of Materials Science and Technology, Heraklion, Crete, Greece

## CONTENTS



*Applied Polymer Rheology: Polymeric Fluids with Industrial Applications*, First Edition.
Edited by Marianna Kontopoulou.

## 7.1 INTRODUCTION

Block copolymers are macromolecules with incompatible block segments of two or several different monomers. Depending on the segment lengths, unfavorable interactions between monomers, and macromolecular architecture, these materials exhibit a great variety of equilibrium phases with long-range order. Block copolymers have received a great deal of attention from both scientific and technological viewpoints. A basic search on the "Web of Science" using the keyword *block copolymer* in the topics entry identifies nearly 10,000 publications for the last decade. It is thus impossible to treat all the aspects of the field, which range from basic synthetic chemistry to material properties through phase behavior, thermodynamics, and dynamics, in a single review. We refer the interested reader to available books and reviews [1–7].

Here we focus on the behavior of block copolymers in external fields and in particular in flow fields of different types, including complex flows that combine shear and elongation deformation fields. We will discuss the linear and nonlinear rheology of block copolymers, the existence of flow-induced transitions, and the possibility of aligning block copolymer domains over large distances, thereby creating near-crystalline materials for modern applications. These different issues are essential for characterizing block copolymers; tuning their mechanical, optical, or electrical properties at the molecular scale; and processing advanced materials. Our presentation encompasses linear diblocks, which are the simplest and most extensively studied block copolymers, and linear multiblocks, which are essential in applications. Our concern to provide the reader with a comprehensive review led us to restrict the essence of our presentation to the case of block copolymer melts. The field of block copolymers solutions is extremely rich, solvent quality being an additional important parameter for tailoring structure, dynamics and rheology [8–11]. We will make some excursions in block copolymer solutions, which will be useful to emphasize the generality of the concepts presented.

This chapter is divided into four parts. We start with a brief introduction on the architecture and composition of block copolymers. Then we provide the reader with key molecular parameters and structural features by presenting the main elements of block copolymer thermodynamics. We continue with a discussion of the linear rheology of the ordered microphases before focusing on non-linear phenomena induced by strong flows. We close with a few remarks on perspectives of the field.

### 7.1.1 Architecture and Composition

Block copolymers encompass a wide class of macromolecules made of one or several incompatible sequences or blocks connected together through covalent bonds. The development of the field has been stimulated by the invention of anionic polymerization techniques that make possible the sequential addition of monomers to reactive carbanion terminated polymer chains [3]. Since the original anionic copolymerization techniques were developed, the invention of a variety of new polymerization techniques in conjunction with the requirements of ever demanding applications have prompted the development of a huge variety of block copolymers with different controlled and well-defined architectures [2, 3, 5]. Diblock copolymers, which are made of two incompatible blocks, are well known and have been widely investigated both theoretically and experimentally, even though they are not necessarily the most relevant for applications. Multiblock copolymers consist of several incompatible blocks arranged in a linear or branched topology or grafted onto a polymeric backbone. Concerning branched macromolecules, it is usual to distinguish miktoarm star copolymers (or heteroarm star copolymers) where the unlike blocks are connected at one junction point and star-block copolymers in which each arm is itself a diblock. Figure 7.1 illustrates typical current architectures that are encountered in fundamental studies and in applications.

The building blocks exhibit a huge variety of compositions and bulk properties. They can be glassy, elastomeric or semicrystalline. Moreover they do not have to be pure homopolymers. Many block copolymers used in applications are gradient or tapered block copolymers whose composition varies gradually along the chain or the blocks from A-rich at one end to B-rich at the other end [12–17]. It is nearly impossible to list the huge variety of block copolymers that have been synthesized and investigated. In Table 7.1, we present some selected examples of synthetic diblock and multiblock copolymers with linear architecture that have been widely used in fundamental studies or in applications and that will be encountered in this review. An extended review of the field is available [29]. The past years have seen a growing interest for the development of novel hybrid block copolymers that combine the outstanding properties of

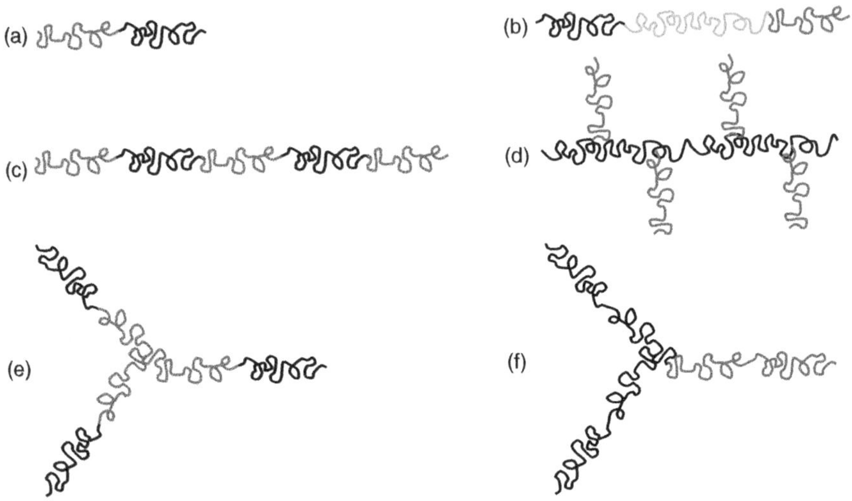


**FIGURE 7.1** Various architectures of block copolymers: (a) linear AB diblocks, (b) linear ABC triblocks, (c) alternate $(AB)_n$ multiblocks, (d) grafted copolymers, (e) star-block copolymers, and (f) miktoarm star copolymers.

**TABLE 7.1 Frequently Used Block Copolymers.**

| Class | Copolymer | Abbrevations | Reference |
|---|---|---|---|
| AB, styrenic | Polystyrene-*b*-poly(methyl methacrylate) | PS-PMMA | [18] |
| | Polystyrene-*b*-polybutadiene | PS-PB | [19] |
| | Polystyrene-*b*-polyisoprene | PS-PI | [20] |
| | Polystyrene-*b*-poly(2-vinylpyridine) | PS-P2VP | [21] |
| AB, Polyolefin | Poly(ethylene)-*b*-poly(ethyl ethylene) | PE-PEE | [22] |
| | Poly(ethylene)-*b*-poly(ethyl propylene) | PE-PEP | [23] |
| ABA | Polystyrene-*b*-polybutadiene-*b*-polystyrene | PS-PB-PS | [3] |
| | Polystyrene-*b*-polyisoprene-*b*-polystyrene | PS-PI-PS | [3] |
| | Polystyrene-*b*-polyethylene-*b*-polystyrene | PS-PE-PS | [3] |
| | Polystyrene-*b*-poly(ethylene-*co*-butylene)-*b*-polystyrene | PS-PEB-PS | [3] |
| ABC | Polystyrene-*b*-polybutadiene-*b*-poly(methyl methacrylate) | PS-PB-PMMA | [24] |
| | Poly(2-vinylpyridine)-*b*-polyisoprene-*b*-polystyrene | P2VP-PI-PS | [25] |
| | Polyisoprene-*b*-polystyrene-*b*-poly(ethylene oxide) | PI-PS-PEO | [26] |
| Amphiphilic | Poly(ethylene oxide)-*b*-poly(propylene oxide)-*b*-poly(ethylene oxide) | PEO-PPO-PEO | [8] |
| | Polystyrene-*b*-poly(acrylic acid) | PS-PAA | [21] |
| | Polystyrene-*b*-poly(ethylene oxide) | PS-PEO | [28] |

synthetic polymers with the structural richness of polypeptides or protein-like polymers [30].

### 7.1.2 Thermoplastic Elastomers

The presence of several incompatible blocks covalently linked inside the same macromolecule has important consequences with respect to the phase behavior of block copolymers. The most fascinating and useful feature of block copolymers is their capacity to microphase-separate beyond some temperature into equilibrium morphologies characterized by long-range order. The microphase separation, also known as the order–disorder transition, is associated with a dramatic change of the viscoelastic properties. This feature, which will be analyzed in more detail in Section 7.2, has been exploited to develop thermoplastic elastomers that constitute an important class of industrially relevant block copolymers [3]. The market of thermoplastic elastomers represents a turnover of several billion euros a year and grows rapidly as these materials replace progressively conventional vulcanized elastomers.

Thermoplastic elastomers (TPE) are based on multiblock copolymers with two crystallizable or high-$T_g$ glassy end blocks (the minor components) and a low-$T_g$ midblock (the major component). The blocks are sufficiently incompatible to microphase separate at low temperature. The minor components then form spherical or cylindrical microdomains embedded in the continuous phase of the major component. In a range of temperatures between the glass-transition temperature of the rubbery midblock and the glass-transition or melting temperature of the end blocks, the latter are glassy or crystalline and play the role of physical cross-links that connect the continuous elastomeric phase. When the material is heated, the microdomains soften; the network looses its strength and, above the temperature of microphase separation, the material flows like a polymer melt. TPEs thus combine the processing characteristic of thermoplastics at high temperature with the physical performance of vulcanized rubbers at the low service temperature. In practice, the development of a specific TPE must satisfy several requirements to adapt the temperature of microphase separation to the processing conditions, to control the morphology of the phase-separated microdomains, and ultimately to obtain the properties that are desired. In some instances, this has led chemists and engineers to design and produce commercially a variety of block copolymers that associate complex architectures with the existence of gradients of composition [7, 13, 31].

Thermoplastic elastomers are of different types. Those based on styrenic block copolymers were among the first studied [3]. They still represent one half of the TPE industry. Many of them have the basic structure polystyrene-*b*-polybutadiene-*b*-polystyrene or polystyrene-*b*-polyisoprene-*b*-polystyrene. The second generation of styrenic block copolymers has a hydrogenated midblock, ethylene–butylene (SEBS) or ethylene–propylene (SEPS). They are intended for use whenever enhanced UV resistance, high service temperature, and processing stability are essential. The recent years have seen the development of styrenic-based copolymers made of more than three blocks, with the motivation to optimize the linear and nonlinear mechanical properties of the final materials [31, 32]. Polyolefin-based thermoplastics are also widely used with the advantage of having a better chemical stability than the styrenic products [3]. The recent development of controlled radical polymerization has opened the route to novel materials containing acrylic monomers [33]. Other important groups of thermoplastic elastomers are based on segmented multiblock copolymers made from random or alternate segments of rigid and soft segments. Examples include polyurethane block copolymers made from polyester or polyether diols and diisocyanate compounds and polyamide/polyether block copolymers [34].

## 7.2 THERMODYNAMICS OF BLOCK COPOLYMERS

### 7.2.1 Formation of Ordered Structures

The stability of homopolymer mixtures depends on a compromise between the entropy of mixing of the chains and the enthalpy penalty associated with unfavorable contacts between monomers. The entropy of mixing, which varies as $1/N$ for two polymer chains with N being the degree of polymerization, is quite low so that it may not overcome the enthalpy of mixing, which depends on temperature as $1/T$. This explains why ordinary mixtures of different polymers often phase separate macroscopically at room temperature, even though their monomers are homogeneously mixed. Phase separation also occurs in block copolymers, but because the incompatible blocks are covalently bonded, the relevant spatial length scale remains comparable to the length of the blocks leading to segregation into microdomains of the size of the blocks. The transition from the disordered state at high temperature, where the monomers belonging to different blocks are molecularly mixed, to the phase-separated state where the blocks segregate into microdomains, is known as the microphase separation or the order-disorder transition (ODT).

Experimentally, the microphase separation leads to the formation of ordered structures of different types. This is illustrated in Figure 7.2 for a polystyrene-*b*-polybutadiene-*b*-poly(methyl methacrylate) copolymer that forms a lamellar morphology at the particular composition investigated. The lamellae consist of microdomains of polystyrene, polybutadiene, and poly(methyl methacrylate), which are regularly stacked. The regions where the smectic-like order is nearly perfect constitute grains that, in the absence of any specific treatment, have a limited spatial extension, generally smaller than a few micrometers. The boundaries between the different grains are associated with the presence of specific topological defects, which play an important role in the equilibration dynamics of the ordered phases after preparation [36]. The deformation and displacement of the defects in the microstructure are at the origin of the unusual relaxation and flow behavior of block copolymers as will be discussed in Sections 7.3 and 7.4.

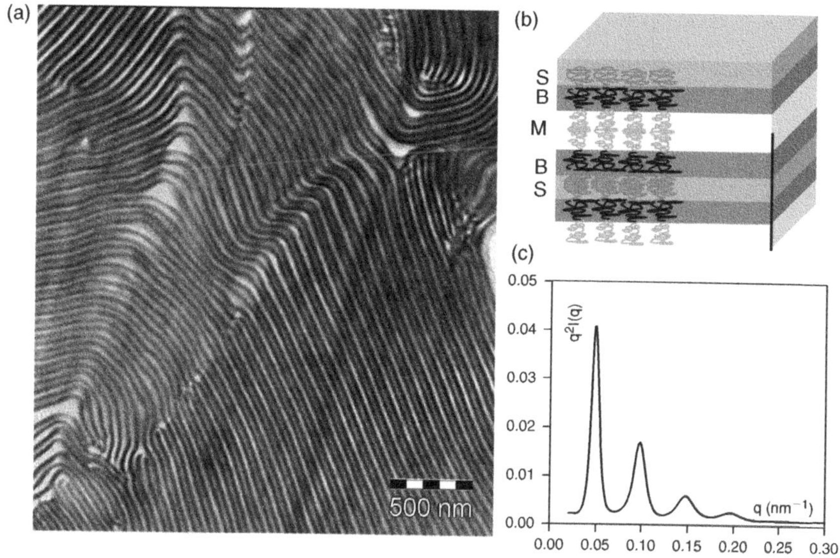


**FIGURE 7.2** Microphase separation in a polystyrene-*b*-polybutadiene-*b*-poly(methyl methacrylate) triblock copolymer [35]. The polystyrene block has a number average molar mass of 37 kg/mol. The copolymer is characterized by the volume fractions of the different blocks: $f_S = 0.25$, $f_B = 0.23$, and $f_M = 0.42$. (a) A transmission electron microscopy observation of a thin microtomed section of a copolymer film prepared by solvent casting and stained with $OsO_4$. *Black*, polybutadiene; *gray*, polystyrene; *white*, poly(methyl methacrylate). The microdomains are separated by narrow interfaces, indicative of strong segregation. (b) The arrangement of the chains inside the microstructure. (c) The diffraction pattern obtained by small-angle X-ray scattering. The long period of the lamellar structure is 132 nm.

### 7.2.2 The Microphase Separation or Order–Disorder Transition

Because most current theories concern AB or ABA block copolymers, we will focus on the case of diblocks in this section to enlight the basic physics at work in block copolymers. The phase behavior of AB copolymers is characterized by three main parameters: the volume fraction, $f$, of the A component (often referred to as its composition), the A/B Flory–Huggins interaction parameter, $\chi$, and the total degree of polymerization $N$. When the interaction parameter, $\chi$, is negligible—i.e., the temperature is high—the blocks are miscible and the melt is disordered. If the parameter $\chi$ is increased by lowering the temperature, the number of unfavorable contacts between A and B monomers tends to decrease, which ultimately leads to microphase separation as discussed above.

The local segregation of the blocks into microdomains involves a loss of translational and configurational entropy of the chains. Since the entropic and

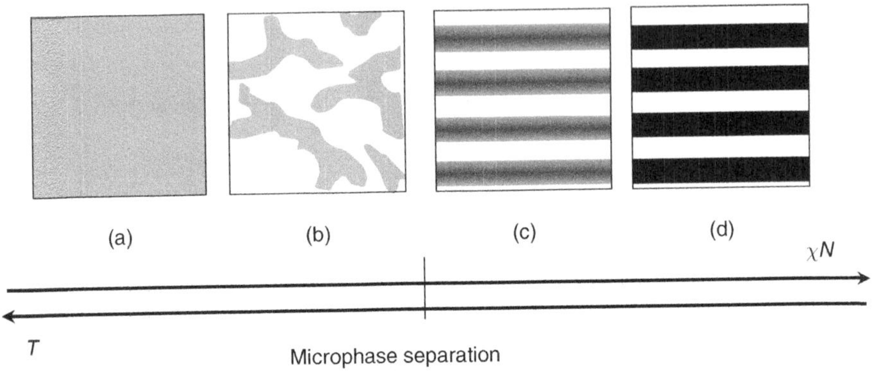


**FIGURE 7.3** The microstructure of an AB diblock copolymer melt as a function of $\chi N \sim 1/T$ and $T$. White, A; black, B. The melt is disordered in panels (a) and (b). The microphase separation or order–disorder transition occurs between panels (b) and (c). Panels (c) and (d) correspond to the weak and strong segregation limit, respectively. The figure does not account for the presence of fluctuations, corresponding to the mean-field limit.

enthalpic contributions to the free energy scale as $1/N$ and $\chi$, respectively, the product $\chi N$ controls the phase behavior of the system. When $\chi N << 1$ the interactions are sufficiently weak compared to the entropic effects that the melt is disordered and the individual chains remain Gaussian. There exist however correlations at a length scale corresponding to the radius of gyration of the chains $R_g \sim a^2 N$, with $a$ being a characteristic segment length, which are revealed by scattering experiments [37]. As $\chi N$ is increased above some critical value, the competition between enthalpic and entropic effects produces the microphase transition. The literature refers to the vicinity of the transition as the weak segregation limit (WSL). In this regime, it is generally considered that the individual chains remain Gaussian and that the composition profile is sinusoidal. By contrast, when $\chi N$ is very large, there exist nearly pure A and B microdomains separated by very narrow interfaces. This is typically the situation presented in Figure 7.2 for the case of triblock copolymers. The literature refers to such a situation as the strong segregation limit (SSL). These different microstructures are illustrated schematically in Figure 7.3.

***7.2.2.1 Strong Segregation Limit*** In the SSL regime, the interaction energy associated with unfavorable contacts between A and B is localized in the interfacial regions between A and B. Therefore, the system tries to minimize the surface area per copolymer chain. However, with the constraints of incompressibility, the chains adopt an extended configuration that generates an entropic penalty. The competition between these two opposing effects, the minimization of the interfacial energy and the stretching of the chains, constitutes the basis of the theoretical description of diblock copolymers in the SSL. Since the very first pioneering studies in the middle 1970s [38, 39] the field has stimulated intense experimental and theoretical research. One of the most influential works is that of Semenov, who developed a systematic computation method and derived the phase co-existence boundaries of several microphases with simple geometries [40].

It was shown that the stretching contribution to the free energy per chain ($F_{el}$) in the SSL has the same scaling as for a Gaussian chain [40, 41]. For the case of a lamellar microphase, the stretching contribution is thus $F_{el}/kT \sim D^2/Na^2$, where $k$ is the Boltzmann constant, $T$ is the absolute temperature, and $D$ is the lamellar period. This energy is balanced by the interfacial energy, $F_i \sim \gamma\sigma$, where $\gamma \sim kT\chi^{1/2}/a^2$ is the interfacial tension and $\sigma \sim Na^3/D$ is the area per chain. The minimization of the total free energy with respect to $D$ leads to a predicted lamellar periodicity, $D \sim aN^{2/3}\chi^{1/6}$. Such lamellar phases are obtained for relatively symmetric diblock copolymers. For more asymmetric copolymers it becomes energetically more favorable for phases with curved interfaces to form. In the SSL regime, the phase boundaries are independent of $\chi N$, which precludes the possibility of transitions between different ordered states (OOT).

***7.2.2.2 Weak Segregation Limit*** The WSL limit refers to the vicinity of the order–disorder transition. The phase diagram for weakly segregated diblocks AB was first computed by Leibler within the Landau mean-field approximation [42]. The theory defines the order parameter $\psi(\mathbf{r}) = <\rho(\mathbf{r}) - f>$, where $\rho(\mathbf{r})$ is the local number density of monomers A and the angle brackets denote a spatial average over the system. The mean-field free energy is expressed in terms of a Landau–Ginzburg expansion, which is valid near second-order or weakly first-order phase transitions. The presentation of the theory is well beyond the scope of this chapter, and we refer the reader to the original paper [42] and to reviews for a detailed description [1–4]. The resulting phase diagram, which involves $f$ and $\chi N$ as control parameters, is given in Figure 7.4a. The theory predicts a weakly first-order transition between the disordered and BCC phases, a critical point between the disordered and lamellar phases ($f = 0.5$; $\chi N = 10.5$), and weakly first-order transitions between solid phases, which can be accessed by varying the temperature in contrast to the case of the SSL. The theory also provides an expression for the structure factor in the disordered phase that has been widely used to interpret X-rays and neutron scattering experiments on diblock copolymers.

The seminal theory of Leibler has stimulated a profusion of experimental research and theoretical developments. Figure 7.4b,c shows the experimental phase diagrams [20, 43] of two model diblocks, which, at least qualitatively, exhibit the main features predicted by the theory. In addition to the lamellar (lam), hexagonal (hex) and BCC phases, the experiments have revealed the existence of a gyroid phase (gyr) and a perforated lamella phase (PL), which were not accounted for in the initial theory. Experimental investigations suggest that the PL structure is a long-lived metastable phase [44]. The gyroid phase is stable and has a bicontinuous cubic symmetry based on a tripod arrangement of channels. It has stimulated a vivid controversy before it was unambiguously

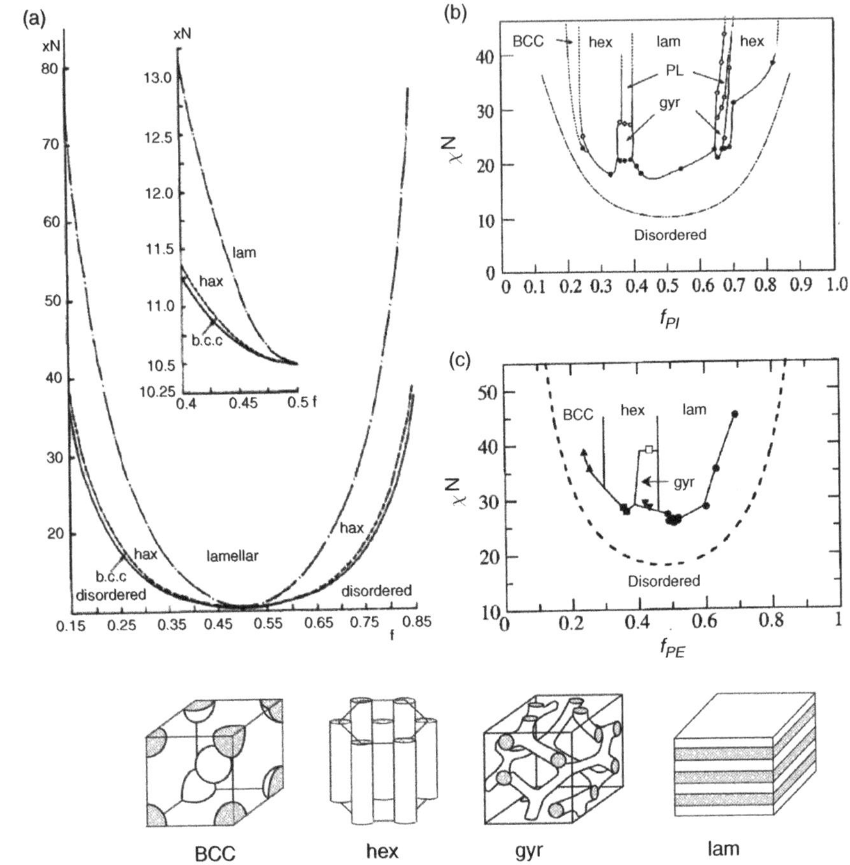


**FIGURE 7.4** (a) Theoretical phase diagram for diblock copolymers in the weak segregation limit. Reprinted with permission from [42]. (b) Experimental phase diagram for PS-PI diblocks. Reprinted with permission from [20]. (c) Experimental phase diagram for PE-PEE diblocks. Reprinted with permission from [43]). The stable phases are represented at the bottom of the figure: body-centered cubic phase (BCC), hexagonal packed cylinders (hex), gyroid (gyr) and lamellar phase (lam), from left to right.

assigned [45]. More recent works have identified a new network morphology of orthorhombic symmetry in a very small region of the phase diagram of polystyrene-*b*-polyisoprene diblocks [46], as well as spherical microdomains with face-cubic symmetry in poly(ethylene oxide)-*b*-polybutadiene melts [47]. On the theoretical side, the effects of fluctuations have been subsequently incorporated in the original WSL theory [48]. Later on, self-consistent field theory (SCFT) was extended as a theoretical framework that unifies the weak and strong segregation limit [49]. The SCFT equations can be solved to compute the solution of an amazing variety of related problems [4].

### 7.2.3 Beyond Diblock Copolymer Melts

The phase behavior of multiblock copolymers poses new and interesting challenges. The case of triblocks of ABA type has stimulated a lot of interest due to their importance in the development of thermoplastic elastomers [3, 7]. The nature and the sequence of phases observed in many ABA copolymers are qualitatively similar to those in AB diblocks [43, 50]. There exist however interesting subtleties for very asymmetric copolymers [51]. ABC triblock copolymers exhibit a broad panel of morphologies as a result of a larger number of control parameters—namely, three distinct interaction parameters and two independent volume fractions. The block sequence also matters when the interaction parameters are very different [25]. For instance, polystyrene-*b*-polybutadiene-*b*-poly(methyl methacrylate) copolymers were found to self-assemble into many different complex structures such as spheres on spheres [52] or helices on cylinders [53] embedded in a continuous matrix formed by the majority component, and many other morphologies like the knitting pattern [54]. The phase diagram of polyisoprene-*b*-polystyrene-*b*-poly(ethylene oxide) has been extensively explored leading to the discovery of a succession of three different network morphologies [26]. The resulting phase diagram has been recently analyzed using SCFT techniques [55]. The phase behavior of branched and grafted copolymers has been analyzed along the same lines in relation with the specific topology of these macromolecules [29].

In most applications, neat block copolymers are never used as one-component systems but instead in association with another component. Blending block copolymers with a solvent, linear homopolymers, other block copolymers, or solid particles constitutes a powerful way to tune the phase behavior of block copolymers [29]. The simplest situation consists in adding a solvent into a block copolymer. This has two main effects. First it renormalizes the segment-segment interactions [56]. When the solvent is nonselective or weakly selective, the interactions are simply screened leading to a shift of the order–disorder transition toward larger $\chi N$. When the solvent is selective, the incompatibility of the blocks is enhanced due to their different affinities with the solvent. In addition, the unequal affinity of the solvent with the different blocks leads to unequal partitioning of the solvent in the microdomains. This effect can be accounted for using the renormalized effective volume fraction of the blocks. The renormalization of the segment–segment interactions together with that of the block volume fraction provide two simple conceptual tools from which the sequence of phases as a function of temperature and concentration can be anticipated qualitatively from the phase diagram of the neat copolymer [56]. The change of morphology induced by a solvent has important implications in regard to the preparation of copolymer films in solution-casting processes where long-lived morphologies nonrepresentative of the melt state can be formed during evaporation [36]. Similar concepts can be used to rationalize changes of morphologies in block copolymer–homopolymers blends [57–59], block copolymer mixtures [60] or block copolymer nanocomposites [61].

## 7.3 RHEOLOGY AS DIAGNOSTIC TOOL: LINEAR VISCOELASTICITY

### 7.3.1 Disorder-to-Order Transitions

In this section we discuss the main features of the linear viscoelastic response of simple flexible block copolymer melts, such as AB diblocks and ABA triblocks. A common method used in polymer science to obtain rheological information across length and time scales consists of rescaling the dynamic frequency sweep measurements collected at different temperatures onto a master curve using the time-temperature superposition principle (tTS). In the case of block copolymers, much like in any heterogeneous system (simply put, with more than one microscopic time), tTS fails and a single master curve cannot be obtained at low frequencies [1, 2]. However, this turns out to be an advantage since the temperature beyond which tTS fails is associated with the order–disorder transition (ODT) temperature [62]. Typical experimental results are shown in Figure 7.5a, which demonstrates the great sensitivity of rheology to ODT for a poly(ethylenepropylene)-*b*-poly(ethylethylene) copolymer (PEP–PEE). The storage (elastic) modulus $G'$ is shown because it is much more sensitive than the loss (viscous) modulus $G''$. An alternative representation of the linear viscoelastic response, which is very sensitive to the ODT, is the Cole–Cole plot, where $G'$ is plotted against $G''$ (Fig. 7.5b). The results shown in Figure 7.5, which are taken from the classic paper of Rosedale and Bates [63], convincingly support the possibility of quantifying the phase diagram and rheology of block copolymers. Although this was not the first rheological study on this important problem [64, 65], it ignited a great activity in the field. An alternative experimental test consists of running a simple temperature scan at a fixed low frequency and heating rate, leading to the accurate identification of the ODT. A representative example is depicted in Figure 7.5c for a mixture of polystyrene-*b*-polyisoprene diblock copolymers (PS–PI), where the ODT is probed for various blend compositions and found to depend linearly on the average degree of polymerization (Fig. 7.5d) [64].

For completeness, it is worth noting that ordered block copolymers bare analogies with semicrystalline polymers, keeping in mind that, although the latter reflects true first-order transition, the former is a weakly first-order transition. In this respect, the true ODT is the so-called equilibrium ODT [67]. The kinetics of ordering of block copolymers has been studied in great detail using combination of shear rheology and time-resolved small-angle X-ray scattering (SAXS) experiments [66, 68–71]. In general, the results are well described by the Avrami equation that also applies to the kinetics of crystallization.

The linear viscoelastic behavior of a wide range of copolymers has been studied systematically and a lot of information has been collected. Briefly, the high-frequency (short time scale) response of the storage, $G'$, and loss, $G''$, moduli is essentially the same for both disordered and ordered phases [62]. At intermediate frequencies, the viscoelastic response of entangled copolymers is characterized by a rubbery plateau region [63], with an elastic plateau modulus

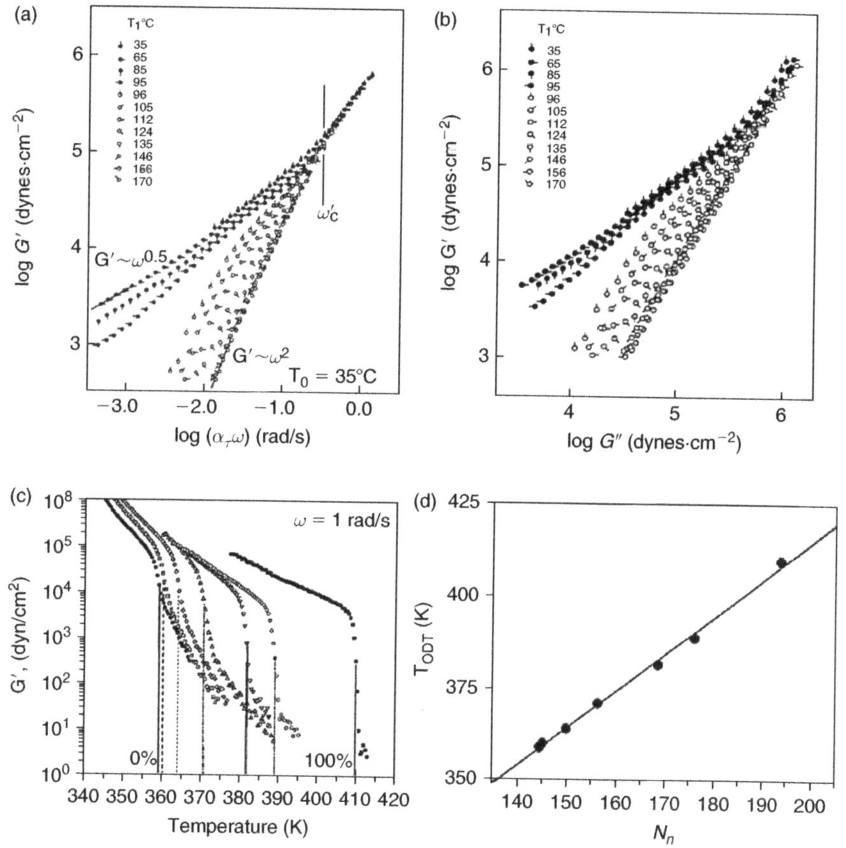


**FIGURE 7.5** (a) Reduced frequency plot (tTS with shift factor $\alpha_T$) of storage modulus $G'$ for a PEP–PEE diblock, over a wide temperature range across the ODT. Reprinted with permission from Ref. 63. (b) Respective Cole–Cole representation of the data in the form of $G'$ against the loss modulus $G''$. Reprinted with permission from Ref. 63. (c) Temperature dependence of $G'$ obtained at $\omega = 1$ rad/s and low strain amplitude while heating two PS–PI diblock copolymers and their blends. The composition of the large diblock is, from left, 0, 1.2, 11, 23.9, 48.8, 64, and 100%. *Vertical lines*, the different ODTs. Reprinted with permission from Ref. 66. (d) Respective ODT temperatures against the average degree of polymerization, $N_n$, for different blends. Reprinted with permission from Ref. 66.

$G_N^0 \approx kT/N_e a^3$, where $N_e$ is the number of statistical segments of length $a$ between entanglements [72]. Unentangled copolymers exhibit Rouse-like behavior [63, 73, 74]. At low frequencies, the response depends on temperature. In the disordered phase well above the ODT, the moduli drop as the frequency is reduced and the terminal regime is reached. The moduli vary as $G' \sim \omega^2 \ll$

$G'' \sim \omega$, conforming to Maxwell's model of a simple viscoelastic liquid. The molecular aspects of the rheology in the disordered state have been studied in great detail [75]. Above but near the ODT, composition fluctuations are significant and affect both the structure (i.e., the phase diagram) and the dynamics, imposing an extra barrier to fluid motion [76, 77]. The rheological signature is a slowing down of the terminal response with the usual scaling for a viscoelastic liquid being observed at much lower frequencies. By properly quantifying the contribution of pretransitional fluctuations to the stress it is possible to determine the spinodal temperature [77]. There is a lot of experimental evidence pointing to the coupling of fluctuations and local relaxation in homogeneous block copolymers under shear near the ordering transition [78]. Below the ODT, terminal behavior is not observed. In the following, we discuss in more detail the rheological properties of different ordered phases of block copolymers in the low frequency regime.

### 7.3.2 Lamellar Phase

This is by far the most extensively studied phase. The linear complex moduli of polycrystalline lamellar phases are nearly equal with a weak power law dependence, $G' \sim G'' \sim \omega^{1/2}$, which has been observed experimentally [63, 68, 79] and predicted theoretically [80–82]. This peculiar response is believed to be a consequence of a distribution of domain orientations, which yields a macroscopic response intermediate between that of a liquid ($G' \sim \omega^2$ and $G'' \sim \omega$) and a solid ($G' \sim \omega^0$) [80]. It remains unclear whether this response represents the actual low-frequency limit. Terminal behavior is expected at frequencies lower than the inverse of the relaxation time of the defects but it has not been observed in experiments. Solid-like response can also be anticipated in analogy to the behavior of other textured materials like smectic-A liquid crystals [83–85]. The latter have a true solid-like behavior dominated by the dynamics of the grains and edge dislocations. Defects of the smectic structure of characteristic size $\xi$ form a network with a low-frequency plateau modulus of $G_{\text{defect}} \approx kT/\xi^3$ extending over a wide frequency window. We are not aware of such observations in lamellar diblocks, but since typical values of $G_{\text{defect}}$ should be $<1$ Pa, they may simply be undetectable experimentally. In more complicated block copolymers, such as triblocks or multiblocks, lamellar connectivity is achieved via blocks of different macromolecules, leading to solid response at low frequencies [73]. This prompts for a quantitative treatment of the collective effects of multiple grains, which is still missing.

### 7.3.3 Cubic Phases

Cubic structures result usually from BCC spherical or bicontinuous gyroid phases. The viscoelastic response of BCC spheres phase is the same as that of the gyroid phase, indicating that the rheology is essentially determined by the three-dimensional crystal-like order of these phases [86, 87]. Both phases behave like viscoelastic solids with an elastic plateau ($G' \sim \omega^0$ and $G' \gg G''$) followed by a terminal regime at very low frequencies [87]. The value of the storage modulus at the elastic plateau is $G_0 \approx AkT/d^3$, where A is a constant and $d$ is the domain spacing, which is proportional to the sphere diameter [88, 89]. $G_0$ depends on the proximity to ODT: $G_0 d^3/kT \sim \exp[-c(T/T_{\mathrm{ODT}} - 1)]$ with $c$ being a constant [88]. Block copolymer cubic structures are thus analogous to ordinary crystalline solids. However, since their domain spacing is much larger than the one of molecular or ionic crystals—i.e., on the order of tens to hundreds of nanometers compared to tens of angstroms—their moduli are smaller by several decades. The onset of terminal relaxation at low frequencies depends on the degree of perfection of the crystalline order [87]. A hypothetical perfectly annealed cubic phase free of defects would exhibit a elastic response in the limit $\omega \to 0$, while grain defects in real systems facilitate long-time relaxation at higher frequencies.

### 7.3.4 Cylindrical Phase

Intermediate behavior between true viscoelastic solid ($G' \sim \omega^0$) and weak solid with power law of ½ ($G' \sim \omega^{1/2}$) has been reported in hexagonally packed cylindrical phases where the moduli vary like $G' \sim G'' \sim \omega^{1/3}$, with $G''$ slightly higher than $G''$ [88]. This behavior reflects the specific arrangement of the cylinders that have viscous liquid-like response along the direction parallel to the long-axis but that resist like elastic solid along the perpendicular direction. Theoretical predictions for the low-frequency response of hexagonal cylinder phases rather suggest a scaling exponent of ¼ [80].

Figure 7.6 summarizes the low-frequency scaling behaviors of $G'$ that have been reported for various diblock copolymers in the disordered and

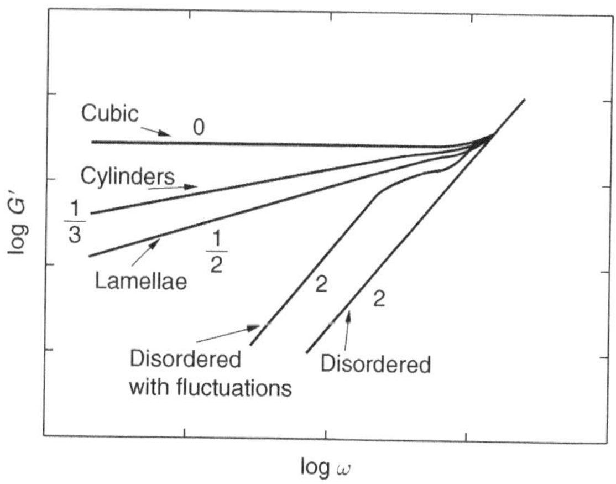


**FIGURE 7.6** The experimental scaling of the storage modulus $G'$ at low frequencies for disordered and various ordered mesophases of diblock copolymers.

ordered phases. For completeness, it is important to note that these distinctive viscoelastic signatures can serve in using rheology as a means not only to probe ODT but also order-to-order transitions (OOT) and anticipating different ordered morphologies, very often in combination with direct structural investigations [86, 90, 91].

### 7.3.5 Other Topologies

In closing, we iterate that the work reviewed here focuses on linear block copolymers. The role of macromolecular architecture—i.e., block sequences like triblocks, multiblocks, and gradient copolymers or branching like star and graft block copolymers—has been explored in length with respect to the different aspects of the rheological response discussed above (mainly linear viscoelasticity, ODT, and OOT). The key findings are qualitatively similar to those reported above [92–94]. The observed quantitative differences are attributed to the architecture and often rationalized theoretically [90, 95, 96]. Along the same lines, linear mixtures of block copolymers with linear homopolymers or other copolymers [71, 97–101] or block copolymers with lengths of different flexibility and or tendency to crystallize [28] have been investigated and continue to pose interesting scientific challenges.

## 7.4 FLOW-INDUCED PHENOMENA

### 7.4.1 Ubiquity of Flow-Induced Phenomena in Block Copolymers

The capacity of external fields in inducing structural changes such as alignment and melting has been known for a long time. In the particular case of block copolymers, the early work of Keller and co-workers [102] and Skoulios and Hadziioannou [103, 104] opened the route for systematic comprehensive studies of flow-induced phenomena. The key ingredient is the interplay between flow field and structure, which constitutes a very active field of experimental and theoretical research, essentially because of the richness of the ordered block copolymer mesophases and of their potential applications [62, 105–107]. In brief, shear flows can cause microphase-separated block copolymers to align in large domains with long-range translational order and various orientations depending on the experimental conditions. The behavior of block copolymers under flow is fundamental to the design and the processing of these materials. When a high degree of flow alignment is achieved, it is possible to fabricate highly organized anisotropic materials with interesting properties that have numerous potential applications in conventional and advanced technologies [108]. Moreover, since the resistance to flow strongly depends on the orientation of the microstructure with respect to the axis of deformation, the coupling between flow and microstructure greatly influences the linear and nonlinear rheology of block copolymers [74, 109]. The present discussion will be limited to shear and extensional flow effects, mainly in lamellar structures, which have been the most extensively studied experimentally.

### 7.4.2 Alignment of Lamellae-Forming Diblocks under Oscillatory Shear

A convenient way of probing shear-induced phenomena consists of applying large-amplitude oscillatory shear (LAOS) [105, 106]. Experimentally LAOS often constitutes a necessary measurement for detecting the limits of linear viscoelastic response. In any ordered phase, the nonlinear response observed beyond the linear regime reflects a complex behavior upon increasing the amplitude of the applied strain: softening, hardening, transition to liquid-like and eventually complete fluidization. The solid-liquid transition, or yielding, is typical of many soft matter materials such as glasses, crystals, and ordered morphologies. The critical strain and stress corresponding to this transition are known as the yield strain and yield stress. Their precise determination remains an open debate in the literature and is beyond the scope of this presentation. The interested reader is referred to the literature. [110–113]. Nevertheless, one can again draw the analogy to liquid crystals, and in particular smectics A for which a linear regime is clearly observed and where $G'$ appears to scale in the post-yield regime with the inverse of strain amplitude. This is confirmed for some but not all of the diblock ordered structures, the block entanglements being identified as the main reason for the discrepancy. In some cases, a stronger $G'$ dependence on amplitude is attributed to phase transition. In the fully fluidized state, the ratio of slopes of $G''$ and $G'$ versus strain amplitude is about 2 [113].

In past years many studies have focused on diblock copolymer melts under large amplitude oscillatory shear flows. Two main orientations have been identified (Figure 7.7a): (1) the parallel orientation where the normal to the lamellae is oriented along the shear gradient, and (2) the perpendicular orientation where the normal is along the vorticity direction. The selection of one particular orientation depends primarily on the frequency, $\omega$, of the oscillatory strain. Three frequency regimes have been defined and conceptualized [74, 79, 105, 116–121]. These regimes are limited by the characteristic frequencies $\omega_d$ and $\omega_c$ (Fig. 7.7b). $\omega_c$ is of the order of the inverse of a single chain relaxation time, whereas $\omega_d$ is a characteristic frequency associated with the relaxation of defects associated with grain boundaries (Section 7.3.2). In the low and high frequency regimes ($\omega < \omega_d$ and $\omega > \omega_c$), shearing leads to parallel orientation, whereas in the intermediate frequency regime ($\omega_d < \omega_c$), the shear stable alignment is perpendicular. A third possible orientation, the transverse orientation where the normal is oriented along the velocity direction (Fig. 7.7a), has been reported. It generally occurs in combination with other orientations during transient states [118–120, 122, 123] and/or in the presence of specific defects like kink bands [124, 125]. In one situation, where the chain dynamics was dominated by the presence of entanglements, pure transverse alignment has been reported [126].

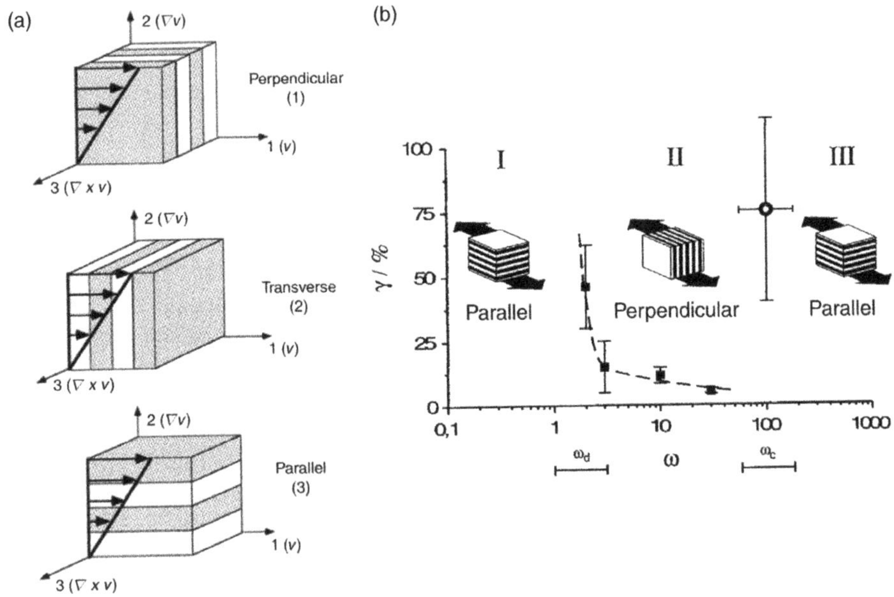


FIGURE 7.7 (a) Different orientations associated with shear-induced alignment of lamellar structures and (b) their occurrence in the strain-frequency parameter space. Reprinted with permission from (a) Ref. 114 and (b) Ref. 115.

All orientations were found to be linearly stable under LAOS, but the perpendicular orientation was stable within a larger range of parameters than the parallel one [127].

### 7.4.3 Selection of Lamellae Orientation by LAOS

A typical experimental method for aligning block copolymers consists of running a dynamic long time sweep at constant frequency and large strain amplitude. It is generally accepted that optimum, albeit not necessarily perfect, alignment leads to lowest viscoelastic moduli. The moduli drop and become constant when complete alignment is reached [62, 128]. Neutron and SAXS experiments, often performed in situ in large-scale facilities (synchrotron SAXS), birefringence measurements, and ex situ TEM observations have also contributed a lot to our present understanding of flow-orientation coupling. For a given diblock copolymer, not all orientations are systematically observed. For instance, the regime of parallel orientation at low frequencies $\omega < \omega_d$ is not systematically observed, which has stimulated some debate in the literature. Actually, the selection of a particular state of orientation is a complex process that involves the interplay of many parameters—namely, material properties, temperature, and experimental conditions like strain amplitude and frequency.

Chain distortions and molecular relaxations, which reflect molar mass effects, seem to play a central role in the development of parallel alignment at high frequencies [118]. A large contrast in the viscoelastic properties of the blocks, which is a consequence of difference in glass-transition temperatures, often referred to as dynamic asymmetry, also favors the parallel orientation at high and low frequencies, since the strain is localized in the softer blocks, which lowers the overall stress [74, 79, 117, 129]. On the other hand, at intermediate frequencies perpendicular alignment can be observed. For nearly symmetric diblocks, parallel to perpendicular alignment transition was reported with increasing frequency [79, 130]. Note that defects, whose motion at low frequencies is largely influenced by entanglements, also affect lamellar alignment [80].

The thermo-mechanical history of the copolymer plays an important role. Parallel orientation is achieved at low frequency only when the melt is annealed at temperature below the order–disorder temperature ($T_{\mathrm{ODT}}$) before shearing [131]. Shearing as the melt is cooled from above $T_{\mathrm{ODT}}$ gradually leads to uniform perpendicular alignment, while parallel alignment is preferred when the shearing is applied once in the ordered state [79]. In the same vein, the proximity of the temperature to $T_{\mathrm{ODT}}$ is another crucial parameter [79, 119–121]. The processing steps before shear application can also induce a particular orientation and/or specific trapped defects that will greatly influence the outcome of subsequent shearing experiments [124].

Concerning the experimental conditions, it has been found that the strain amplitude not only influences the rate and the degree of alignment but also contributes to the selection of the direction of alignment [114, 115, 119–121]. Typically, in the frequency range where the transition from parallel to perpendicular alignment takes place, an increase of the strain amplitude leads to parallel alignment [114]. In conclusion, a given orientation can be accessed through different pathways depending on the initial conditions [120]. The kinetics of orientation depends of the microscopic processes at the origin of orientation. At least four different mechanisms of flow-induced alignment are frequently mentioned in the literature [120]: selective melting, defect migration, grain rotation, and selective creation of lamellae from poorly ordered region. However, a complete and predictive understanding of shear-induced alignment and the microscopic mechanisms at work is still lacking.

### 7.4.4 Steady Shear Flows of Ordered Diblocks

While the abovementioned studies have focused on flow alignment dynamics of block copolymer melts subjected to reciprocating shear flows at fixed amplitude, comparatively less attention has been paid to block copolymers in steady shear flow. In addition, given that the strain amplitude and the frequency have appeared to be important parameters in oscillatory flow alignment, new phenomena are expected to occur in strong flow conditions, where large strains accumulate during shearing. In general, in steady shear flow experiments well below the ODT, a behavior similar to that in LAOS is found [132]: Perpendicular alignment is observed at low shear rates, and parallel alignment at high shear rates. The direction of alignment flips when the *steady* shear flow conditions are changed. As temperature is increased however, the degree of perpendicular alignment obtained at low shear rates decreases, unlike the behavior in large amplitude oscillatory flow where high degrees of perpendicular alignment are achieved at low frequencies even close to the ODT. This suggests that flow alignment in steady shear is more complex than that in oscillatory shear flow, perhaps owing to an increased likelihood for defect formation [106].

Under steady shear, the lamellae contract, especially when they are in the parallel orientation. This effect is due to elastic chain distortion during shear deformation. It depends on the rate of shear and molecular weight, as was shown in an experimental study with polystyrene-*b*-pol(ethylenepropylene) copolymers [133] and also predicted theoretically [134]. The elastic distortion of chains in lamellae domains is often reflected as kink bands and tilting [135].

Although the behavior of block copolymers in solvent does not constitute the main focus of this review, it is interesting to note that the alignment of diblock copolymer solutions under reciprocating shear flow has also been observed and studied [136–141]. The results are globally consistent with the behavior of melts except that parallel alignment is never observed at very low frequencies. The orientations in diblock solutions also flips from perpendicular to parallel when the shear rate is increased but the parallel alignment at high shear rates is associated with highly non-linear elasticity [139].

### 7.4.5 Role of Block Complexity

Notwithstanding the conceptual importance of diblock copolymers, practical applications of block copolymers rely on triblock or multiblock molecular architectures, which involve different possibilities of bridging and looping of the internal blocks. The most comprehensive studies of flow alignment of multiblock copolymers have focused on ABA or ABC triblock and ABABA pentablock copolymer melts [73, 142–146]. Although triblock copolymers share common features with diblocks, they exhibit a greater sensitivity to experimental conditions, and several contradictory results have been reported [73, 128, 142–144]. ABC copolymers have a higher degree of complexity than AB and ABA copolymers due to their connectivity. In ABC copolymers, the midblocks B necessarily bridge the end blocks A and C, which belong to two different microdomains. By contrast, in ABA copolymers the midblocks can adopt a loop or a bridge conformation so that they belong either to the same or to different microdomains. The new and important result is that connectivity and chain conformation play an important role in determining lamellar alignment of multiblock copolymers. The effect of the connectivity in triblock copolymers has been studied for the case of polystyrene-*b*-polybutadiene-*b*-poly (methyl methacrylate) copolymers dissolved in weakly selective solvents [144]. The perpendicular orientation is observed at the lowest accessible shear rates, but a highly defective parallel orientation forms at shear rates just above the inverse of the relaxation time of individual chains. The lamellar alignment reversibly flips from one orientation to another. At much higher shear rates, a significant fraction of lamellae is destroyed by the flow and reorient in transverse orientation upon flow cessation. These phenomena are associated with peculiar variations of the viscosity with the shear rate. In pentablock copolymers [145, 146], the only stable arrangement is the low-frequency perpendicular orientation. A transition to a disordered state occurs at high frequency. Upon flow cessation, the molecules reorder into a well-defined stable transverse orientation. This behavior has been interpreted in relation with the pentablock architecture that imposes a strong mechanical coupling between the block layers and prevents the microdomains from sliding relative to each other, thus destroying the parallel orientation. Similar arguments have been developed for other multiblock copolymers [147–149]. The effect of bridge and loops on the nonlinear response of layered materials and the dynamical consequences of possible bridge-to-loop conversion under flow have been analyzed theoretically [150].

Shear alignment has been also observed in other block copolymer morphologies, and we mention here a few indicative examples. Unidirectional shearing of the hexagonal cylinder phase of diblock or triblock copolymers has been reported to induce alignment of the cylinders in the shear direction. The distance from ODT and the shear intensity are the important control parameters for the degree of alignment and possible phase transformation [151]. Almdal et al. [152] and Okamoto et al. [153] have explored the effect of LAOS on the orientation of BCC phases of poly(ethylethylene)-*b*-poly(ethylenepropylene) and polystyrene-*b*-poly(ethylenepropylene) melts, respectively. Both studies report strong preferential orientation of a particular lattice plane, although the orientation behaviors of the two cases are quite different. Shin et al. subsequently demonstrated that the main control parameter for BCC phase alignment is the difference in the mechanical properties of the matrix and the dispersed phase [154].

### 7.4.6 Shear-Induced Disorder–Order, Order–Order, and Order–Disorder Transitions

It is well known that the imposition of shear flow on weakly ordered systems like colloids and surfactants can have profound effects on the phase transitions. In block copolymers, shear flows suppress fluctuations in the disordered phase (anisotropically), thereby making the system more mean-field-like and causing an increase of the ODT temperature, $T_{\mathrm{ODT}}$. It is thus possible to induce disorder–order transitions by shearing a disordered copolymer melt. Experiments with diblocks and triblocks copolymer melts have confirmed the possibility of inducing ordering by shearing the disordered phase near the ODT [155]. The corresponding shift of the ODT temperature, $T_{ODT}$, is proportional to the square root of the shear rate, $\Delta T_{ODT} \sim \dot{\gamma}^{1/2}$, in qualitative agreement with theoretical predictions [156, 157]. However, the coupling between flow and fluctuations is not unique and can change qualitatively with shear rate. For instance, Balsara et al. observed both shear-induced ordering and shear-induced disordering in block copolymer melts above their $T_{ODT}$ by changing the shear rate [158]. This provides an additional tool to tune the phase behavior and the properties of block copolymer materials.

A related question concerns the evolution of the shear-induced phase on shear cessation. The exact answer, which depends on the thermodynamic state of the quiescent sample at the probed temperature, can shed light into the mechanisms of the order–disorder transition. This has been recognized as an important problem and addressed in the literature. For instance, investigations by Bates and co-workers on asymmetric PEP–PEE copolymers found that LAOS could stabilize an hexagonal cylindrical phase above the equilibrium ODT [159]. Upon shear cessation, the hexagonal cylinder to disorder transition was found to proceed over time via intermediate undulating cylinder and BCC spherical phases, (Fig. 7.8).

Disordering has been reported for the BCC sphere phase in melts of polystyrene-*b*-polyisoprene copolymers subjected to steady shear [112]. The final state depends on the applied stress. At low stresses, the BCC lattice is not affected by the flow. As the stress is increased, there is a critical stress above which the structure is disrupted and the long-range order is lost. At still higher stresses, the system behaves as if it were disordered, both structurally and rheologically. If the shear stress is lowered below the critical stress, the BCC structure reforms much like after thermal disordering.

Shear flows have also the capacity to induce order–order transitions when one ordered phase is transformed into another one. For instance, in an investigation of polystyrene-*b*-polyisoprene ordered phases, it has been found that the cubic gyroid structure subjected to LAOS transforms reversibly to an cylindrical phase with hexagonal order [160]. As for the oriented hexagonal perforated lamellar phase HPL, it is metastable and transforms into the gyroid phase upon shearing.

**FIGURE 7.8** Oriented cylinders in hexagonal order-to-disorder transition of PEP–PEE block copolymer upon cessation of oscillatory shear. Reprinted permission from Ref. 159.

### 7.4.7 Extensional and Other Complex Flows

The pioneering work of Keller and co-workers [102] who extruded polystyrene-*b*-polybutadiene-*b*-polystyrene copolymers to produce near-single crystals stimulated many experimental studies with flow fields involving extensional components. Uniaxial deformations are important due to their industrial relevance as well as their ability to effectively align ordered block copolymers [106, 161]. In practice it is much more efficient to align block copolymer microstructures than shear deformation. For example, triblock copolymer lamellar structures under planar elongation show more effective orientation in the flow direction (and none in the perpendicular) than under shear deformation [162]. An elongational deformation was recently imposed on polystyrene-*b*-polyisoprene-*b*-polystyrene triblock copolymer melts below the order–disorder temperature where polystyrene segregates into spherical microdomains arranged in BCC symmetry at equilibrium [161]. Strain hardening was probed in combination with simultaneous SAXS and TEM structural observations. It was found that under elongation the spherical domains transformed into cylinders with short-range order.

The development of in situ rheophysical instrumentation has contributed a lot to obtaining useful data and to our current understanding of this field. For example, using elongational flow optorheometry (EFOR)—i.e., combining an elongational rheometer with polarimetry [163], Kotaka and co-workers investigated systematically the elongational flow-induced morphology of block copolymers in the melt state [164]. Strain hardening was observed in the disordered state at low Hencky strain rates and strain softening at high rates [165]. The situation was far more complex in the ordered state. For example, the cylindrical morphology was elongated in two different directions, parallel and perpendicular with respect to the average cylinder axis. In the former case, the elongational viscosity exhibited ultimate strain softening behavior, whereas in the latter case it showed strain hardening.

Tensile tests constitute a common tool for investigating the mechanical properties of ordered block copolymers in the solid state where one of the phases is glassy or semicrystalline. This subject is central to many applications of thermoplastic elastomers. The relation between symmetry of the ordered phases (cubic, cylindrical, or lamellar) and the mechanical properties is of particular importance. The planar extensional deformation of polystyrene-*b*-poly(ethylene-*co*-butylene)-*b*-polystyrene copolymers containing either pre-oriented spherical or cylindrical polystyrene spheres has been studied in detail [166]. The deformation of the BCC phase is affine and reversible. By contrast, the deformation of hexagonal phase is nonaffine. The cylindrical domains do change their domain spacing but eventually break, as shown in Figure 7.9. Breaking proceeds via undulation of the cylinders, and then their breakup into small ellipsoids. The undulated cylinders confirmed the mechanism of buckling or micronecking observed earlier in block copolymers under tensile deformation [167].

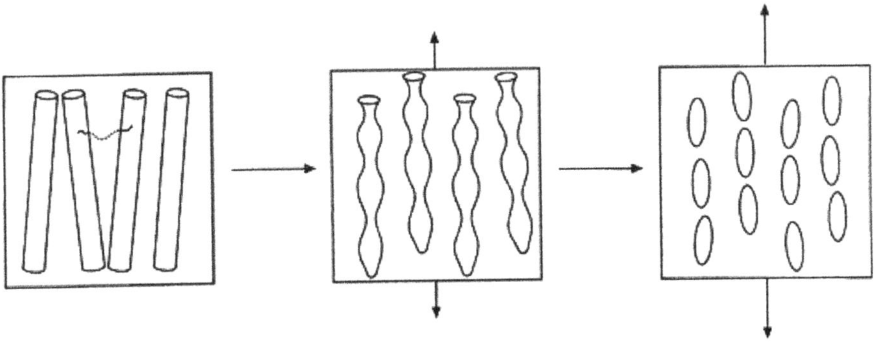

**FIGURE 7.9** The transformation of the hexagonal phase of glassy polystyrene cylinders in a rubbery matrix of poly(ethylene-*co*-butylene) under nonaffine planar extension. From left to right: undeformed, undulated cylinders (micronecking regime) and broken cylinders. Note that the domain spacing remains unchanged. Reprinted with permission from Ref. 166.

Other comprehensive studies of the mechanical properties of solid copolymers concern polystyrene-*b*-polyisoprene-*b*-polystyrene and polystyrene-*b*-polybutadiene-*b*-polystyrene copolymers. Using a combination of conventional mechanical tests, synchrotron SAXS experiments, and transmission electron microscopy, Thomas and co-workers studied in great detail the microscopic mechanisms of deformation of these styrenic TPEs with lamellar and cylindrical morphologies [168, 169]. They also demonstrated the importance of parameters like the symmetry of the morphology, the quality of alignment, the nature of the defects and their degree of annealing, and, in near-single crystal materials, the orientation of the sample with respect to the direction of extension. This study was further extended to layered silicate and lamellar triblock nanocomposites [170].

Engineering processes used to process films, fibers, and coatings involve flows that are still more complex with a combination of shear and extensional deformation. Roll casting has proved to be a reliable method for producing films with long-range order starting from block copolymer solutions [171]. For the case of extrusion, it has been reported that diblock copolymers orient in parallel and perpendicular directions or become disordered, depending on the shear rate and annealing [172, 173]. Still more puzzling, the extrusion of pentablock solutions at high shear rates produces a stable transverse orientation [174].

## 7.5 CLOSING REMARKS AND OUTLOOK

Rheology is an indispensable tool for understanding the viscoelastic properties, controlling the material's response and its final properties, and processing different nanostructures. It is therefore fundamental to the evolution of the field and has played a key role in recent achievements and developments that have emerged. Our initial objective concerned the phase behavior and the dynamical properties of block copolymers under external fields. Up to now we have essentially considered the effect of flow fields which by far have been most extensively studied. This should not obliterate the fact that other external fields have been used to manipulate ordered and disordered phases of block copolymers and microfabricate tailor-made materials. Recently the use of temperature gradients or electric fields have appeared as extremely promising in this context.

The gradient zone technique well known in metallurgy has been extended by Hashimoto and co-workers to create nearly single crystals of block copolymers of macroscopic size [175]. The technique consists of applying a moving temperature gradient field during the ordering process. Lamellae of cylinders orient parallel to the temperature gradient over long distances and form single grain materials. The technique has been applied to different block copolymers with different morphologies [176].

Several studies have given evidence for the role of external electric fields in aligning block copolymer ordered structures [177, 178]. Both lamellar and cylindrical diblocks can be aligned parallel to static electric fields. Experiments in the melt are generally limited by the high viscosities of structured copolymers. These limitations can be removed by using block copolymer solutions in weakly selective solvents [179]. The mechanism often invoked to explain alignment is based on a dielectric effect, but this relatively weak effect seems to require high electric fields exceeding the dielectric breakdown of the films [180]. It has been proposed that the presence of free lithium ions introduced during the anionic synthesis of the copolymers can induce morphological changes and phase transitions [180, 181]. For polystyrene-*b*-poly(methyl methacrylate) diblocks, another possible mechanism considers that the formation of lithium-poly (methyl methacrylate) complexes enhances the dielectric contrast thereby reducing the strength of the critical electrical field [182].

Up to now we have essentially focused on block copolymers in bulk. Both shear and extensional fields have also been successfully applied in orienting block copolymer thin films and creating a variety of nanostructures. The field was recently reviewed comprehensibly by Marencic and Register [183]. Finally, we have to emphasize that interactions between block copolymers and surfaces associated with important confinement produce new morphologies with outstanding ordering properties [184–186]. The subject greatly influences emerging areas such as block copolymer lithography for realizing small microelectronic components [187, 188] or nanopatterning in thin films [189–191].

In conclusion, the field of block copolymers is huge and the applications seem to be unlimited. This richness encompasses amazing physicochemical properties and interesting technological challenges which are demonstrated in a great number of publications and patents. Here we have tried to highlight some of these fascinating properties.

## REFERENCES


1. I.W. Hamley, *The Physics of Block Copolymers*, Oxford University Press, Oxford, 1998.
2. N. Hadjichristidis, S. Pispas, and G. Floudas, *Block Copolymers: Synthetic Strategies, Physical Properties & Applications*, Wiley-VCH, New-York, 2003.
3. G. Holden, H.R. Kricheldorf, and R.P. Quirk, eds., *Thermoplastic Elastomers*, Carl Hanser Verlag, Berlin, 2004.
4. G.H. Fredrickson, *The Equilibrium Theory of Inhomogenous Polymers*, Oxford University Press, Oxford, 2006.
5. K. Matyjaszewski, Y. Gnanou, and L. Leibler, eds., *Macromolecular Engineering*, Wiley-VCH, Weinheim, 2007.
6. G.H. Fredrickson and F.S. Bates, *Annu. Rev. Mater. Sci.*, 26, 501 (1996).
7. A.-V. Ruzette and L. Leibler, *Nat. Mat.*, 4, 19 (2005).
8. P. Alexandridis and B. Lindman, *Amphiphilic Block Copolymers*, Elsevier, Amsterdam, 2000.
9. I.W. Hamley, *Block Copolymers in Solution: Fundamentals and Applications*, Wiley, New York, 2005.
10. J.-F. Gohy, *Adv. Polym. Sci.*, 190, 65 (2005).
11. J. Rodriguez-Hernandez, F. Checot, Y. Gnanou, and S. Lecommandoux, *Prog. Polym. Sci.*, 30, 691 (2005).
12. Tsukahara, Y., N. Nakamura, T. Hashimoto, and H. Kawai, *Polym. J.*, 12, 455 (1980); T. Hashimoto, Y. Tsukahara, and H. Kawai, *Polym. J.*, 10, 699 (1983); T. Hashimoto, Y. Tsukahara, K. Tachi, and H. Kawai, *Macromolecules*, 16, 648 (1983).
13. K. Knoll and N. Niessner, *Macromol. Symp.*, 132, 231 (1998); R. Adhikari, G.H. Michler, T.A. Huy, et al., *Macromol. Chem. Phys.*, 204, 488 (2004).
14. P. Hodrokoukes, G. Floudas, S. Pispas, and N. Hadjichristidis, *Macromolecules*, 34, 650 (2001).
15. S. Jouenne, J.A. Gonzalez-Leon, A.-V. Ruzette, et al., *Macromolecules*, 40, 2432 (2007).
16. M.M. Mok, S. Pujari, W.R. Burghardt, et al., *Macromolecules*, 41, 5818 (2008).
17. M.M. Mok, J. Kim, C. L. H. Wong, et al., *Macromolecules*, 42, 7863 (2009).
18. T.P. Russell, R.P. Hjelm Jr., and P.A. Seeger. *Macromolecules*, 23, 890 (1990).
19. R.J. Roe, M. Fishkis, and J.C. Chang, *Macromolecules*, 14, 1091 (1981).
20. A.K. Khandpur, J.S. Farster, F S. Bates, et al., *Macromolecules*, 28, 8796 (1995).
21. M.F. Schulz, A.K. Khandpur, F.S. Bates, et al., *Macromolecules*, 29, 2857 (1996).
22. J. Zhao, B. Majumdar, M.F. Schulz, et al., *Macromolecules*, 29, 1204 (1996).
23. J.H. Rosedale, F.S. Frank, K. Almdal, et al., *Macromolecules*, 28, 1429 (1996).
24. R. Stadler, C. Auschra, J. Beckmann, et al., *Macromolecules*, 28, 3080 (1995).
25. Y. Mogi, H. Kotsuji, Y. Kaneko, et al., *Macromolecules*, 25, 5408 (1992); S.P. Gido, D.W. Schwark, E.L. Thomas, and M. do Carmo Gonçalves, *Macromolecules*, 26, 2636 (1993).
26. T.S. Bailey, C.M. Hardy, T.H. Epps III, and F.S. Bates, *Macromolecules*, 35, 7007 (2002).

27. L. Zhang and A. Eisenberg, *Science*, 268, 727 (1995).
28. G. Floudas and C. Tsitsilianis, *Macromolecules*, 30, 4381 (1997).
29. V. Abetz and P.F.W. Simon, *Adv. Polym. Sci.*, 189, 125 (2005).
30. A. Carlsen and S. Lecommandoux, *Curr. Opin. Colloid Interface Sci.*, 14, 329 (2009).
31. S. Jouenne, J.A. Gonzalez-Leon, A.-V. Ruzette, et al., *Macromolecules*, 41, 9823 (2008).
32. F.S. Bates, G.H. Fredrickson, D. Hucul, and S.F. Hahn, *AICHE J.*, 47, 762 (2001).
33. A.H.E. Müller and K. Matyjaszewski, eds., *Acrylic Copolymer Controlled and Living Polymerizations: From Mechanisms to Applications*, Wiley-VCH, Weinheim, 2009.
34. S. Fakirov, ed., *Handbook of Condensation Thermoplastic Elastomers*, Wiley-VCH, 2005.
35. C. Fleury, *Structuration et rhéologie des solutions de copolymères triblocs SBM*, PhD thesis, Université Pierre et Marie Curie, Paris, 2001.
36. L. Corte, K. Yamauchi, F. Court, et al., *Macromolecules*, 36, 7695 (2003).
37. P.G. de Gennes, *Scaling Concepts in Polymer Physics*, Cornell University Press, New York, 1979.
38. E. Helfand, *Macromolecules*, 8, 552 (1976).
39. E. Helfand and Z.R. Wasserman, *Macromolecules*, 9, 879 (1976).
40. A.N. Semenov, *Soviet Physics JETP*, 61, 70 (1985).
41. P. G. de Gennes, J. Phys. (Paris), 37, 1443 (1976); S. Alexander, J. Phys. Paris, 38, 983 (1997).
42. L. Leibler, *Macromolecules*, 13, 1602 (1980).
43. A.J. Ryan, S.-M. Mai, J.P.A. Fairlough, et al., *Phys. Chem. Chem. Phys.*, 3, 2961–2971, (2001).
44. D.A. Hajduk, H. Takenouchi, M.A. Hillmyer, et al., *Macromolecules*, 30, 3788 (1997).
45. D.A. Hadjuk, P.E. Harper, S.M. Gruner, et al., *Macromolecules*, 27, 4063 (1994).
46. M.I. Kim, T. Wakada, S. Akasaka, et al., *Macromolecules*, 42, 5266 (2009).
47. Y.-Y. Huang, J.-Y. Hsu, H.-L. Chen, and T. Hashimoto, *Macromolecules*, 40, 406 (2007).
48. G.H. Fredrickson and E. Helfand, *J. Chem. Phys.*, 87, 697 (1987).
49. M.W. Matsen and F. Bates, *Macromolecules*, 29, 1091 (1996).
50. M.W. Matsen, *J. Chem. Phys.*, 113, 5539 (2000).
51. M.W. Hamersky, S.D. Smith, A.O. Gozen, and R.J. Spontak, *Phys. Rev. Lett.*, 95, 168306 (2005).
52. U. Breiner, U. Krappe, T. Jacob, et al., *Polym. Bull.*, 40, 219 (1998).
53. U. Breiner, U. Krappe, V. Abetz, and R. Stadler, *Macromol. Chem. Phys.*, 198, 1051 (1997); U. Krappe, R. Stadler, and I. Voigt-Martin, *Macromolecules*, 28, 4558 (1995).
54. U. Breiner, U. Krappe, E.L. Thomas, and R. Stadler, *Macromolecules*, 31, 135 (1998).
55. J. Qin, F.S. Bates, and D.C. Morse, *Macromolecules*, 43, 5128 (2010).

56. K.J. Hanley, T.P. Lodge, and C.-I. Huang, *Macromolecules*, 33, 5918 (2000).
57. D.J. Kinning, K.I. Winey, and E.L. Thomas, *Macromolecules*, 21, 3502 (1988).
58. T. Hashimoto, H. Tanaka, and H. Hasegawa, *Macromolecules*, 23, 4378 (1990); H. Tanaka, H. Hasegawa, and T. Hashimoto, *Macromolecules*, 24, 240 (1991).
59. P. Falus, H. Xiang, M.A. Borthwick, et al., *Phys. Rev. Lett.*, 93, 145701 (2004).
60. F. Chen, Y. Kondo, and T. Hashimoto, *Macromolecules*, 40, 3714 (2007).
61. M.R. Bockstaller, R.A. Mickiewicz, and E.L. Thomas, *Adv. Mat.*, 17, 13331 (2005).
62. R.H. Colby, *Curr. Opin. Coll. Interface Sci.*, 1, 454 (1996).
63. J.H. Rosedale and F.S. Bates, *Macromolecules*, 23, 2329 (1990).
64. C.D. Han and J. Kim, *J. Polym. Sci. Polym. Phys.*, 25, 1741 (1987).
65. C.D. Han, J. Kim, and J.K. Kim, *Macromolecules*, 22, 383 (1989).
66. G. Floudas, D. Vlassopoulos, M. Pitsikalis, et al., *J. Chem. Phys.*, 104, 2083 (1996).
67. G. Floudas, T. Pakula, G. Velis, et al., *J. Chem. Phys.*, 108, 6498 (1998).
68. G. Floudas, T. Pakula, E.W. Fischer, et al., *Acta Polym.* 45, 176 (1994).
69. H.H. Winter, D.B. Scott, W. Gronski, et al., *Macromolecules*, 26, 7236 (1993).
70. J.L. Adams, D.J. Quiram, W.W. Graessley, and R.A. Register, *Macromolecules*, 29, 2929 (1996).
71. B. Stuhn, A. Vilesov, and H.G. Zachmann, *Macromolecules*, 27, 3560 (1994).
72. M. Doi and S.F. Edwards, *The Theory of Polymer Dynamics*, Oxford, New York, 1986.
73. B.L. Riise, G.H. Fredrickson, R.G. Larson, and D.S. Pearson, *Macromolecules*, 28, 7653 (1995).
74. S.S. Patel, R.G. Larson, K.I. Winey, and H. Watanabe, *Macromolecules*, 28, 4313 (1995).
75. S. Choi and C.D. Han, *Macromolecules*, 37, 215 (2004).
76. G.H. Fredrickson and E. Helfand, *J. Chem. Phys.*, 89, 5890 (1988).
77. R.G. Larson and G.H. Fredrickson, *Macromolecules*, 20, 1897 (1987).
78. L.A. Archer and G.G. Fuller, *Macromolecules*, 27, 7152 (1994).
79. K.A. Koppi, M. Tirrell, F.S. Bates, et al., *J. Phys. II (France)*, 2, 1941 (1992).
80. M. Rubinstein and S.P. Obukhov, *Macromolecules*, 26, 1740 (1993).
81. K. Kawasaki and A. Onuki, *Phys. Rev. A.*, 42, 3664 (1990).
82. T.A. Witten, L. Leibler, and P.A. Pincus, *Macromolecules*, 23, 824 (1990).
83. R.G. Larson, K.I. Winey, S.S. Patel, et al., *Rheol. Acta*, 32, 245 (1993).
84. R.H. Colby, C.K. Ober, J.R. Gillmor, et al., *Rheol. Acta*, 36, 498 (1997).
85. A. Wewerka, K. Viertler, D. Vlassopoulos, and F. Stelzer, *Rheol. Acta*, 40, 416 (2001).
86. J. Zhao, B. Majumdar, M.F. Schulz, et al., *Macromolecules*, 29, 1204 (1996).
87. M.B. Kossuth, D.C. Morse, and F.S. Bates, *J. Rheol.*, 43, 167 (1999).
88. C.Y. Ryu, M.S. Lee, D.A. Hajduk, and T.P. Lodge, *J. Polym. Sci. Polym. Phys.*, 35, 2811 (1997).
89. J.M. Sebastian, C. Lai, W.W. Graessley, and R.A. Register, *Macromolecules*, 35, 2707 (2002).
90. G. Floudas, N. Hadjichristidis, H. Iatrou, et al., *Macromolecules*, 27, 7735 (1994).

91. K. Kimishima, T. Koga, and T. Hashimoto, *Macromolecules*, 33, 968 (2000).
92. N.Y. Ryu and T.P. Lodge, *Macromolecules*, 32, 7190 (1999).
93. N. Sakamoto, T. Hashimoto, C.D. Han, and N.Y. Vaidya, *Macromolecules*, 30, 1621, (1997).
94. M.M. Mok, S. Pujari, W.R. Burghardt, et al., *Macromolecules*, 41, 5818 (2008).
95. G. Floudas, S. Pispas, N. Hadjichristidis, et al., *Macromolecules*, 29, 4142 (1996).
96. S.T. Milner, *Macromolecules*, 27, 2333 (1994).
97. M.W. Matsen, *Macromolecules*, 28, 5765 (1995).
98. G. Floudas, N. Hadjichristidis, M. Stamm, et al., *J. Chem. Phys.*, 106, 3318 (1997).
99. S. Koizumi, H. Hasegawa, and T. Hashimoto, *Macromolecules*, 27, 6532 (1994).
100. A.D. Vilesov, G. Floudas, T. Pakula, et al., *Macromol. Chem. Phys.*, 195, 2317 (1994).
101. R.A. Mickiewicz, E. Ntoukas, A. Avgeropoulos, and E.L. Thomas, *Macromolecules*, 41, 5785 (2008).
102. M.J. Folkes, A. Keller, and F.P. Scalisi, *Colloid Polym. Sci.*, 251, 1 (1973).
103. G. Hadziioannou, A. Mathis, and A. Skoulios, *Colloid Polym. Sci.*, 257, 136 (1979).
104. G. Hadziioannou, C. Picot, A. Skoulios, et al., *Macromolecules*, 15, 263 (1982).
105. U. Wiesner, *Macromol. Chem. Phys.*, 198, 3319 (1997).
106. I.W. Hamley, *J. Phys.: Condens. Matter*, 13, R643 (2001).
107. H.D. Ceniceros, G.H. Fredrickson, and G.O. Mohler, *J. Comput. Phys.*, 228, 1624 (2009).
108. J.W. Park, J. Cho, and E.L. Thomas, *Soft Matter*, 4, 739 (2008).
109. K. Luo and Y. Yang, *Macromolecules*, 35, 3722–3730 (2002).
110. D. Bonn and M.M. Denn, *Science*, 324, 1401 (2009).
111. P. Coussot, H. Tabuteau, X. Chateau, et al., *J. Rheol.*, 50, 975 (2006).
112. J.M. Sebastian, C. Lai, W.W. Graessley, et al., *Macromolecules*, 35, 2700 (2002).
113. M. Cloitre, in *Microgel-Based Materials*, Wiley-VCH, 2010.
114. V.K. Gupta, R. Krishnamoorti, J.A. Kornfield, and S.D. Smith, *Macromolecules*, 29, 1359 (1996).
115. D. Maring and U. Wiesner, *Macromolecules*, 30, 660 (1997).
116. K.A. Winey, S.S. Patel, R.G. Larson, and H. Watanabe, *Macromolecules*, 26, 4373 (1993).
117. Y. Zhang, U. Wiesner, and H.W. Spiess, *Macromolecules*, 28, 778 (1995).
118. Z.-R. Chen, J.A. Kornfield, S.D. Smith, et al., *Science* 277, 1248 (1997).
119. V.K. Gupta, R. Krishnamoorti, Z.-R. Chen, et al., *Macromolecules*, 29, 875 (1996).
120. Z.-R. Chen, A.M. Issaian, J.A. Kornfield, et al., *Macromolecules*, 30, 7096 (1997).
121. Z.-R. Chen and J.A. Kornfield, *Polymer*, 39, 4679 (1998).
122. S. Okamoto, K. Saijo, and T. Hashimoto, *Macromolecules*, 27, 5547 (1994).
123. B.S. Pinheiro, D.A. Hajduk, S.M. Gruner, and K.I. Winey, *Macromolecules*, 29, 1482 (1996).
124. D.L. Polis and K.I. Winey, *Macromolecules* 29, 8180 (1996).
125. D.L. Polis and K.I. Winey, *Macromolecules*, 31, 3617 (1998).
126. Y. Zhang and U. Wiesner, *J. Chem. Phys.*, 103, 4784 (1995).

127. P. Chen and J. Vinals, *Macromolecules*, 35, 4183 (2002).
128. S. Stangler and V. Abetz *Rheol. Acta*, 42, 569 (2003).
129. G.H. Fredrickson, *J. Rheol.*, 38, 1045 (1994).
130. R.M. Kannan and J.A. Kornfield, *Macromolecules*, 27, 1177 (1994).
131. Y. Zhang, U. Wiesner, Y. Yang, et al., *Macromolecules*, 29, 5427 (1996).
132. K. Luo and Y. Yang, *Macromolecules*, 35, 3722 (2002).
133. L. Qiao, A.J. Ryan, and K.I. Winey, *Macromolecules*, 35, 3596 (2002).
134. D.R. Williams and F.C. MacKintosh, *Macromolecules*, 27, 7677 (1994).
135. L. Qiao and K.I. Winey, *Macromolecules*, 33, 851 (2000).
136. N.P. Balsara, B. Hammouda, P.K. Kesani, et al., *Macromolecules*, 27, 2566 (1994).
137. H. Wang, P.K. Kesani, N.P. Balsara, and B. Hammouda, *Macromolecules*, 30, 982 (1997).
138. H. Wang, M.C. Newstein, A. Krishnan, et al., *Macromolecules*, 32, 3695 (1999).
139. J.L. Zryd and W.R. Burghardt, *Macromolecules*, 31, 3656 (1998).
140. K. Mortensen, E. Theunissen, R. Kleppinger, et al., *Macromolecules*, 35, 7773 (2002).
141. S. Kitade, N. Ochiai, Y. Takahashi, et al., *Macromolecules*, 31, 8083 (1998).
142. T. Tepe, D.A. Hadjuk, M.A. Hillmyer, et al., *Rheol.*, 41, 1147 (1997).
143. T.J. Hermel, L. Wu, S.F. Hahn, et al., *Macromolecules*, 35, 4685 (2002).
144. E. Di Cola, C. Fleury, P. Panine, and M. Cloitre, *Macromolecules*, 41, 3627 (2008).
145. M.E. Vigild, C. Chu, M. Sugiyama, et al., *Macromolecules*, 34, 951 (2001).
146. A. Phatak, C. W. Macosko, F. S. Bates, and S. F. Hahn, J. Rheol. 49, 197 (2005).
147. L. Wu, T.P. Lodge, and F.S. Bates, *Macromolecules*, 37, 8184 (2004).
148. L. Wu, T.P. Lodge, F.S. Bates, *Macromolecules*, 39, 294 (2006).
149. Y. Matsumiya, M. Matsumoto, H. Watanabe, et al., *Macromolecules*, 40, 3724 (2007).
150. V. Ganesan and G.H. Fredrickson, *J. Rheol.*, 45, 161 (2001).
151. F.A. Morrison and H.H. Winter, *Macromolecules*, 22, 3533 (1989).
152. K. Almdal, K.A. Koppi, and F.S. Bates, *Macromolecules*, 26, 4058 (1993).
153. S. Okamoto, K. Saijo, and T. Hashimoto, *Macromolecules*, 27, 3753 (1994).
154. G. Shin, N. Sakamoto, K. Saijo, et al., *Macromolecules*, 33, 9002 (2000).
155. A. Nakatani, F.A. Morrison, J.F. Douglas, et al., *J. Chem. Phys.*, 104, 1589 (1996).
156. M.E. Cates and S.T. Milner, *Phys. Rev. Lett.*, 62, 1856 (1989).
157. M. Muthukumar, in *Flow-Induced Structure in Polymers*, ACS, Washington, DC, 1995.
158. N.P. Balsara and H.J. Dai, *J. Chem. Phys.*, 105, 2942 (1996).
159. F.S. Bates, K.A. Koppi, M. Tirrell, et al., *Macromolecules*, 27, 5934 (1994).
160. R. Eskimergen, K. Mortensen, and M.E. Vigild, *Macromolecules*, 38, 1286 (2005); J.-H. Ahn and W.-C. Zin, *Macromolecules*, 33, 641 (2000).
161. W.-K. Lee, H.D. Kim, and E.Y. Kim, *Curr. Appl. Phys.*, 6, 718 (2006).
162. H.L. Lee, R.A. Register, D.A. Hajduk, and S.M. Gruner, *Polym. Eng. Sci.*, 36, 1414 (1996).
163. G.G. Fuller, *Optical Rheometry of Complex Fluids*, Oxford, New York, 1994.

164. T. Kotaka, A. Kojima, and M. Okamoto, *Rheol. Acta*, 36, 646 (1997).
165. T. Kotaka, M. Okamoto, A. Kojima, et al., *Polymer*, 42, 3223 (2001).
166. C. Daniel, I.W. Hamley, and K. Mortensen, *Polymer*, 41, 9239 (2000).
167. T. Pakula, K. Saijo, H. Kawai, and T. Hashimoto, *Macromolecules*, 18, 1294 (1985).
168. Y. Cohen and E.L. Thomas, *Macromolecules*, 36, 5265 (2003); Y. Cohen, R.J. Albalak, B.J. Dair, et al., *Macromolecules*, 33, 6502 (2000).
169. C.C. Honeker, E.L. Thomas, R.J. Albalak, et al., *Macromolecules*, 33, 9395 (2000); C.C. Honeker, and E.L. Thomas, *Macromolecules*, 33, 9407 (2000).
170. Y.-H. Ha and E.L. Thomas, *Macromolecules*, 35, 4419 (2002).
171. M.A. Villar, D.R. Rueda, F. Ania, and E.L. Thomas, *Polymer*, 43, 5139 (2002).
172. H. Leist, K. Geiger, and U. Wiesner, *Macromolecules*, 22, 1317 (1999).
173. E.S. Carreras, J.-M. Piau, N. El Kissi, et al., *J. Rheol.*, 50, 803 (2006).
174. T. Harada, F.S. Bates, and T.P. Lodge, *Macromolecules*, 36, 5440 (2003).
175. T. Hashimoto, J. Bodycomb, Y. Funaki, and K. Kimishima, *Macromolecules*, 32, 952 (1999); J. Bodycomb, Y. Funaki, K. Kimishima, and T. Hashimoto, *Macromolecules*, 32, 2075 (1999).
176. K. Mita, H. Tanaka, K. Saijo, et al., *Macromolecules*, 40, 5923 (2007); K. Mita, M. Takenaka, H. Hasegawa, and T. Hashimoto, *Macromolecules*, 40, 5923 (2007).
177. K. Amundson, E. Helfand, X. Quan, et al., *Macromolecules*, 27, 6559 (1994).
178. J. DeRouchey, T. Thurn-Albrecht, R. Kolb, and T.P. Russell, *Macromolecules*, 37, 2538 (2004).,
179. A. Böker, H. Elbs, H. Hänsel, et al., *Phys. Rev. Lett.*, 89, 135502 (2002); K. Schmidt, H.G. Schoberth, M. Ruppel, et al., *Nature Mat.*, 7, 142 (2008).
180. Y. Tsori, F. Tournilhac, D. Andelman, and L. Leibler, *Phys. Rev. Lett.*, 90, 145504 (2003).
181. Y. Tsori, F. Tournilhac, and L. Leibler, *Macromolecules*, 36, 5873 (2003).
182. J.-Y. Wang, T. Xu, J.M. Leiston-Belanger, et al., *Phys. Rev. Lett.* 96, 128301 (2006).
183. A.P. Marencic and R.A. Register, *Annu. Rev. Chem. Biomol. Eng.*, 1, 277 (2010).
184. K. Ch. Daoulas, M. Müller, M.P. Stoykovich, et al., *Langmuir*, 24, 1284 (2008).
185. X. Zhang, B.C. Berry, K.G. Yager, et al., *ACS Nano*, 2, 2331 (2008).
186. D.O. Shin, B.H. Kim, J.-H. Kang, et al., *Macromolecules*, 42, 1189 (2009).
187. M. Park, C. Harrison, P.M. Chaikin, et al., *Science*, 276, 1401 (1997).
188. C. Tang, E.M. Lennon, G.H. Fredrickson, et al., *Science*, 322, 429 (2008).
189. J.W. Park, J. Cho, and E.L. Thomas, *Science*, 321, 939 (2008).
190. R. Ruiz, H. Kang, F.A. Detcheverry, et al., *Science*, 321, 936 (2008).
191. C.C. Chao, T.C. Wang, R.M. Ho, et al., *ACS Nano*, 4, 2088 (2010).

# CHAPTER 8

# REACTIVE SYSTEMS AND THERMOPLASTIC VULCANIZATES

PHILIPPE CASSAGNAU,[1] GRÉGORY MARTIN,[2] and CLAIRE BARRÈS[3]
[1] Université de Lyon, Ingénierie des Matériaux Polymères, Villeurbanne, France
[2] Hutchinson S.A., Centre de Recherche, Chalette-sur-Loing, France
[3] Université de Lyon, INSA de Lyon, Ingénierie des Matériaux Polymères, Villeurbanne Cedex, France

## CONTENTS



*Applied Polymer Rheology: Polymeric Fluids with Industrial Applications*, First Edition.
Edited by Marianna Kontopoulou.

## 8.1 INTRODUCTION

Why is rheological analysis so intimately associated to reactive processing? In the field of reactive processing, rheology can be viewed as the interface between chemistry and flow modeling [1]. Rheological behavior, and especially viscoelastic properties, are very sensitive to small variations of the molecular weight and can be used as a tool to study the kinetics of bulk polymerization processes, both for uncross-linked and cross-linked systems. In addition, rheology is generally viewed as a tool for identifing suitable processing conditions in reactive extrusion. On the other hand, modeling the changes of the viscoelastic properties during a chemical reaction is of great importance in the prediction of the flow behavior during melt processing, when the nature of flow is strongly associated with the change in properties induced by this reaction. Rheology is also a tool for investigating and controling the production of new materials involving a drastic modification of the viscoelastic properties during processing. Thermoplastic vulcanizates (TPVs), which are an example of such a reactive material, consist of a blend of two polymeric components, one of which undergoes cross-linking during processing.

By definition, the process of dynamic vulcanization involves cross-linking of an elastomer (the major phase) during its melt mixing with a thermoplastic (the minor phase). This most often results in a final TPV morphology where the elastomer phase is fully vulcanized and finely dispersed in the thermoplastic matrix. Actually, such thermoplastic–elastomer blends present some specific and very interesting properties, depending on their morphology, as they can combine the elasticity of the cross-linked elastomeric phase with the processability of the thermoplastic. Over the years, many studies have provided very useful information regarding the highly important correlation between this unique polymer blend morphology and the final mechanical properties. Thus the key to adjusting and optimizing TPV properties is to exhaustively control the morphology development. This can be quite complex, especially considering the drastic change of the viscoelastic properties of the elastomeric phase during processing, which ultimately affects its dispersion within the thermoplastic. Coran and Patel [2] showed that a large number of elastomers and thermoplastics can in theory be combined to produce different TPVs, but in practice the best chemical and physical properties for these materials have been obtained from ethylene-propylene-diene-monomer (EPDM) and polypropylene (PP) blends.

Obviously, the final morphology of such blends directly depends on the composition and compounding route, which clearly play a primary role on the final properties of TPVs [3–8]. The initial morphology of such blends before the cross-linking step is also a subject of great interest. It must be stable enough so that it can withstand the drastic increase of the elastomeric phase viscosity during cross-linking, which induces complex morphology evolution and stabilization mechanisms. In addition, the nature of the cross-linking agent and the extent of cross-linking of the elastomeric phase are important factors because they modify the structure, the viscosity, and the elasticity of the elastomeric phase and thus the final morphology and properties of the blends [9–20]. Dynamic cross-linking (cross-linking under processing conditions) can also induce phase inversion or at least phase co-continuity, which must be taken into consideration during the processing of such blends [21–24]. All these different aspects of phase morphology developments in TPVs have been recently reviewed by Radusch [25] and Naskar [26].

From this short and nonexhaustive literature review, it can be concluded that the morphology development in TPV processing is strongly related to the variation of the viscoelastic properties of the rubber phase upon cross-linking during processing. Thus the first section of this chapter focuses on the evolution of rheological properties that takes place during cross-linking of EPDM and other polyethylene-based polymers.

On the other hand, some questions in TPV processing are still open. For example: What is the deformation of a cross-linked droplet in a thermoplastic phase, which behaves as a viscous liquid? What about the balance of the coalescence and breakup mechanisms with the extent of the cross-linking reaction of one or both phases? And finally, can we suggest a general framework for TPV processing by studying the effects of blending combined with the cross-linking aspect? The second section will try to address these questions by describing (1) the deformation of cross-linked droplets in a model system, and (2) the dispersion mechanisms of a cross-linked phase during dynamic vulcanization. We will finally conclude with the rheology of industrially relevant TPVs.

## 8.2 VISCOELASTIC PROPERTIES OF RUBBERS

### 8.2.1 Variation of Rheological Properties during Cross-linking

Linear viscoelasticity has been extensively used to characterize the kinetics of formation of a polymer network and more specifically to determine the sol-gel transition. The sol-gel phase transition or gel point occurs during a random process of subunit association into larger and larger molecules. Clusters of polymer will be formed as the cross-linking reaction progresses. A complete review of the rheology of systems undergoing liquid–solid transition was published by Winter and Mours [27]. Nevertheless, their approach for identifying the transition cannot be applied to elastomers such as EPDM. Actually EPDMs, even when plasticized, do not show a well-defined terminal relaxation zone, and the criterion for gelation cannot be applied. Although EPDMs cannot be described in the framework of Winter and Mours, the evolution of the viscoelastic properties of some other polyethylene-based peroxide-cured systems has been studied [28–31]. The relationships between rheology and reaction kinetics derived from these publications are useful toward TPV understanding and development.

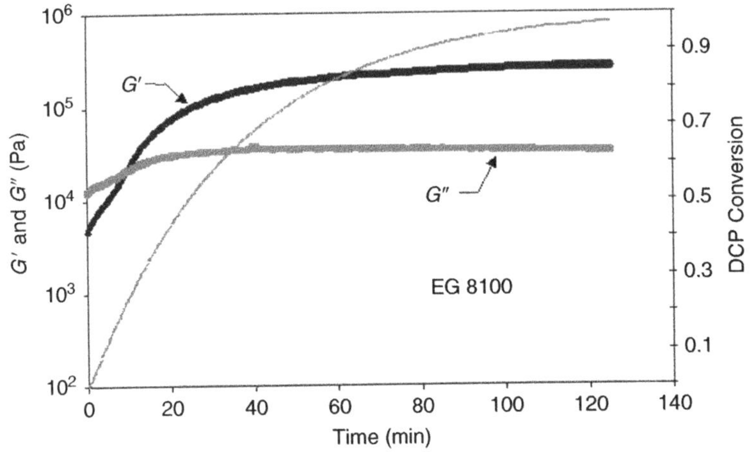


**FIGURE 8.1** Variation of the complex shear modulus and DCP conversion versus time, at $T = 145°C$, $\omega = 1$ rad/s, dicumyl peroxide = 1 wt % during static cross-linking of an ethylene–octene copolymer (Engage™ 8100 by Dow Chemical). Reprinted with permission from Ref. 32.

When the cross-linking reaction occurs, the molecular chain mobility is strongly affected and the storage and loss moduli increase with time (Fig. 8.1). Assuming self-similarity of the relaxation time spectrum ($G(t)\ \alpha\ t^{-n}$) for a material at the sol-gel transition, the frequency dependence of the shear moduli can then be obtained [27]. $G'$ and $G''$ are then described by the same frequency dependence ($G'(\omega) \propto G''(\omega) \propto \omega^n$) near the percolation threshold, where $n$ is the relaxation exponent. Moreover, the loss tangent is frequency-independent and given by $\tan\delta = \frac{G''(\omega)}{G'(\omega)} = \tan\left(\frac{n\pi}{2}\right)$, which represents the necessary condition for the occurrence of a gel point in the system. It was shown [27] that the gel point of a cross-linking polymer coincides with the $G'$-$G''$ crossover only if the power law on both storage and loss moduli followed the power law $G' \propto G'' \propto \omega^n$ with $n = 1/2$. Such a result was observed by Cassagnau et al. [28] and Espinasse et al. [29] for copolymers of ethylene and vinyl acetate (EVA) cross-linked by exchange reactions. However, Scanlan and Hicks [30] and more recently Li and Kontopoulou [31] observed a critical exponent close to $n \approx 0.3$ for ethylene copolymers, cross-linked by radical chemistry. We can conclude from these results that the network formation at the gel point depends on the chemical route used to cross-link the polymer chains. This dependence was also observed for the recovery of elastic properties (compression set experiments) of cross-linked EVA [33].

The choice of conditions, i.e. dynamic (inside the chamber of a batch mixer for instance) versus static is also crucial. Original results obtained recently by Msakni et al. [32] for copolymers of ethylene and octene cross-linked by radical chemistry showed that the dynamically cross-linked samples were reprocessable, which means that the final microstructure was totally different from that obtained under static conditions. Indeed, the rheological behavior of the dynamically cross-linked samples was characteristic of materials with complex molecular structures. Furthermore, these samples were soluble, and the size exclusion chromatography analysis revealed a broad molecular weight distribution with a significant tail of high molecular weights situated around $10^6$ g/mol. This melt can be imagined as a continuum of multiscale clusters whose sizes range from that of a free precursor chain to larger clusters of high molecular weights. The balance between reaction and mixing efficiency is the controlling parameter for the development of these microstructures under dynamic conditions.

The same conclusions were drawn by Fortelny and Krulis [34] with PP containing dynamically cross-linked EPDM blends, in which they observed a significant drop in the loss tangent (at 0.1 rad · $s^{-1}$) between 20 and 30 wt % of EPDM. This may be interpreted as the occurrence of percolation of the cross-linked EPDM domains, but solvent extraction experiments showed that the gels contained in those blends disintegrated within days. It was thus proposed that long-living entanglements between cross-linked inclusions were responsible for both the rheological and dissolution results. The rheological behavior and reprocessability of dynamically cross-linked blends may therefore be explained in the same manner, because a continuous dynamically cross-linked phase apparently consists of entangled finite cross-linked domains.

### 8.2.2 Viscoelastic Properties of Rubber Networks

When the cross-linking reaction is complete, the traditional methods for analyzing rubber networks are equilibrium swelling measurements, mechanical testing and finally linear viscoelasticity. The analysis of such properties has been used extensively to demonstrate the existence of different network features and has enabled the development of the first theoretical models like the affine model [35]. According to this model the shear equilibrium modulus is given by:

$$G_e^{\text{affine}} = \nu RT = \frac{\rho RT}{\overline{M}_c} \tag{8.1}$$

where $\nu$ is the density of elastically active chains, $\rho$ the density, R the gas constant, $T$ the absolute viscosity and $\overline{M}_c$ the average molar mass of these chains. An elastic chain is defined as a chain attached to the network at each of its two ends.

The phenomenological model developed by Langley [36] and Dossin and Graessley [37] involves an additional term accounting for the topological contributions. According to the entanglement interpretation of the topological contributions, a portion of the restrictions on configurational rearrangements of macromolecules becomes permanently trapped when a network is formed and therefore is able to contribute to the equilibrium elasticity. The modulus can be expressed as:

$$G_e^{\text{Langley-Graessley}} = (\nu - h\mu)RT + G_e^{\max} T_e \tag{8.2}$$

where $h$ is an empirical parameter between 0 and 1; $T_e$ is the fraction of trapped entanglements in the network—i.e. the fraction of the maximum concentration of topological interactions that contributes to the modulus; and $G_e^{\max}$ is the maximum topological contribution. It is expected to be very close to $G_N^0$, the plateau modulus of the uncross-linked, high molar mass polymer.

Upon review of the relevant literature it is striking that, despite the intensive use of EPDM in industrial applications, its viscoelastic analysis is not very developed. Actually, EPDM is a complex polymer whose microstructure has often been said to contain "microgels", i.e., microcrystalline regions, that make rheological analysis even more complicated. Furthermore, most of the time EPDM networks also contain a large amount of plasticizer and are thus swollen by the extractable component, which acts as a solvent. This is not taken into account unless corrections are introduced for the plasticizer into both contributions to the modulus. Consequently, on one hand, the network contribution $(\nu - h\mu)$ $RT$ must be corrected by $\theta^{1/3}$, where $\theta$ is the polymer volume fraction of the compound. On the other hand, the entanglement contribution is also altered by the presence of oil, and in that case the effect is that of a solvent in an entangled, uncross-linked polymer [38]. The terminal viscoelastic parameters of EPDM in the presence of plasticizer were derived from the bulk polymer ones with a correction based on the free volume theory. This correction is of the form $\theta^{2.25}$ as shown in different polymer systems [39, 40], in agreement with the groundbreaking theoretical work by Daoud et al. [41]. The shear modulus of the network in presence of a plasticizer can be finally expressed as:

$$G_e = (\nu - h\mu)RT\theta^{\frac{1}{3}} + G_e^{\max} T_e \theta^{2.25} \tag{8.3}$$

The parameters $\nu$, $\mu$, and $T_e$ have been calculated (Table 8.1) for cross-linked EPDM samples [42] using the theoretical relations established by Pearson and Graessley [43, 44].

Table 8.1 displays the variation of the major structural parameters—namely, $\nu$ and $T_e$—with increasing concentration of a cross-linker (resol), expressed in phr. Logically, the cross-link density, $\nu$, and the probability of trapped entanglements, $T_e$, which is linked to the cross-link density, increase with the cross-linker fraction. Simultaneously, the probability of forming dangling chains decreases. However, whereas $\nu$ follows an almost linear variation with the cross-linking agent fraction, $T_e$ exhibits a dramatic increase with resol phr while it remains lower than about a fifth of the reference concentration. This was expected, considering that beyond a certain network density, the long range chain mobility is highly hindered (60% of trapped entanglements) and consequently the possibility for a chain to "escape" from getting involved in cross-links with other chains largely decreases. Furthermore, this study showed that the $\nu/\mu$ ratio increases with increasing the curing agent concentration and tends

TABLE 8.1 Properties of Plasticized EPDM at Different Extents of cross-linking[a]

| Samples | % w/w of resol | Gel fraction | Tan $\delta$ | $\mu$ (mol.m$^{-3}$) | $\nu$ (mol.m$^{-3}$) | $G_e$ (Pa) | *Te* | $E_\infty$ (Pa) | $\tau_0$ (min) | *m* |
|---|---|---|---|---|---|---|---|---|---|---|
| **REF** | 2.43 | 0.998 | 0.01 | 64 | 124 | $3.3 \times 10^5$ | 0.85 | $6.3\ 10^5$ | 0.09 | 0.067 |
| **RES1/4** | 0.62 | 0.989 | 0.07 | 21 | 39 | $9.8 \times 10^4$ | 0.63 | $2.6\ 10^5$ | 13 | 0.070 |
| **RES1/7** | 0.35 | 0.962 | 0.12 | 9.6 | 17 | $5.8 \times 10^4$ | 0.42 | $1.2\ 10^5$ | 104 | 0.170 |
| **RES1/10** | 0.25 | 0.896 | 0.18 | 3.90 | 6.5 | $3.4 \times 10^4$ | 0.21 | $7.0\ 10^4$ | 512 | 0.184 |
| **RES1/30** | 0.08 | 0.700 | 0.40 | 0.80 | 1.25 | $2.0 \times 10^3$ | 0.04 | $4.0\ 10^3$ | 8000 | 0.348 |

[a] $\tan \delta = G''/G'$ is the loss tangent at equilibrium, $\mu$ is the density of chemical cross-linking bridges, $G_e$ is the equilibrium elastic modulus. $E_\infty$, $\tau_0$, and $m$ are the fitted parameters of the Chasset–Thirion equation (8.4) for the shear stress relaxation of plasticized EPDM networks (strain = −0.25). Reprinted with permission from Ref. 42.

toward 2, which is the theoretical value for a perfect tetrafunctional network. This is an indication of the network evolving toward a more and more perfect structure with increasing cross-link density and concomitantly decreasing dangling chain occurrence.

On the other hand as far as industrial applications are concerned, the relaxation mechanisms in the long time limit are of great importance because they define the elastic properties and creep behavior of cross-linked elastomers. Among numerous other experimental procedures, the compression set test is commonly carried out. It reflects the elastic recovery properties of elastomeric materials. Discrepancies between theory and experiment still remain in the long time relaxation range and the role of chain entanglements, network defects such as dangling chains, and network heterogeneities, is certainly a key issue. Experimentally, Chasset and Thirion [45] postulated in the 1960s that an excellent representation of the isothermal relaxation modulus data at long times, $t$, for many rubbers can be given by:

$$E(t) = E_{\infty}\left[1 + \left(\frac{t}{\tau_0}\right)^{-m}\right] \tag{8.4}$$

where $E_{\infty}$ is the equilibrium modulus, and $m$ and $\tau_0$ are material parameters.

Ferry [46] suggested that the molecular mechanism responsible for the long time relaxation process is the diffusion of dangling chain ends in the presence of entanglements. The introduction of polymer dynamics concepts allowed some progress in the molecular theories developed for such vulcanizates. A molecular interpretation of the long-time relaxation of elastomers, based on results by De Gennes [47] for the reptation of a single branched chain with topological constraints, was first given by Curro and Pincus [48]. Their model is based on the retraction of pendant chains in a cross-linked network with topological constraints (entanglements). It predicts that a polymer network containing a random distribution of dangling chain ends leads to a relaxation modulus having a power law dependence with time [48, 49], as in the phenomenological Chasset and Thirion equation [45]. The theory by Curro and Pincus predicts that the model parameter, $m$, is proportional to the cross-link density, which was found to be in agreement with some experimental data on natural rubber. Furthermore, we demonstrated in our previous work [32] that the elastic properties after prolonged action of compression stresses strongly depend on the microstructure of the network for an identical cross-linking density. As EVA has many more $CH_2$ potential reactive sites than CH-$COOCH_3$ potential sites, the elastic recovery of EVA cross-linked by random radical reaction is better than the elastic recovery of samples cross-linked by the random exchange reaction of cross-linking. This defines a probability $P(n)$ of having a dangling end of $n$ units lower than $x_o$ in a radical process ($x_o$ is the molar fraction: [CH-$(COOH)_3$]/[$CH_2$]) compared to the cross-linking process by exchange reaction.

The characteristic parameters $E_{\infty}$, $\tau_0$, and $m$ of the Chasset-Thirion equation (8.4) were determined for each EPDM specimen shown in Table 8.1 [42] using

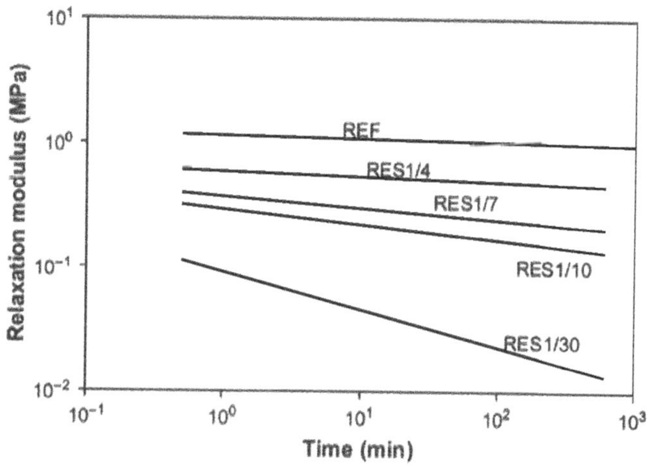


**FIGURE 8.2** Variation of the relaxation modulus of the samples shown in Table 1 measured over 10 hours at 100°C, under 0.25 compressive strain. Reprinted with permission from Ref. 42.

the relaxation modulus data of Figure 8.2. The variation of these parameters with the cross-linking density is very consistent with the assumed role of the dangling chains. As shown in the work by Curro and Pincus, which refers to natural rubber with zero soluble fraction, the relaxation of a dangling chain (i.e. a chain linked to the network by only one of its extremities) is assimilated to the reptation of star-like molecules, whose relaxation times evolve exponentially with their molar mass. So, from a qualitative point of view, our results demonstrate that for imperfect networks such as our lightly cross-linked samples, the long time relaxation mechanisms are largely emphasized, supposedly due to the presence of long dangling chain ends. However, the exponent m from Chasset-Thirion equation cannot be compared to published data, and the analysis of the cross-link dependence of $m$ cannot be carried out with the Curro and Pincus approach. Indeed, the decrease of $m$ when cross-link density increases is opposite to the proportional relation derived by these authors in their theory. The major reason for this discrepancy is certainly that EPDM networks, which contain both plasticizer and soluble polymer fractions, do not fit into the assumptions of the Curro and Pincus theory.

## 8.3 MORPHOLOGY DEVELOPMENT

Generally speaking, the final morphology of an immiscible polymer blend depends on the type of flow, the blend composition, interfacial tension and viscoelasticity of the components. Taylor [50] and Grace [51] defined two dimensionless parameters, the capillary number and the viscosity ratio, shown in (8.5) and (8.6), respectively, that enable the prediction of morphology in the melt state.

$$Ca = \frac{\sigma R}{\alpha} \tag{8.5}$$

where $\sigma$ represents the shear stress, $R$ the droplet radius and $\alpha$ the interfacial tension of the blend.

$$p = \frac{\eta_d}{\eta_m} \tag{8.6}$$

where $\eta_d$ and $\eta_m$ are the viscosity of the dispersed phase and the viscosity of the matrix, respectively. A critical value of the capillary number exists, above which breakup of the deformed droplets of the dispersed phase is expected. However, it has been demonstrated that, in shear flow, no breakup of the deformed droplets can take place for blends with $p > 4$. In this case, the critical capillary number remains constant. For capillary numbers under the critical value, a stable ellipsoidal shape of the droplet is obtained. In most of studies discussed in the literature the blend components used are generally model, Newtonian fluids, whereas in usual processing conditions, polymers are viscoelastic fluids. It is therefore necessary to consider the effect of elastic forces on the droplet deformation. Some authors have been reported the effects of elasticity on morphology development. Mighri et al. [52, 53] and Lerdwijitjarud et al. [54] demonstrated the influence of elasticity on the critical capillary number using Boger model fluids. They demonstrated that matrix elasticity helps the droplets deformation, whereas droplet elasticity resists the droplet deformation. Nevertheless limited results have been reported on deformation of a cross-linked droplet, i.e., a droplet with strong permanent elasticity. In this case, it is expected that the elastic properties will overcome the effect of viscosity ratio on the blend morphology. Before engaging on a discussion on actual systems, it is useful to present results obtained for model systems, under well-controlled conditions.

### 8.3.1 Model System

***8.3.1.1 Shear Deformation of a Cross-linked Droplet*** In one of our previous studies [55] we showed the influence of different levels of permanent elasticity of the dispersed phase on droplet deformation in a PDMS/EVA blend (EVA phase is the cross-linked one). The EVA cross-link density (i.e., the permanent elasticity) was inferred from $\tan\delta = G''/G'$ as this parameter is representative of the cross-linking reaction progress. At the gel point of the EVA network, we assumed that $\tan\delta = 1$ so that below the gel point $\delta > 1$ (no cross-linked EVA: $\tan\delta \approx 77$ at $\omega = 0.1\ \mathrm{rad\,s^{-1}}$) and $\tan\delta < 1$ beyond the gel point. Results of droplet deformation ($\lambda_{d\text{-}1}$; see Ref. 55 for the definition of the deformation in the case of droplet), including droplet images, are presented in Figure 8.3 as a function of strain amplitude at various extents of cross-linking. As one might expect, droplet deformation decreases with increasing extent of

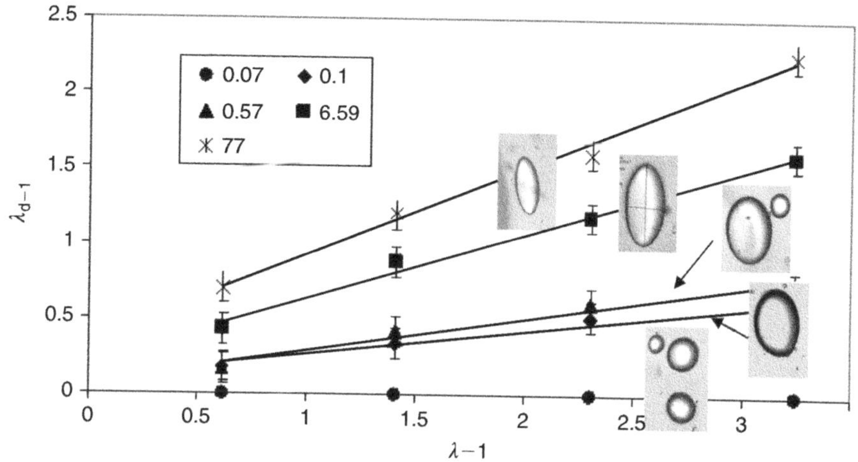


**FIGURE 8.3** Droplet deformation versus matrix deformation as a function of the extent of cross-linking. Pure EVA (tan $\delta = 77$) deformation has been added for comparison purposes. Values in the legend correspond to the value of tan $\delta$ of the cross-linked phase. The pictures correspond to an applied strain amplitude $(\lambda - 1) = 1.41$.

cross-linking since both the elasticity and viscosity of the dispersed phase increase with the level of cross-linking. In addition, it is observed that droplet deformation increases linearly as a function of the applied deformation ($\lambda < 3.5$) regardless the extent of the cross-linking. It can also be seen in Figure 8.4 that whatever the strain amplitude is, the resulting normalized droplet deformation is the same. This observation is valid for each extent of cross-linking. Furthermore as shown in Figure 8.4, if one considers the normalized droplet deformation $(\lambda_d - 1)/(\lambda - 1)$ as a function of tan $\delta$, no major effect on droplet deformation at the gel point can be seen. On the other hand, deformation of cross-linked droplets is possible when tan $\delta$ is held between 0.57 and 0.1. Finally, a drastic decrease in droplet deformation is observed for tan $\delta \leq 0.07$, which corresponds to an insoluble polymer fraction $> 0.8$. Beyond this value, droplet deformation cannot be measured optically, regardless of the applied strain. However in spite the fact that the droplet deformation cannot be measured in this simple flow field, the droplets might still deform under real processing conditions.

***8.3.1.2 Droplet Relaxation*** Following a deformation experiment, such as the one described above, the deformed droplet relaxes into a spherical equilibrium shape. According to Luciani and co-workers [56], the relaxation time of the deformed droplets can be calculated with the following equation:

$$\tau_d = \frac{R_0 \eta_m}{\alpha} \frac{(2p+3)(19p+16)}{40(p+1)} \tag{8.7}$$

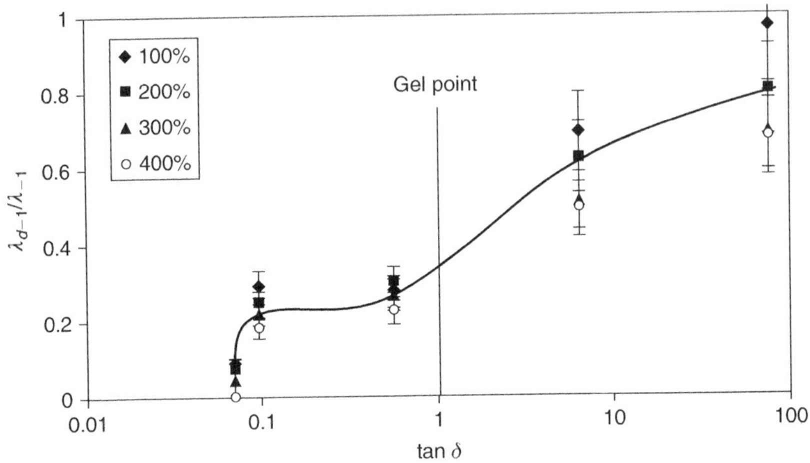


**FIGURE 8.4** Normalized droplet deformation as a function of tan $\delta$ at different strains (100, 200, 300 and 400). Pure EVA deformation has been added for comparison purposes. The vertical line at tan $\delta = 1$ corresponds to the gel point. The dashed vertical bar signifies the optical measurement accuracy. Reprinted with permission from Ref. 55.

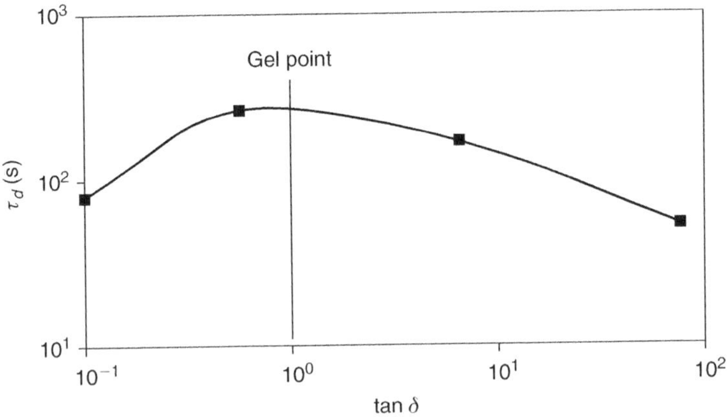


**FIGURE 8.5** Relaxation time of cross-linked EVA droplets as a function of tan $\delta$. Reprinted with permission from Ref. 55.

For uncross-linked EVA droplets, the theoretical relaxation time is $\tau_d = 56$ s. From the deformation experiments carried out on pure EVA droplets, the experimental relaxation time was found to be approximately 52 s, which is close to theoretical value. Furthermore assuming that the interfacial tension of the blend does not change with cross-linking, the relaxation $\tau_d$ should increase as the viscosity ratio increases, according to (8.7). As expected, and as can be seen in Figure 8.5, the relaxation time of the cross-linked droplets increases linearly as tan $\delta$ decreases. Beyond the gel point (tan $\delta = 0.57$), the experimental relaxation time decreases very rapidly. Below this value of tan $\delta$, as the cross-linked droplets no longer deform, it is not possible to perform relaxation experiments. The sudden fall in relaxation time of deformed cross-linked droplets above the gel point can be explained by the effect of elastic forces that combine with the interfacial forces to make the droplet recover a spherical shape. Indeed, elastic forces generated by the deformed droplet tend to cause it to retract, and as a consequence the droplet recovery becomes faster. It is interesting to note that the maximum relaxation time corresponds to the gel point.

***8.3.1.3 Morphology Control*** From the previous paragraphs the following question arises: Can we achieve a desired morphology, either nodular or fibrillar, by controlling the extent of cross-linking of the dispersed phase?

In the PDMS–EVA example discussed above, the capillary number of PDMS–nonreactive EVA blends was found to be in the range $2.6 \leq Ca \leq 5.2$ [55]. Considering the viscosity ratio of the PDMS–EVA blend before the cross-linking reaction, the critical capillary number was found to be $Ca_{cr} \approx 0.47$. The cross-linking of the dispersed phase drastically modifies the value of the critical capillary number, which quickly exceeds the critical value of 4. From a theoretical point of view, EVA fibers that have been cross-linked beyond the gel point cannot break. Consequently we can assume that under steady shear flow, substantial deformation of the EVA particles is expected as long as the gel point has not been reached. According to Huneault et al. [57], theoretical breakup times of EVA fibers under shear and quiescent conditions can be calculated. A steady shear flow at 0.1 $s^{-1}$ over 120 s generated fibers having an average diameter of 10 $\mu$m; the theoretical breakup time in quiescent conditions was about 7 min. Such breakup time is too short compared to the cross-linking kinetics, as a curing time of 1 h is required to obtain a high degree of cross-linking. Therefore, it is necessary to create the EVA fibers during the curing process and maintain the shear rate at sufficiently low levels to prevent the fibers from breaking up during the shearing process. However, the deformation of the EVA phase must be substantial since cross-linking gradually limits it with curing time. For instance, a shear treatment of 0.05 $s^{-1}$ during the 1-h curing period at 160°C leads to a stable fibrillar morphology, as shown in Figure 8.6a. It is important to note that the cross-linking level must be sufficient to fix the fibrillar morphology, which means that tan $\delta$ must be $<0.1$. In fact it was experimentally observed that cross-linked EVA fibers can break into rather spherical droplets in spite of a cross-linking density corresponding to tan $\delta = 0.57$.

On the other hand, a stable nodular morphology can be obtained in the same manner. Actually a nodular morphology can be generated (Fig. 8.6b) by choosing an appropriate shear rate allowing short breakup times of the EVA fibers. High shear rates lead to fine nodular morphology, whereas low shear rates lead to coarse nodular morphology. Knowing that the cross-linking

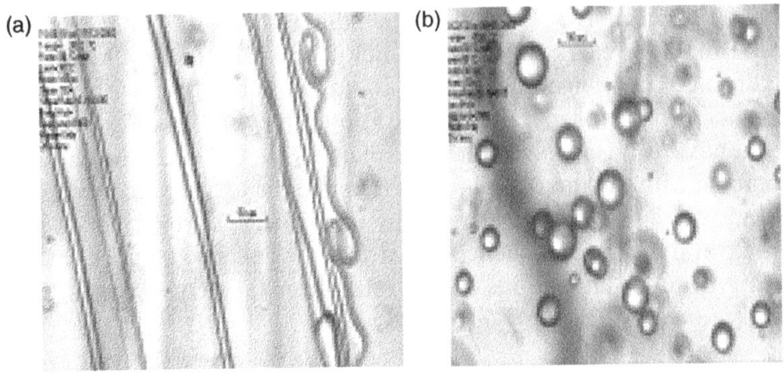

**FIGURE 8.6** Morphology control through cross-linking of the dispersed phase. The fibrils and droplets are stable under shearing. Reprinted with permission from Ref. 55.

kinetics are much slower than the breakup of the EVA fibers under shear flow, a fixed nodular morphology is easily generated.

To conclude, it has been demonstrated that the deformation of a cross-linked droplet decreases when the cross-linking level increases. However, no significant effect was noted near the gel point. Furthermore it has been shown that deformation of a cross-linked droplet is possible just above the gel point. Actually, a drastic decrease in droplet deformation is observed only for $\tan\delta \leq 0.07$, which corresponds to an insoluble polymer fraction $>0.85$. Finally, a fixed fibrillar morphology can be obtained by adjusting the shear rate so that the cross-linking reaction occurs before the breakup of the dispersed phase fibers. In the same manner, a fixed nodular morphology can be generated if fiber relaxation occurs before the cross-linking level becomes too high. In other words, the morphology of the dispersed phase can be controlled through the cross-linking extent, which depends on the processing control. A dimensionless number can be defined as the ratio between the time cross-linking, $t_c$, and the time of fiber breakup, $t_b$. The time of cross-linking is the time to reach a gel fraction $>0.8$.

$$\lambda_{DC} = \frac{t_c}{t_b} \tag{8.8}$$

If $\lambda_{DC} \ll 1$, a fibrillar morphology is expected. On the contrary, if $\lambda_{DC} \gg 1$, a nodular morphology will be obtained. At $\lambda_{DC} \approx 1$ the morphology will not be well defined and generally a coarse morphology is observed.

### 8.3.2 Application to Industrial Systems

Ethylene-propylene-diene (EPDM)–polypropylene (PP) blends are surely the most extensively used elastomer–thermoplastic blends on the market. The reasons for their impressive processing and mechanical properties have been the subject of great interest, having many relevant consequences on the industrial production of thermoplastic vulcanizates (TPVs) [2, 24, 58, 59].

As cited in the introductory part, numerous studies have been performed on the morphology development of TPVs. However, due to the complexity of the involved mechanisms that are, indeed, tightly linked to the synthesis of any TPV, it is of importance to separate the effect of blending from the actual chemical cross-linking aspect. For that purpose, we studied in a previous publication [60], the impact of shearing on both the thermoplastic and the precross-linked EPDM phases (cross-linked EPDM samples described in Table 8.1). More precisely, the main morphologies of blends consisting of PP and pre-cross-linked EPDMs having various cross-link densities, as well as with uncross-linked and dynamically cross-linked EPDM were investigated.

Under our experimental conditions of shearing in a laboratory batch mixer, it appeared possible to fragment and homogeneously disperse EPDM networks having a gel fraction as high as 0.7 into a thermoplastic matrix, whatever the proportion of the PP–EPDM phases (80/20 or 30/70). Above this critical value, the precross-linked EPDM networks were not dispersed efficiently into fine particles (see morphologies in Fig. 8.7c–f) leading to coarse morphologies with remaining fragments that were quite large (order of magnitude of 200 $\mu$m with a broad distribution). These acted as strong imperfections during mechanical property testing. Wu et al. [61] came to the same conclusion in the case of dynamically vulcanized chlorinated butyl rubber–polyethylacrylate TPV as they observed a drastic modification of the extrudate surface for gel fractions $>0.8$.

As far as the mechanisms of morphology development are concerned, although the Rayleigh–Taylor instabilities can be involved in such immiscible blend morphology development, another dominant mechanism such as erosion has to be taken into account, depending on the EPDM cross-link density and PP–EPDM ratio. Actually, two mechanisms of erosion are mainly observed for EPDM, depending on gel fraction. For cross-linking density corresponding to gel fractions $g_{EPDM} < 0.7$, a particle collision-coalescence-separation type of erosion proposed by Bhadane et al. [62, 63] can be assumed. However, we suppose from the observed mechanical properties that submicron size of the dispersed phase cannot be achieved from precross-linked EPDM even at low cross-link density. At higher cross-link density ($g_{EPDM} > 0.7$) another mechanism of erosion, which is time dependent, can be suggested. This mechanism is only pure mechanical erosion by stress breakup of EPDM domains. At low concentration of the EPDM phase (20%), the stress is applied through the viscosity of the molten PP while at higher concentration of EPDM (70%) pure mechanical breaking (grinding process) induces a rough but effective breaking up of the EPDM fragments.

Actually these results are in agreement with our previous work [55] discussed earlier, which showed that droplet deformation and relaxation of a cross-linked sample drastically changed and vanished at a gel fraction around 0.8. Such qualitative results have been also reported for other systems; e.g., Fenouillot and Perier-Camby [64] observed that, for a thermoplastic–thermoset blend with

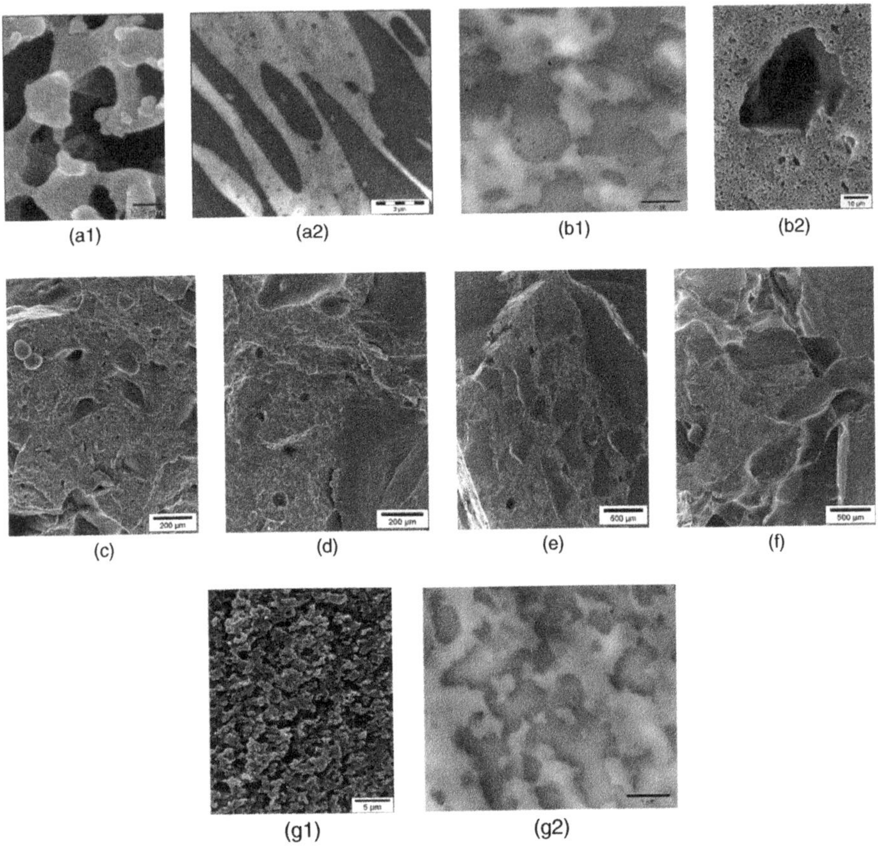


**FIGURE 8.7** SEM and TEM micrographs of PP–EPDM blends (30/70) after 10 min of mixing. $g_{EPDM}$ is the gel fraction of the EPDM phase. The EPDM phase was selectivly etched for SEM analysis. As the EPDM phase was stained, it appears dark in TEM analysis. (a1) SEM of uncross-linked blend; (a2) TEM of uncross-linked blend; (b1) TEM of blend containing EPDM:RES1/30 – $g_{EPDM} = 0.7$; (b2) TEM of blend containing EPDM sample RES1/30 – $g_{EPDM} = 0.7$; (c) blend containing EPDM sample REF1/10, $g_{EPDM} = 0.896$; (d) blend containing EPDM sample REF1/7, $g_{EPDM} = 0.962$; (e) blend containing EPDM sample REF1/4, $g_{EPDM} = 0.989$, (f) blend containing EPDM sample REF, $g_{EPDM} = 0.998$); (g1) SEM of dynamically cross-linked blend ($g_{EPDM} \sim 1$); and (g2) TEM of dynamically cross-linked blend. Reprinted with permission from Ref. 60.

a gel fraction between 0.4 and 0.7, it was possible to make long and thin elastomeric fibers, which could potentially break up into droplets, within the thermoplastic matrix. Earlier, De Loor et al. [65, 66] showed that the equilibrium between breakup and coalescence of an in situ cross-linked dispersed phase was shifted toward coalescence, whereas at a gel fraction of the dispersed phase >0.6 the classical mechanisms were largely modified and replaced by a "coagulation" mechanism. Furthermore, it was shown that, for the fully cross-linked dispersed phase, the final morphology remained remarkably stable during successive extrusion steps. To sum up, all the results of the literature agree with the same order of magnitude of the gel fraction of the dispersed phase ($g \sim 0.7$) as the critical value for drastic modification of breakup and coalescence mechanisms.

Finally, the cross-linking of the EPDM phase in molten PP under mixing (i.e., dynamic cross-linking) leads to a very fine and homogeneous morphology with a complex and cohesive EPDM structure. The advantage of cross-linking under shearing, compared to the fragmentation and dispersion of precross-linked EPDM in presence of a thermoplastic phase, appears clearly vital in the synthesis of cohesive TPVs. From our results it can be concluded that the cross-linking time (time needed to reach $g_{EPDM} = 0.7$) must be longer than the time of mixing (time necessary to get the morphology by a type of erosion). Furthermore, it appears from this study and conclusions that a gradient of cross-linking density is required to have the best mechanical properties.

This discussion leads to the following question: Why do dynamically cross-linked PP–EPDM blends (TPVs) have so good and specific mechanical properties? As generally admitted and observed, the time for the cross-linking reaction of the EPDM phase must be longer than the characteristic time of mixing. According to the present results, the time of cross-linking is the time necessary to reach a gel fraction of the EPDM phase close to 0.7 and the time of mixing is the time needed to obtain the morphology observed in uncross-linked blends. Consequently, the best route from a processing point of view should be to add the cross-linking system just after the blending stage of both thermoplastic and elastomeric phases. However, this conclusion is a simplified view of the complexity of the morphology development during TPV synthesis. Furthermore, Litvinov [67] based on proton NMR relaxation experiments suggested that the polypropylene-rich chain fragments of EPDM and the amorphous PP phase form a thin interfacial layer, which is the source of physical junctions (or entanglements) at the EPDM–PP interface. It is then clear that such an interface cannot be reached when precross-linked EPDM phase is used, as it yields elastic domains without any mobility of the EPDM chains. However, such behavior can be expected for low density cross-linking ($g_{EPDM} < 0.7$) of the EPDM phase as some chains and more precisely the shortest ones are free to reptate to the interface.

## 8.4 TPV RHEOLOGY

Generally speaking, the rheological properties of TPVs are of importance for both melt processing and shaping operations. However, the rheological behavior of TPVs depends on numerous interrelated parameters, such as type of constituent polymers, phase concentration, compatibilizers, and morphology modifications through the cross-linking step. Cross-linking inevitably leads to a drastic increase of viscosity of the elastomeric phase upon reaction, inducing a complex viscosity evolution and stabilization to the final products. In addition, the nature of the cross-linking agent and the extent of reaction in the elastomeric phase are very important as they modify the structure, the viscosity and the elasticity of the elastomeric phase, thus the blend morphology and therefore the final rheology. Finally, the effect of the proportion of plasticizer and its diffusion within each phase are just as important. Paradoxically, despite this high complexity, only few research studies have been reported on the rheology of TPVs. Radhesh Kumar et al. [68] and Rajesh Babu et al. [69] reviewed in the introductory part of their papers the main factors contributing to the melt viscoelastic properties of TPVs.

It is surprising that the shear rheological behavior of TPVs can be viewed as analogous to those of filled polymers composites. From a linear viscoelasticity point of view, TPVs can be described as viscoelastic solids, since the storage modulus is larger than the loss modulus over the entire frequency range. TPVs exhibit a pseudo-plastic (shear thinning) behavior with an apparent yield stress, due to a network structure of the dispersed cross-linked domains. As a result, the zero shear viscosity tends to infinity. Babu et al. [69] postulated a dependence of viscoelastic properties on the particle size and interactions, whereas Li and Kontopoulou [31] proved that the rheological behavior depends also on the cross-link density of the rubber phase. Generally, the rheological behavior of TPVs can be well fitted by a simple power law of the absolute complex viscosity:

$$|\eta^*| = k\omega^n \tag{8.9}$$

where $k$ is the consistency index and $n$ is the shear thinning index. However, similarly as composite polymers, TPVs do not obey the Cox-Merz rule [70]. Furthermore, the non-linearity region (Payne effect) of TPV samples is shorter than that of uncross-linked blends [71, 72]. Through strain sweep experiments, Rajesh Babu [69] showed that the onset of nonlinearity shifted to lower deformation, whereas the initial storage modulus was regained to large extent. Finally, Wu et al. [61] showed that the formation of a co-continuous phase in TPV processing results in periodic distortions and melt fracture at the die exit of the extruder, making it difficult to obtain a smooth extrudate surface. These flow instabilities were also observed, depending on processing conditions, such as temperature and residence time. With increasing extrusion temperatures the surface of the extrudates becomes gradually smooth. The effect of temperature on nonlinear properties was discussed by Leblanc [71] from large amplitude harmonic experiments. He concluded that oil-free thermoplastic regions (PP in this study) become possible at high temperature in the hardest TPV grades. Consequently, the viscoelastic behavior of these TPV reflects more the role of the thermoplastic phase, i.e., larger linear domain.

In conclusion, only general trends can be deduced on the rheology of TPVs, as their formulations are complex and their material structure depends on numerous interrelated parameters, including processing operating conditions.

Moreover, the addition of fillers is an integral part of industrial TPV formulation. Our previous work showed that nano-fillers can play a dominant role in the morphology development in polymers blends [73–76]. On the other hand, the final properties of PP–EPDM co-continuous blends filled with fumed silica have shown amazing results especially regarding the relaxation and strain recovery behavior encountered during compression set experiments [77].

## 8.5 CONCLUSIONS

TPVs, and especially their rheology, are one of the most fascinating research topics in polymer blending. TPVs present some specific and very interesting properties depending on their morphology, as they can combine the elasticity of the cross-linked elastomeric phase with the processability of the thermoplastic. The key to adjusting and optimising their properties is to control their morphology development on in situ cross-linking of the rubber phase. Rheology is then a relevant technique to study such a complex system. The present chapter is aimed at illustrating the key issues in controlling the TPVs development, from the fundamental study of the deformation of a cross-linked droplet to the morphology development in real life system. It is very interesting that, as a main result, all the literature results agree with the same order of magnitude of the gel fraction of the dispersed phase ($g \sim 0.7$–$0.8$) as the critical value for drastic modification of break-up and coalescence mechanisms. Consequently, and as proved in real life systems, the cross-linking time (time needed to reach $g_{EPDM} = 0.7$–$0.8$) must be longer than the time of mixing (time necessary to get the morphology by an erosion mechanism).

## REFERENCES

1. P. Cassagnau, V. Bounor-Legaré, and F. Fenouillot, *Intern. Polym. Process.*, 22, 218 (2007).
2. A.Y. Coran and R.P. Patel., *Thermoplastic Elastomers Based on Dynamically Vulcanized Elastomer-Thermoplastic Blends in Thermoplastic Elastomers*, Munich, Vienna, New York, Hanser-Garner Publications, 1996.
3. C. Joubert, P. Cassagnau, A. Michel, and L. Choplin, *Polym. Eng. Sci.*, 42, 2222 (2002).
4. X. Wang, J. Sun, and R. Huang, *J. Appl. Polym. Sci.*, 42, 2268 (2006).
5. S. Shafei Sararudi, H. Nazockdast, and A.A. Katbab, *Rubber Chem. Technol.*, 77, 847 (2004).
6. E. Prut, N.A. Erina, J. Karger-Kocsis, and T.I. Medintseva, *J. Appl. Polym. Sci.*, 109, 1212 (2008).
7. M. Van Duin and A.V. Machado, *Polym. Degr. Stab.*, 90, 340 (2005).
8. A.V. Machado, M. Van Duin, and J.A. Covas, *Material Science Forum*, 514, 838 (2006).

9. M.D. Ellul, A.H. Tsou, and W. Hu, *Polymer*, 45, 3351 (2004).
10. E. Prut, T. Medintseva, and V. Dreval, *Macromol. Symp.*, 233, 78 (2006).
11. A.K. Jain, A.K. Nagpal, R. Singhal, and N.K. Gupta, *J. Appl. Polym. Sci.*, 78, 2089 (2000).
12. A. Mousa, *Int. J. Pol. Mat.*, 54, 619 (2005).
13. M. Hernandez, J. Gonzalez, C. Albano, et al., *Polym. Bull.*, 50, 205 (2003).
14. B. Ohm, R. Annicelli, T. Jablonowski, and R. Mazzeo, *Rubber World*, 226, 33 (2002).
15. Z. Li and M. Kontopoulou, *Polym. Eng. Sci.*, 49, 34 (2009).
16. W. Wang, Q. Wu, and B. Gu, *Polym. Eng. Sci.*, 43, 1798 (2003).
17. H.W. Xiao, S.Q. Huang, and T. Jiang, *J. Appl. Polym. Sci.*, 92, 357 (2004).
18. T. Zaharescu, R. Setnescu, S. Jipa, and T. Setnescu, *J. Appl. Polym. Sci.*, 77, 982 (2000).
19. T. Marinovic, Z. Susteric, I. Dimitrievski, and Z. Veksliz, *Kautschuk Gummi Kunststoffe*, 51, 189 (1998).
20. A. Verbois, P. Cassagnau, A. Michel, et al., *Polym. Intern.*, 53, 523 (2004).
21. N. Mekhilef and H. Verhoogt, *Polymer*, 37, 4069 (1996).
22. A.V. Machado and M. Van Duin, *Polymer*, 46, 6575 (2005).
23. C. Joubert, P. Cassagnau, A. Michel, and L. Choplin, *Polym. Eng. Sci.*, 42, 2222 (2002).
24. P.A. Bhadane, N. Virgilio, D.B. Favis, et al., *AIChE J.*, 52, 3411 (2006).
25. H.J. Radusch, in *Micro- and Nanostrutured Multiphase Polymer Blend Systems*, Taylor&Francis Group, Boca Raton London New York, 2006.
26. K. Naskar, *Rubber Chem. Technol*, 80, 504 (2007).
27. H.H. Winter and M. Mours, *Adv. Polym. Sci.*, 134, 135 (1997).
28. P. Cassagnau, V. Verney, M. Bert, and A. Michel, *Polymer*, 34(1), 124 (1993).
29. I. Espinasse, P. Cassagnau, M. Bert, and A. Michel, *J. Appl. Polym. Sci.*, 54, 2083 (1994).
30. J.C. Scanlan and M.J. Hicks, *Rheol. Acta*, 30(5), 412 (1991).
31. Z. Li and M. Kontopoulou, *Polym. Eng. Sci.*, 49, 34 (2009).
32. A. Msakni, P. Chaumont, and P. Cassagnau, *Polym. Eng. Sci.*, 46, 1531 (2006).
33. C. Joubert, A. Michel, L. Choplin, and P. Cassagnau, *J. Polym Sci, Part B: Polym Phys*, 41, 1779–1790 (2003).
34. Z. Krulis and I. Fortelny, *Eur. Polym J.*, 33, 513–518 (1997).
35. P.J. Flory, *Principles of Polymer Chemistry*, Cornell University Press, Ithaca, NY, (1953).
36. N.R. Langley, *Macromolecules*, 1, 368 (1968).
37. L.M. Dossin and W.W. Graessley, *Macromolecules*, 12, 123 (1979).
38. M. Ponsard-Fillette, C. Barrès, and P. Cassagnau, *Polymer*, 46, 10256 (2005).
39. G. Marin, E. Menezes, V.R. Raju, and W.W. Graessley, *Rheol. Acta*, 19, 462 (1980).
40. J. Gimenez, P. Cassagnau, and A. Michel, *J. Rheol*, 44, 527 (2000).
41. M. Daoud, J.P. Cotton, B. Farnoux, et al., *Macromolecules*, 8, 804 (1975).
42. G. Martin, C. Barres, P. Cassagnau, et al., *Polymer*, 49, 1892 (2008).

43. D.S. Pearson and W.W. Graessley, *Macromolecules*, 11, 528 (1978).
44. D.S. Pearson and W.W. Graessley, *Macromolecules*, 13, 1001 (1980).
45. R. Chasset and P. Thirion, in *Proceedings of the Conference on Physics of Non-Crystalline Solids*, North-Holland publishing company, Amsterdam, PAYS-BAS, 1965.
46. J.D. Ferry, *Viscoelastic properties of polymers*, 3rd ed. Wiley, New York, 1980.
47. P.G. De Gennes, *Scaling Concepts in Polymer Physics*, Cornell University Press, Ithaca, NY, 1979.
48. J.G. Curro and P. Pincus, *Macromolecules*, 16, 559 (1983).
49. J.G. Curro, D.S. Pearson, and E. Helfand, *Macromolecules*, 18, 1157 (1985).
50. G.I. Taylor, *Proc. R. Soc. London Ser. A*, 146, 501 (1934).
51. H.P. Grace, *Chem. Eng. Commun.*, 14, 225 (1982).
52. F. Mighri, A. Ajji, and P.J. Carreau, *J. Rheol.*, 41, 1183 (1997).
53. F. Mighri, P.J. Carreau, and A. Ajji, *J. Rheol.*, 42, 1477 (1998).
54. W. Lerdwijitjarud, A. Sirivat, and R.G. Larson, *Polym. Eng. Sci.*, 42, 798 (2002).
55. Y. Deyrail and P. Cassagnau, *J. Rheology*, 48, 505 (2004).
56. A. Luciani, M.F. Champagne, and L.A. Utracki, *J. Polym. Sci.*, 35, 1393 (1997).
57. M.A. Huneault, M.F. Champagne, and A. Luciani, *Polym. Eng. Sci.*, 36, 1694 (1996).
58. A.L. Da Silva, M. Tavares, D. Politano, et al., *J. Appl. Polym Sci.*, 66, 2005 (1997).
59. M. Van Duin, *Macrom. Symp.*, 233, 11 (2006).
60. G. Martin, C. Barres, P. Cassagnau, et al., *Eur. Polym. J.*, 45, 3257 (2009).
61. J. Wu, Q. Pan, and G. Huang, *J. Mater. Sci.* 42, 449 (2007).
62. P.A. Bhadane, M. Champagne, and M.A. Huneault, *Polymer*, 47, 2760 (2006).
63. P.A. Bhadane, M. Champagne, M.A. Huneault, et al., *J. Polym. Sci. Part B. Polym. Phy.*, 44, 1919 (2006).
64. F. Fenouillot and H. Perier-Camby, *Polym. Eng. Sci.*, 44, 625 (2004).
65. A. De Loor, P. Cassagnau, A. Michel, and B. Vergnes, *Intern. Polym. Process*, 9, 211 (1994).
66. A. De Loor, P. Cassagnau, A. Michel, and B. Vergnes, *J. Appl. Polym. Sci*, 3, 1675 (1994).
67. V.M. Litvinov, *Macromolecules*, 39, 8727 (2006).
68. C. Radhesh Kumar, S.V. Nair, K.E. George, et al., *Polym. Eng Sci.*, 43, 1555 (2003).
69. R.R. Babu, N.K. Singha, and K. Naskar, *J. Appl. Polym. Sci.*, 113, 3207 (2009).
70. P. Steeman and W. Zoetelief, *58th ANTEC* Conference proceedings, Vol 1 3297 (2000).
71. J.L. Leblanc, *Rheo Acta*, 46, 1013 (2007).
72. P. Ezzati, I. Ghasemi, M. Karrabi, and H. Azizi, *Iranian Polym. J.*, 17, 265 (2008).
73. L. Elias, F. Fenouillot, J.C. Majesté, and P. Cassagnau, *Polymer*, 48, 6029 (2007).
74. L. Elias, F. Fenouillot, J.C. Majesté, et al., *Polymer*, 49, 4378 (2008).
75. L. Elias, F. Fenouillot, J.C. Majesté, et al., *J. Polym. Sci.: Polym. Phys.*, 46, 1976 (2008).
76. F. Fenouillot, J.C. Majesté, and P. Cassagnau, *Polymer*, 50, 1333 (2009).
77. G. Martin, C. Barres, P. Cassagnau, et al., *Mater. Chem. Phys.*, 113, 889 (2009).

# CHAPTER 9

# STRUCTURE AND RHEOLOGY OF POLYMER COMPOSITES CONTAINING THERMOTROPIC LIQUID CRYSTALLINE POLYMERS

JIASONG HE

Beijing National Laboratory for Molecular Sciences (BNLMS), Key Laboratory of Engineering Plastics, Joint Laboratory of Polymer Science and Materials, Institute of Chemistry, Chinese Academy of Sciences, Beijing, China

## CONTENTS



*Applied Polymer Rheology: Polymeric Fluids with Industrial Applications*, First Edition.
Edited by Marianna Kontopoulou.

## 9.1 INTRODUCTION

A liquid crystalline polymer (LCP) is a polymeric material that, under suitable conditions of temperature, pressure, and concentration, exists as a liquid crystalline (LC) mesophase [1]. Depending on the conditions for the formation of their liquid crystalline states, LCPs are classified into lyotropic and thermotropic. Poly(*p*-phenylene terephthalamide) (PPTA) is a typical example of a lyotropic LCP, which can be processed to high-strength, high-modulus fibers by a solution spinning technique. Most thermotropic LCPs are processed and molded into structural parts of different shapes in their molten state, by means of conventional processing techniques for thermoplastics, such as extrusion and injection molding. Thermotropic LCPs, offering high mechanical strength and stiffness as well as good thermal stability have been studied extensively and have gotten more applicable over the past two decades. For LC molecules in the nematic phase—one of three basic types and the lowest level of mesogenic orders—within a microscopic volume element, the axes of all molecules are oriented in a specific direction. In this state, the intermolecular forces are very small, so molecules can easily pass by each other. Similar to LC molecules, thermotropic LCP melts in the nematic phase exhibit very unusual rheological behavior, which is attributed to their complex microstructure. Compared to lyotropic LCPs, the practical application of thermotropic LCPs as a major or minor component in blends has become widespread, and their rheology is of interest. Thus this chapter will discuss the rheological properties of thermotropic LCPs and their blends and composites, together with the distinct structures of the latter. For the brevity, hereafter the term *thermotropic liquid crystalline polymer* is abbreviated into LCP.

## 9.2 THERMOTROPIC LIQUID CRYSTALLINE POLYMERS

Thermotropic main-chain liquid crystalline polymers, containing mesogenic units in their main chains, are formed by linking together suitable relatively rigid units directly or through appropriate functional groups (i.e. flexible spacers) [1]. Examples of the general molecular structure of main-chain liquid crystalline polymers are available [1]; a simple schematic of some typical structures is shown in Figure 9.1. The rheology of LCPs has been reviewed by Seo [2]. Due to the rigidity of the LCP molecular chains, injection-molded parts made of pure LCPs usually show strong anisotropy, evidenced by higher strength and modulus along the flow direction and weak lateral adhesion in the transverse direction. The resulting parts look like buckled bamboo with an

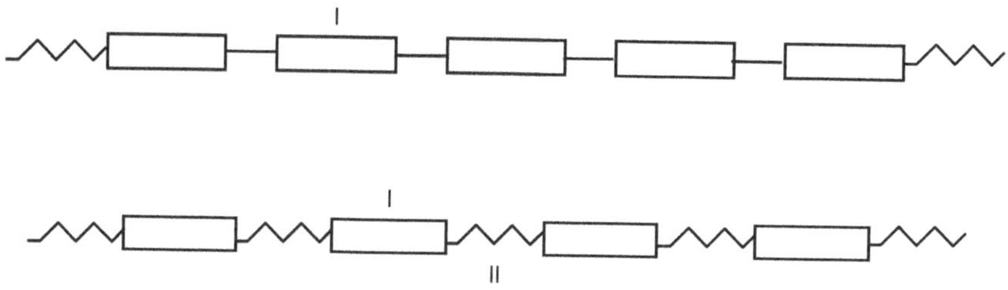


**FIGURE 9.1** Schematic structure of main-chain polymer liquid crystals. I, mesogenic unit; II, spacer.

expanded transverse size. Obviously, this kind of texture is not suitable for real applications. For this reason, to obtain a balanced performance in injection molding, LCPs are usually blended with other thermoplastics, or even fibers to reduce the above-mentioned anisotropy of the final products and widen their applications.

One of the main features of the rheology of LCPs in the nematic phase is the shear thinning behavior at high shear rates, usually $>50\,s^{-1}$, i.e., in the range encountered in extrusion and injection molding. This feature makes LCPs particularly attractive in these applications and has thus attracted extensive research interest leading to the concept of the so-called in situ composite.

## 9.3 IN SITU COMPOSITES CONTAINING LCPs

Although polymer-based composites have generally excellent mechanical properties, the relatively high content of the reinforcing fiber always increases melt viscosity of the composite, resulting in a deterioration in processability. The decrease of the melt viscosity by means of increasing processing temperature may degrade matrix polymers and consequently impair their mechanical performance. In seeking to balancing the flow properties and reinforcing effect, LCPs attracted researchers' attention, because of their strong shear thinning behavior.

In 1987 Kiss [3] proposed a new technique for blending LCPs with polymers in their molten state. This processing method produced polymer blends with decreased melt viscosity, attributed to the shear thinning behavior of the minor LCP component, and a reinforced structure, induced by LCP fibrils generated in situ during melt processing. He called this group of polymer blends in situ composite, and emphasized that thermotropic LCPs had sufficient chain stiffness to give extremely high rigidity and strength to the composite, and yet could be melted. LCPs formed fibers extremely readily, even in injection-molded parts. It is likely that LCPs will form fibrous domains of inclusions in a matrix polymer when melt blended. Given the inherent strength and stiffness of the LCPs, these fibrous inclusions will then act as reinforcement, much like chopped glass fiber. In short, LCPs will act like "meltable chopped glass fibers." These in situ composites are actually easier to process than the neat matrix polymers because of the low viscosity of the LCP component, a very different situation from that found for chopped-glass reinforced composites. This widely cited paper has diffused the boundary between polymer blends and polymer composites. Blends of thermotropic LCPs with thermoplastic polymers are processed and molded by conventional techniques, such as extrusion, injection molding, and spinning with related equipment. However, the resultant products have the characteristic reinforced structure and mechanical behaviors of fiber-reinforced plastics or composites. It is common to use the technical terms of LCP blends and in situ composites interchangeably. LCP blends almost always refer to blends with thermotropic LCPs as the minor portion.

The shear-thinning effect of LCP melts leads to decreased viscosity of LCP-containing polymer blends. This is one of the reasons LCPs are widely used as processing aids to other thermoplastic polymers. Chan and Gao [4, 5] investigated the mechanism causing the lubricating effect in blends of an LCP and high molecular weight polyethylene (HMWPE). They suggested that the highly stretched LCP domains along the flow direction are responsible for lubricating the whole melt system. Using experimental observation and phenomenological models they have concluded that the chain alignment in the elongated LCP domains causes chain alignment and disentanglement in the neighboring HMWPE melt. The high level of alignment of pure PE is attributed to the stronger interaction between the two constituents during flow, caused by molecular alignment and entanglement reduction at the interface. Their results provide a basic understanding for the drastic viscosity reduction found in LCP-containing polymer systems.

It seems that in situ composites possess the advantages of both decreased melt viscosity and increased mechanical properties over conventional chopped fiber-reinforced composites, so that blending LCPs with polymers has been a hot spot of basic and applied research for at least two decades. Tjong published a well-written review on the structure, morphology, mechanical, and thermal characteristics of the in situ composites based on liquid crystalline polymers and thermoplastics [6]. After his discussion about rheology of pure LCPs, his recommendation was to process LCPs above their rheological transition temperature to ensure complete melting of crystallites. According to Tjong, the fibrillation, morphology and dispersion of the LCP dispersed phase in the matrix is affected by the processing conditions, including the rheological characteristics of the matrix, viscosity ratio of the components, LCP content, and interfacial adhesion between the components. Based on all the published results, compared to the expected rheological features of in situ composites, the effect of LCP fibrillation on the rheological characteristics is very important for this group of composites.

Despite the intensive research on the various aspects of in situ composites, the prospect of these composites seems not as bright as previously expected. One of the main reasons is the limited improvement of mechanical performance achieved by in situ generated LCP fibrils reinforcement. For effective reinforcement in composites, reinforcing fillers such as carbon fiber, glass fiber and LCP fibrils should have a large enough intrinsic aspect ratio of their length to diameter. This has been confirmed by theoretical analysis and experimental observation in composite mechanics. Starting from the publication of Kiss' paper, most efforts have been made to incorporate LCP fibrils with aspect ratios as large as possible in polymer melts by means of a variety of processing techniques, including hot drawing, sheet extrusion, and stretching, followed by sheet laminate processing. The key point is to produce long LCP fibrils and to keep their aspect ratio in consequent hot molding. These efforts on fabrication techniques are mainly of industrial interest, rather than having a fundamentally scientific value. However the limited improvement of mechanical performance of LCP blends containing LCP fibrils limits the prospect of these in situ composites.

Studies have also been conducted on in situ ternary LCP–thermoplastics composites—i.e., when a third polymeric component is added to improve the mechanical performances of in situ composites—as reviewed by Tjong [6]. In spite of these efforts, in situ composites have many intrinsic limitations [7]. Carbon–graphite fibers have quite high strengths and moduli up to 7 GPa and 800 GPa, respectively. Glass fibers have as high as 4 GPa of strength and 80 GPa of modulus with a very low price, compared to carbon fibers. Compared to these fibers, LCP fibers, spun under optimized conditions, have higher strengths and moduli than glass fibers, but at a higher cost. Furthermore, when LCPs are blended with thermoplastic polymers to make in situ composites, the formation of fibrils (i.e., fibrillation) in the molten blend should be accomplished during melt processing of the final parts such as extrusion and injection molding. However, in these forming processes the fibrillation conditions are inferior to those encountered in the optimized LCP spinning process. As a result, LCP fibrils generated in situ in the polymer matrix have larger diameters, smaller aspect ratios and lower mechanical properties, compared to the fibers spun from the same LCP. To overcome these limitations the new concept of in situ hybrid composites had been introduced by He et al. [8]. This group of composites, which are reinforced both macroscopically and microscopically, uses the advantages of LCPs, while avoiding their shortcomings.

## 9.4 IN SITU HYBRID COMPOSITES

In principle in situ hybrid composites consist of three components: macroscopic fibers such as glass fiber or carbon fiber, microscopic LCP fibrils generated in situ, and a polymer matrix [8]. This group of in situ hybrid composites has excellent mechanical performance compared to chopped fiber-reinforced polymer composites [7, 8]. Macroscopic fibers act as the main load bearer, while LCP domains function to lubricate the whole system. The LCP fibrils formed subsequently during melt processing contribute to the enhanced mechanical performance by resisting and blocking the propagation of microcracks in the composite. These reinforcements spanning in size two orders of magnitude construct an ideal composite structure. More important, their excellent mechanical performance is accompanied by substantially improved flow properties, attributed to the lubricating role of LCPs. This combination of excellent processing and mechanical behavior is exactly what so many research efforts had been targeting. The studies mentioned above used in situ hybrid composites that consisted only of the three basic components—the matrix polymer, glass fiber, and LCPs—to obtain scientifically clear results. Neither compatibilizers nor interfacial modifiers were used, and the interfacial interaction was not optimized. As concluded by Tjong [6], the hybrids generally offer better processability and mechanical performance over conventional short fiber, whisker or inorganic filler reinforced composites.

The viscosity reduction of a viscous melt system by the addition of a less viscous component is interesting from an application point of view, but of limited scientific interest. It is an expected outcome that the more LCP is added, the more significantly the viscosity will be decreased. Indeed Kulichikhin and co-workers [9] reported that the addition of LCP at concentrations >20 wt % led to a decrease of blend viscosity of glass fiber–reinforced polypropylene composites. In addition to this expected effect, another strange phenomenon was observed, which has been the focus of intense investigation, as reviewed below.

## 9.5 RHEOLOGICAL HYBRID EFFECT

In principle, the melt viscosity of the in situ hybrid composites should be decreased in a similar manner as in in situ composites on addition of LCP, attributed to a lubricating effect caused by the LCP domains. This viscosity decrease is common place in applications where LCPs are used as a lubricant to increase the flow property of the matrix of highly viscous polymers such as polyethersulfone, polysulfone, and polyetheretherketone [7]. However, an unexpected synergistic phenomenon appeared when a polyamide 6 (PA6) matrix of low viscosity was blended with a more viscous LCP as the dispersed phase [10]. After the addition of glass fiber into the molten blend of these two polymers, the resultant composites displayed a very low melt viscosity, which in some cases was even lower than that of the PA6 and binary PA6–LCP blends. This new observation triggered lengthy in-depth investigations. After accumulating sufficient evidence it has been found that this phenomenon is a commonly occurring tendency rather than a special case, encountered in many in situ hybrid composites and has been termed scientifically as a rheological hybrid effect [11]. Its definition is as follows: "a phenomenon in which the melt viscosity of a ternary polymer blend decreases with increasing filler loading, influenced by the minor polymer phase in the blend." Up to now this effect has been observed in several ternary blend or composite systems with added fillers of different shapes and sizes. Case studies of various systems are shown below, where all the rheological data were measured by capillary rheometry.

### 9.5.1 Composites Filled with Micrometer Glass Beads

Spherical glass beads (GB) were added into a binary blend of Nylon 6 (N) and Vectra A950 (a wholly aromatic copolyester of of 73 mol % *p*-hydroxybenzoate (HBA) and 27 mol % 2,6-hydroxynaphthanoate (HNA) synthesized by Hoechst-Celanese) [11]. For GB-filled Nylon 6 composites, the ball-bearing effect of GBs worked at only lower GB loadings, similarly to polycarbonate (PC)–GB composites [12]. But this ball bearing effect did not work at higher GB loadings, as evidenced by the increased melt viscosity of filled Nylon 6 composites, which increased with the increase of GB content over the entire range of shear rate investigated, e. g., $\eta_{\text{app N/GB1(45/55)}} > \eta_{\text{app N/GB1(62/38)}} > \eta_{\text{app N/GB1(95/5)}}$ at shear rate of $115\,s^{-1}$ (Fig. 9.2). However, in the presence of LCP the addition of GB changed the tendency completely. The melt viscosity of N–GB–LCP composites has the order of $\eta_{\text{app N/GB1/LCP(84/5/11)}} > \eta_{\text{app N/GB1/LCP(51/38/11)}} > \eta_{\text{app N/GB1/LCP(34/55/11)}}$. This trend of decreased viscosity with increasing GB filling goes up to 55 wt % of glass bead loading. Moreover, over the entire range of shear rate, the viscosity of N–GB–LCP composites is lower than that of neat Nylon, neat LCP, binary N–LCP blends and N–GB composites. At the shear rate of $1152\,s^{-1}$, the shear viscosities of N–GB–LCP composites containing

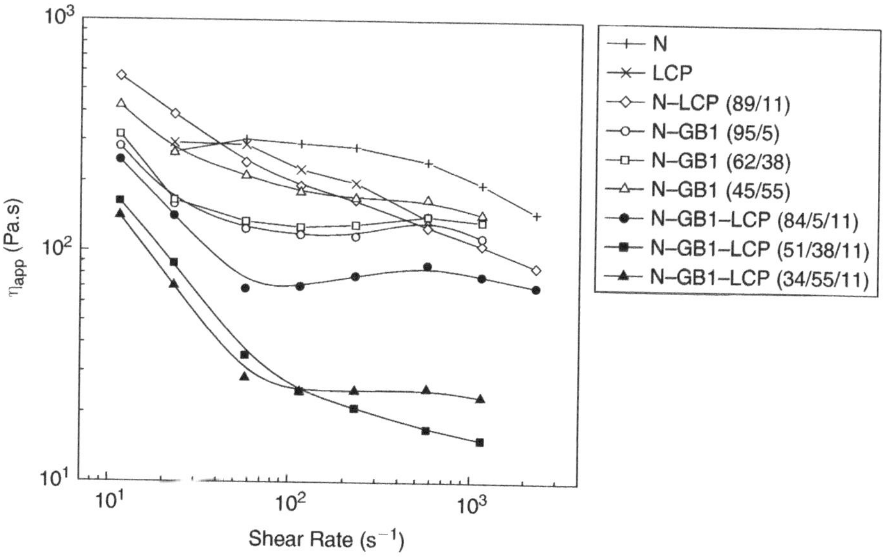


**FIGURE 9.2** Rheological curves of polyamide 6 (N)–glass bead (GB1) composites and N–GB–LCP composites with different GB contents by weight percentage. Reprinted with permission from Ref. 11.

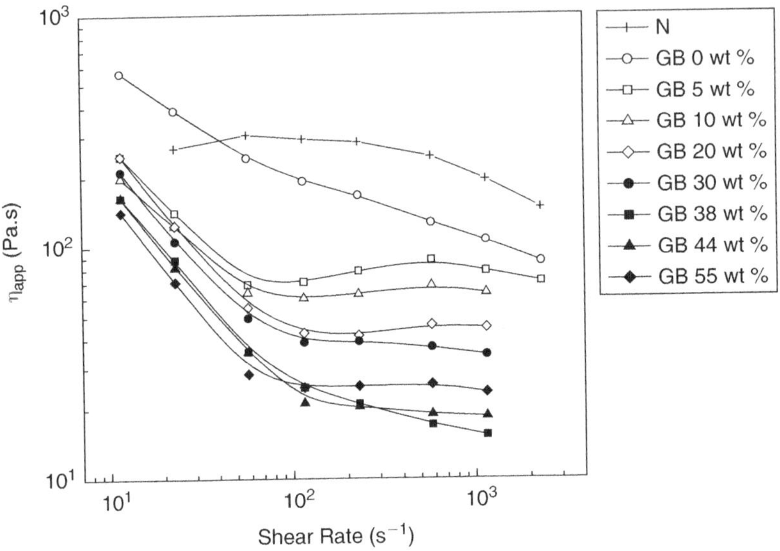


**FIGURE 9.3** Rheological curves of polyamide 6 (N)–GB–LCP ((N + GB)–LCP = 89/11) composites with different GB weight percentages. Reprinted with permission from Ref. 11.

38 and 55 wt % of glass bead are only 8% and 12% of the neat Nylon 6, respectively. Figure 9.3 shows in more detail the successively decreased melt viscosity of Nylon 6–GB–LCP composites with the increase of GB content up to 55 wt %. This rheological hybrid effect was also observed with the addition of glass beads of different diameters.

### 9.5.2 Composites Filled with Micrometer Glass Fibers

Cylindrical glass fibers (GF) were added into binary PC–LCP (Vectra A950) blends for the primary cause of investigating the processability of in situ hybrid composites. It was found that PC–GF–LCP-h 80/15/5 (*h* refers to samples having GF added halfway through the melt blending process) showed a lower viscosity than PC–GF–LCP-h 90/5/5, and PC–GF–LCP-b 90/5/5 (*b* refers to samples having GF added at the beginning of melt blending) showed much lower viscosity than PC–LCP 95–5 blend (Fig. 9.4) [13], hinting a rheological hybrid effect in the presence of the filler. When a polyamide 6 (PA6), which was less viscous than Vectra A950, was chosen as the matrix of PA6–GF–LCP in situ hybrid composites [10], it was found that the addition of glass fiber significantly reduced the melt viscosity of PA6–LCP blends to such an extent that the PA6–GF–LCP composites had viscosities lower than those of neat PA6, neat LCP, and PA6–GF composites as well as the corresponding PA6–LCP blends. As shown in Figure 9.5a, in the whole shear rate range studied, the

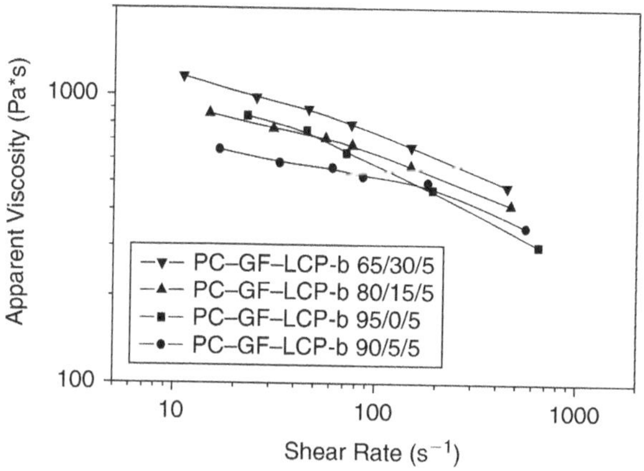


**FIGURE 9.4** Rheological curves of PC–GF–LCP-b in situ hybrid composites at 300°C. Reprinted with permission from Ref. 13.

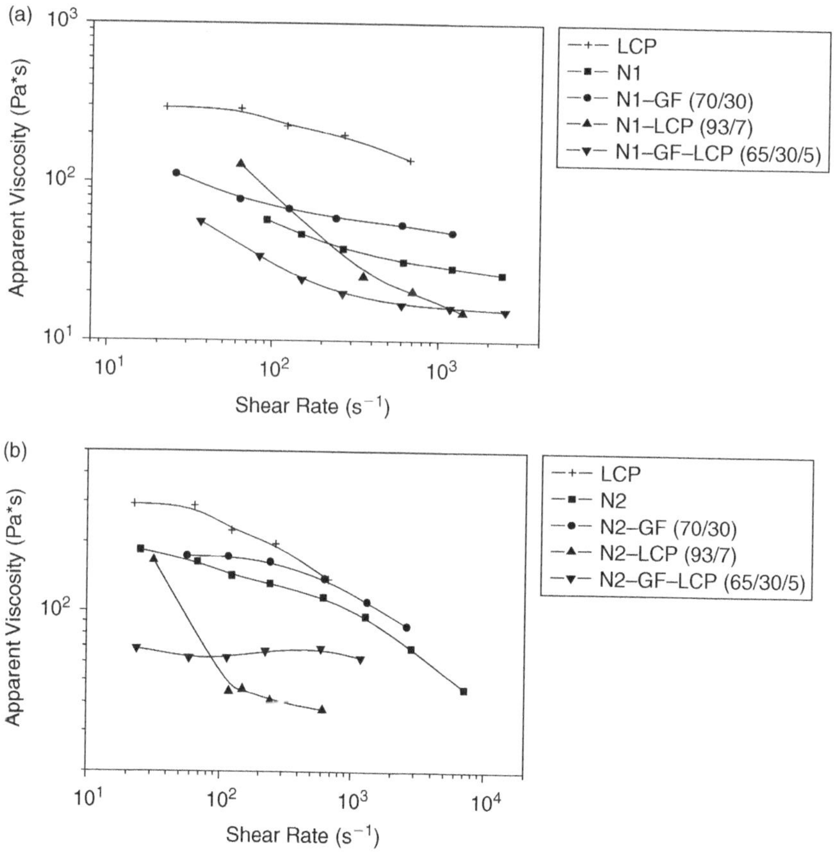


**FIGURE 9.5** Rheological curves of polyamide 6 (N)–GF–LCP composites at 280°C. Reprinted with permission from Ref. 10.

viscosity of N1 (a PA, Akulon K120, DSM)–GF–LCP containing 30% GF was not only much lower than that of N1–GF composite but also lower than the viscosities of the neat resins and N1–LCP blend. At a shear rate of $100\,s^{-1}$, the viscosity of the hybrid composite was only one third of neat N1. Similarly at a shear rate of $85\,s^{-1}$, a N2 (a PA, Akulon K123 GL, DSM)–GF–LCP composite had a lower viscosity than the N2–LCP blend, although the former contained 30 wt % GF (Fig. 9.5b).

### 9.5.3 Composites Filled with Nanometer Silica

Spherical nano-silica filled bisophenol-A polycarbonate (PC)–Vectra A950 blend was another typical example showing a rheological hybrid effect [14]. By taking the relative shear viscosity as the evidence (Fig. 9.6), PC–$SiO_2$ composites with nano-$SiO_2$ content increasing from 1 to 10 wt %, have their relative shear viscosity increased from 1.10 to 1.50 at $70\,s^{-1}$ of shear rate, and from 1.0 to 1.24 at $2600\,s^{-1}$, respectively. All the PC–$SiO_2$ composites have their relative viscosity larger than unity and monotonous increase with increasing nano-$SiO_2$ content at given shear rates. This commonly occurs with filler-loaded polymers. However, due to the lubricating effect of LCP, binary PC–LCP blends with 10.7 wt % LCP have their relative viscosity decreased at different shear rates, such as 0.98 at $70\,s^{-1}$ and 0.84 at $2900\,s^{-1}$, respectively. But in contrast to these,

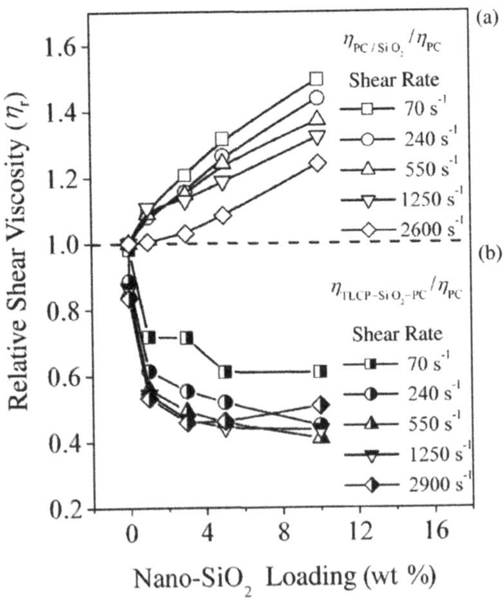


**FIGURE 9.6** Effect of nano-$SiO_2$ loading on the relative shear viscosity of composites at 285°C. (a) Binary PC–$SiO_2$ composites. (b) Corresponding ternary PC–$SiO_2$–LCP composites with 10.7 wt % LCP [14].

the PC–$SiO_2$–LCP composites show a completely different picture. All the points of relative shear viscosity ($\eta_{TLCP/SiO_2/PC}/\eta_{PC}$) are located far below the horizontal line of relative viscosity of unity. At 70 $s^{-1}$ the relative viscosity decreases to 0.72 and 0.61 for PC–LCP–$SiO_2$ 1% and PC–LCP–$SiO_2$ 10%, respectively. Decreases as a function of $SiO_2$ loading are seen at all other shear rates. At 2900 $s^{-1}$, the relative viscosity decreased to 0.53 for PC–LCP–$SiO_2$ 1% and further to 0.46 for PC–LCP–$SiO_2$ 3%, and slightly increased to 0.51 for PC–LCP–$SiO_2$ 10%. Here, the slightly increased viscosity at such high shear rate might be attributed to the shear-thickening behavior of fumed nano-$SiO_2$. In summary, the relative viscosity of these composites decreases with the increase of silica content over a wide range of shear rates in the presence of LCP.

Shown by the above-mentioned composite systems, the lubricating effect of LCP is produced not simply by its content in the composite. The lubricating effect is magnified in the presence of fibers and fillers, i.e., the more the filler that is added, more significant the reduction effect, no matter what the fillers size of micrometer or nanometer and their shape of cylinder or sphere. These findings provide strong evidence of rheological hybrid effect.

### 9.5.4 Contribution of Highly Elongated LCP Domains to Viscosity Reduction of Filled Polymer Melts

During all the studies of the rheological behaviors of these hybrid composites, their morphology was investigated closely to reveal the cause of decreased viscosity. It is getting clearer that the reduced melt viscosity correlates well with the formation of LCP fibrils in these hybrid composites. For N–GB–LCP composites, LCP fibrils are observed in the samples exhibiting viscosity decrease (Fig. 9.7). The better the fibrillation, the more the viscosity reduction [11]. For PC–GF–LCP composites, if LCP forms fibrils, the hybrid composites have improved flowability; if droplets dominate the morphology, the composites have higher viscosity. In the PC–GF–LCP-h 80/15/5 and PC–GF–LCP-b 90/5/5 composites having much decreased viscosity, as discussed previously, more LCP fibrils are clearly seen under polarized optical microscopy (Fig. 9.8) [13]. In a N1–GF–LCP composite, the sample extruded at a shear rate of 230 $s^{-1}$ had a morphology consisting of LCP droplets and fibrils with average length to diameter ratios up to 86, although the viscosity ratio of 5 of the dispersed LCP

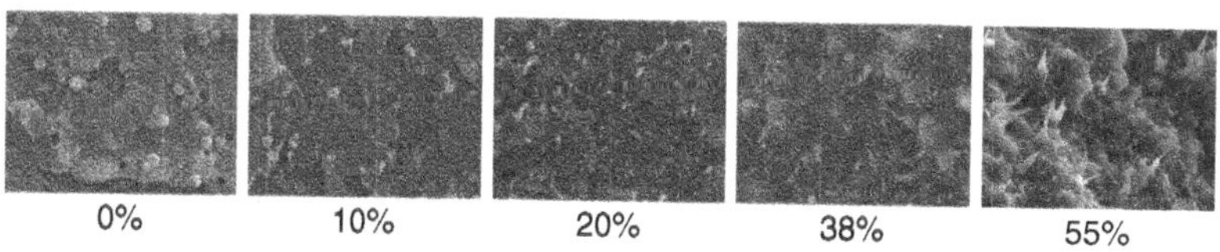


**FIGURE 9.7** Morphology of ternary polyamide 6–GB–LCP composites with different GB contents at shear rate of 115 $s^{-1}$ (LCP content is 11 wt %). Reprinted with permission from Ref. 11.

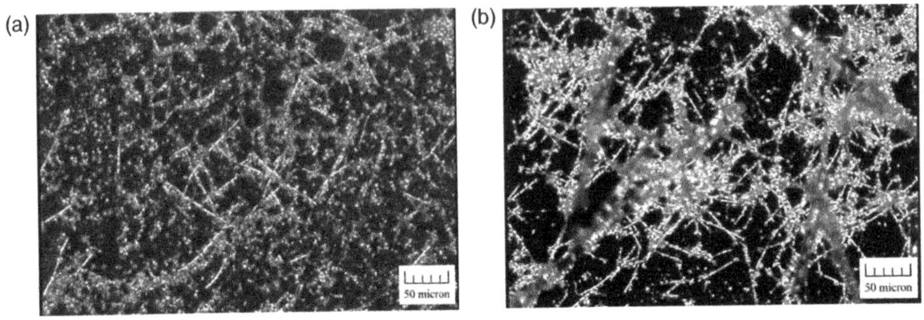


**FIGURE 9.8** POM photos of extruded (a) PC–GF–LCP-h 80/15/5 (b) and PC–GF–LCP-b 90/5/5 in situ hybrid composites. Reprinted with permission from Ref. 13.

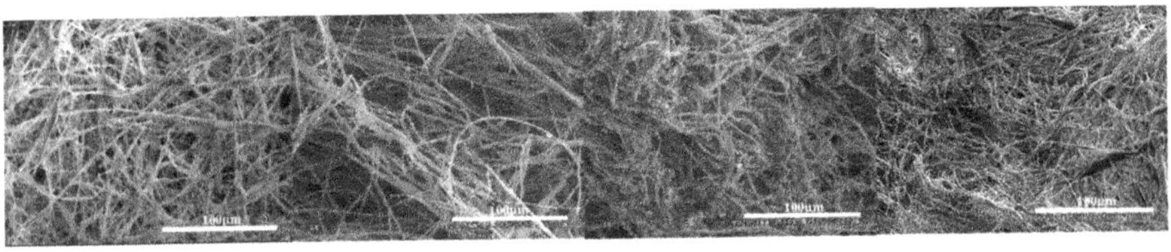

**FIGURE 9.9** SEM pictures of flocculated long LCP fibrils extracted from extrudates of PC–$SiO_2$–LCP composites with varying nano-$SiO_2$ contents. From left to right: PC–LCP–silica 1%, PC–LCP–silica 3%, PC–LCP–silica 5%, and PC–LCP–silica 10%, at $\sim 550\,s^{-1}$ [14].

phase to Nylon N1 matrix phase did not favor the LCP fibrillation without extensive drawing according to commonly accepted rheological considerations [10]. For PC–$SiO_2$–LCP composites, the SEM pictures of LCP extracted from these composites (Fig. 9.9) show that the fibrillation of LCP droplets was greatly enhanced in the presence of nano-$SiO_2$, particularly at high shear rates. A large number of LCP fibrils with quite high aspect ratios were formed at shear rates of 550 and $2900\,s^{-1}$. After selectively dissolving of the matrix PC, not only were unflocculated LCP fibrils collected but also the flocculated LCP fibrils were collected as a mat floating on the top of the extraction solution [14]. This indicates that the presence of nano-$SiO_2$ in PC–$SiO_2$–LCP composites made the deformation and fibrillation of LCP droplets much easier, even at a lower shear rate as $550\,s^{-1}$. The high shear rate led to a great increase in the aspect ratio of LCP fibrils and the quantity of long LCP fibrils as well as to the disappearance of LCP spheres. It is clear that in spite of the fillers size of micrometer or nanometer and their shape of cylinder or sphere, the viscosity reduction in these filler-containing systems correlates well with the deformation and fibrillation of LCP droplets and becomes much more significant when the deformation and fibrillation are greatly enhanced at relatively high shear rates.

Although these LCP fibrils are observed in quenched samples of all these LCP- and filler-containing hybrid composites, actually the lubricating effect is

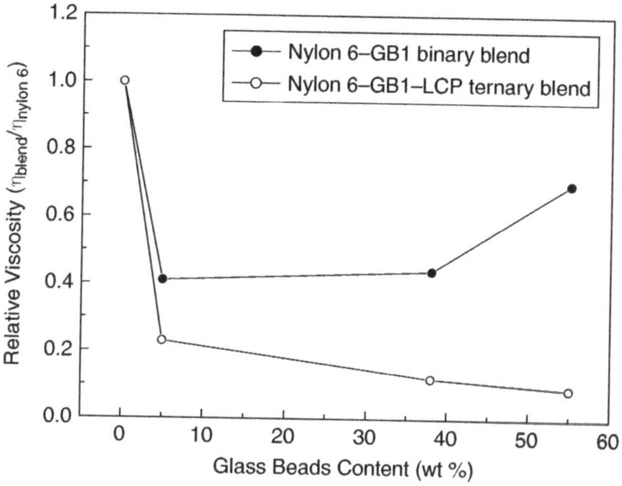


**FIGURE 9.10** Curves of relative viscosity vs. glass beads content for Nylon 6–GB composites and Nylon 6–GB–LCP composites containing 11 wt.% of LCP at shear rate of $57.6\,s^{-1}$. Reprinted with permission from Ref. 11.

not a result of the LCP fibrils themselves in their solid form. Rather it is the LCP domains that are highly stretched along the flow direction that lubricate the whole melt system. For understanding this point we should go back to the critical conclusions reached by Gao for LCP–HMWPE systems [4, 5]. The chain alignment in the elongated LCP domains causes chain alignment and disentanglement in the neighboring matrix polymer melt. The high level of alignment of pure matrix polymer is attributed to the stronger interaction between the two constituents, i.e., LCP and matrix polymer, during flow caused by molecular alignment and entanglement reduction at the interface. The drastic viscosity reduction is a result of the interactions between the two materials after the LCP domains become highly stretched along the flow direction. The long relaxation time of LCP rigid molecular chains makes these highly elongated LCP domains frozen to the form of final fibrils in the cooling process of these composite systems.

Owing to their shear thinning behavior, LCPs can always decrease the viscosity of polymer blends. However, if a comparison is made between the effect of LCP spheres and fibrils, their distinct roles are clearly distinguished. Figure 9.10 shows the viscosity changes in Nylon 6–GB composites without and with LCP [11]. The relative viscosity of Nylon 6–GB composites decreases with increasing GB loading at first, as a result of the ball-bearing effect of glass beads [12]; then increases upon increased loading. However, the relative viscosity of Nylon 6–GB–LCP composites decreases with GB at all loadings. It was found that at 5 wt % GB content, LCP domains were in the form of spheres with the limited viscosity reduction of Nylon 6–GB–LCP composite. However, at 55 wt % GB content, LCP domains were in the form of long fibrils, resulting in a significantly decreased viscosity of the composite. Furthermore, the viscosity of N–GB–LCP 34/55/11 is lower than that of N–GB–LCP 84/5/11, due to the contribution of LCP fibrils at high GB filling. In summary, the highly elongated LCP domains, which appear in the form of fibrils in resultant solidified composites, are much more effective for the viscosity reduction of blends and composites than LCP spheres.

When working with polymer blends, it is always considered that their resultant structure depends on the rheological characteristics of component polymers such as their viscosities, the viscosity ratio, the capillary number and critical capillary number, in addition to the conditions related to processing. The successive influences start from the rheological parameters of component polymers and the processing conditions through the morphology of polyblends and composites, and finally end at performances of resultant materials. However, from the newly found rheological hybrid effect, it is obvious that the morphology of the hybrid polymer composites, especially the morphology of the dispersed LCP domains, reversely has a strong influence on the rheological properties of the material systems. Now the key issue of interest is how to achieve the formation of highly elongated LCP domains, which leads to the highly desirable rheological hybrid effect. For brevity, the phrase *the formation of highly elongated LCP domains* is shortened into “LCP fibrillation.”

## 9.6 THERMODYNAMIC AND HYDRODYNAMIC EFFECTS ON LCP FIBRILLATION

The conditions under which the rheological hybrid effect occurs in filled polymer melts in the presence of LCPs have been investigated for different polymers, loaded with microscale and nanoscale fillers of various shapes. Studies have shown that LCP fibrillation is promoted by filler addition, depending on thermodynamic and hydrodynamic driving forces.

### 9.6.1 Thermodynamic Driving Forces

When fillers such as GF, GB, $SiO_2$, and whiskers are added to the molten polymer blends, the migration and distribution of these fillers in the melt is controlled by interfacial tension. Premphet and co-workers [15] have reported that in filled blends the filler distributes selectively in the polymer phase with which it has the lowest interfacial tension. According to the interfacial values shown in Table 9.1, for PA6–GB–LCP and PC–GB–LCP systems glass beads will migrate thermodynamically to the LCP droplets, while for a polysulfone (PSF)–$CaCO_3$ whisker–LCP system, the $CaCO_3$ whisker will be located in the PSF phase.

The migration of fillers creates different environments and thus affects their function. When the content of spherical particles in the vicinity of LCP droplets

**TABLE 9.1 Interfacial Tension Between Possible Pairs in Ternary Composites**

| Possible Pair | Interfacial Tension (mN/m) |
|---|---|
| PA6-LCP | 3.96 |
| PA6-GB | 142.6 |
| LCP-GB | 125.7 |
| PC-LCP | 2.86 |
| PC-GB | 134.1 |
| LCP-GB | 115.2 |
| PSF-TLCP | 1.37 |
| PSF-$CaCO_3$ whisker | 85.6 |
| TLCP-$CaCO_3$ whisker | 101.8 |

increases to a certain amount, these spherical particles produce hydrodynamic driving forces on the LCP droplets during flow. In the case of fillers migrating in the matrix polymer, they increase its elastic modulus thus changing the flow field and consequently the morphology of the composites. The detailed effects originating from the filler distribution will be discussed in the next paragraphs.

### 9.6.2 Hydrodynamic Driving Forces

The rheological hybrid effect is caused by the addition of fillers into thermoplastics–LCP blends. The migration and distribution of fillers in different regions of these ternary composites produces hydrodynamic driving forces, which act on the dispersed LCP droplets, or on the polymer matrix, thus influencing the local flow field.

Hybrid composites containing spherical particles such as glass beads are among the typical examples that can be easily understood from the point of view of hydrodynamic driving forces. The morphological features of PA6–GB–LCP composites clearly show the influence of hydrodynamic driving forces on LCP fibrillation. As shown in Figure 9.11a, LCP domains in PA6–GB–LCP have curved tails, which result from the rolling of glass beads. The rolling effect is so strong that LCP microfibrils having diameters at the order of magnitude of $10^1$ nm are being pulled out from the circumference of a melt droplet of LCP and twisted into 1-μm-thick fibrils by strong local shearing and extensional action (Fig. 9.11b) [16]. This situation occurs just because of the formation of micro-rollers of glass beads after their accumulation in the vicinity of LCP droplets reaches a certain amount. The extensional flow field formed by these micro-rollers of glass beads exerts strong extensional stresses on LCP coils, resulting in the formation of LCP microfibrils with an average diameter of 30 nm.

The addition of various contents of GB spheres also promoted the in situ fibrillation of LCP in a PC–LCP melt during shear flow [17]. The increase in

(a)

(b)

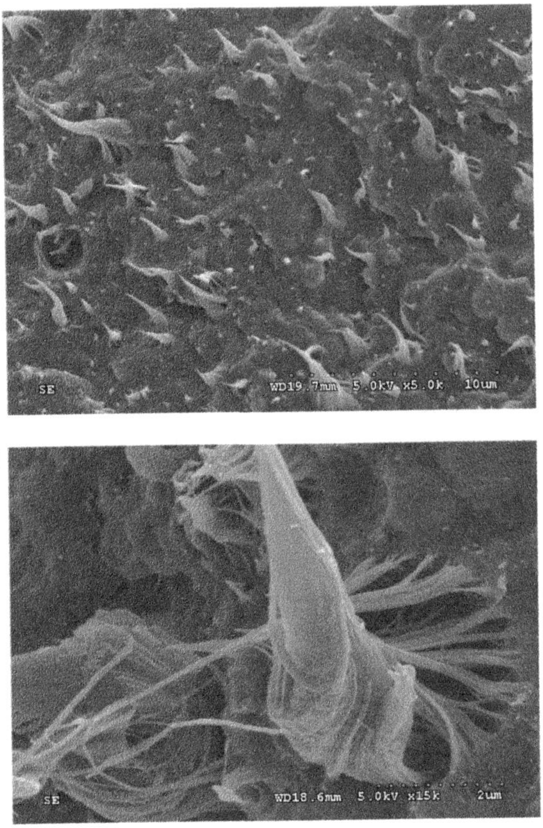


**FIGURE 9.11** (a) SEM micrograph of the fracture surface of a PA6–GB–LCP 51/38/11 (by wt %) ternary composite under lower magnification. (b) Locally magnified morphology of a PA6–GB–LCP 34/55/11 (by wt %) ternary composite at the shear rate of 115.2 $s^{-1}$. Reprinted with permission from Ref. 16.

shear rate also facilitated the deformation and fibrillation of LCP as a result of the decreased viscosity ratio ($p$) and increased capillary number ($Ca$) at higher shear rates. Rheological results showed that neither $p$ nor $Ca$ varied significantly with the addition of GB. But actually the enhancement of LCP fibrillation by adding GB spheres was much more significant than that achieved only by increasing the apparent shear rate. When GB spheres were added to the PC–LCP blend and subjected to the externally applied shear deformation, they rotated along the shear direction with a high local shear between the closely neighboring spheres. This high local shear increased the capillary number effectively and promoted the deformation of LCP even at lower apparent shear rates. Moreover, the presence of rotating GB spheres facilitated the coalescence of LCP droplets, and subsequently the deformation and fibrillation of LCP. It was the hydrodynamic effects of GB spheres that produced the effectively enhanced fibrillation of LCP during processing.

Investigations on the effect of glass bead packing on the LCP fibrillation in PC were conducted with GBs of diameters of 2, 5, 20 μm; GB content at 5, 20, 30 vol %; and shear rates of 20–1260 $s^{-1}$ [18]. The rheological measurements indicated that the GB addition increased the viscosity ratio and seemed unfavorable to LCP fibrillation. However, the morphological observation showed that the LCP fibrillation was promoted by the GB addition and varied with the GB packing. The relationship between GB packing and LCP fibrillation revealed two kinds of hydrodynamic effects of GB promoting the LCP fibrillation: At lower GB packing, the shear flow was enhanced by the high local shear between GBs; and for a high enough GB packing, the shear flow was transformed into an elongational flow, which was more effective for the LCP fibrillation. By adding large amounts of glass beads to PC–LCP blends, very long and perfectly oriented fibrils of LCP were formed in situ in capillary flows [19]. Thermodynamically the different interfacial tensions between the components caused the GB migrate to the dispersed LCP phase. Then the locally high content of GB spheres produced a confined condition, so that the LCP droplets passed through the simultaneously formed micro-capillaries, and were subsequently stretched into fibrils and stabilized.

If the size of spherical particles goes down to the nanometer scale, the hydrodynamic driving forces are still working toward LCP fibrillation. The extra aspect to be considered is the high interfacial area in nanometer particle-filled polymer melts. In the case of PC–nano-silica–LCP, the esterification between PC and LCP molecules during melt blending was significantly reduced in nano-$SiO_2$ filled LCP–PC blends, compared to the unfilled LCP–PC ones [20]. This was a result of the high surface coverage of LCP droplets by nano-$SiO_2$, because the interfacial tension dominated the selective distribution of nano-$SiO_2$ at the PC–LCP interface. The increased shear rate enlarged the viscosity disparity between PC and LCP, thus forcing the nano-$SiO_2$ particles to migrate from the PC–LCP interface to PC matrix, enhancing the mobility of LCP phase, and therefore promoting the coalescence and the subsequent fibrillation of LCP droplets in capillary flow.

When nano-$SiO_2$was added into a polysulfone (PSF)–LCP blend [21], very long and perfectly oriented LCP fibrils were in situ formed in capillary flows by only 5 vol % nano-$SiO_2$. Higher contents of nano-$SiO_2$ increased the elasticity of the matrix sharply, as revealed by dynamic rheology analysis. Then at the entrance of capillary die the entrance angle was decreased and elongational stress increased when the polymer melt flowed through the abrupt contraction, which resulted in LCP fibrillation in PSF–nano-$SiO_2$–LCP system.

In the case of fillers with a platelet-like structure, such as nanoclays, organophilic montmorillonite (OMMT) was used to fill PA6 in the presence of an LCP [22]. Morphological observations showed that the clay platelets had dramatic influences on the dispersion and deformation of the LCP phase. The LCP droplets were deformed into fibrils at clay contents up to 5 and 7 wt %. The morphological evolution of LCP in the hybrid composites, especially at 7 wt % of clay loading, was consistent with the prediction based on the micro-rhelogy parameters such as the viscosity ratio of the dispersed phase to the matrix and the ratio of capillary number to the critical capillary number. This enhanced fibrillation of LCP droplets was attributed to the role of nanoclay particles as a compatibilizer to improve the interfacial adhesion and to suppress the interfacial slip between LCP and polyamide phases in the melt, so that the shear stress was effectively transferred to the dispersed LCP phase.

Other fillers, such as $CaCO_3$ whiskers have been added into polysulfone (PSF)–LCP blends [23]. In contrast to the spherical or ellipsoidal droplets of LCP formed in binary PSF–LCP blends, the LCP fibrillation was promoted by the introduction of whisker particles in all ternary composites at the shear rates studied; and the more whiskers that were added, the larger the aspect ratios of LCP fibrils that were obtained. Although the decreased viscosity ratio achieved by the whisker addition might favor the LCP fibrillation, the calculation of the capillary number showed that the change of viscosity ratio was not the predominant factor controlling the LCP fibrillation in the hybrid composites. The results measured from the capillary entrance pressure drop showed that the introduction of whiskers increased the elongational stress at the entrance of the capillary die. The analysis of the flow field confirmed that the presence of rigid whiskers altered the flow field and enhanced the vortex in the converging flow area at the entry of the capillary die. It is this vortex enhancement that resulted in the increase of the elongational stress and promoted the fibrillation of LCP droplets in the whisker-filled PSF–LCP melts.

Further studies used whiskers of calcium carbonate and aluminum borate as the filler in PSF–LCP blends [24]. LCP mainly formed spheres and less-deformed ellipsoids in the binary PSF–LCP blend at the shear rates studied. However, after adding the whisker into the binary blend, LCP fibrils with large aspect ratios were generated in all the hybrid composites; the more whiskers that were added, the larger the aspect ratio of LCP that was obtained. The addition of longer aluminum borate whiskers produced the most pronounced LCP fibrillation effect. With the whisker added, the elongational viscosity ratio decreased gradually and relatively limited, indicating this ratio was not the dominant factor influencing the LCP fibrillation. In contrast to this, the elasticity of composites increased more with increasing content and aspect ratio of whiskers and thus resulted in vortex enhancement and generated the extra elongational stress in the converging flow area at the entry of capillary, thus promoting the formation of LCP fibrils with larger aspect ratio. This reveals that the vortex enhancement was the predominant factor promoting LCP fibrillation. For a clear view, all the results of the multiscale effect of LCP fibrillation are summarized in Table 9.2.

Except for the above-discussed driving forces, the topological factor in these filler-loaded LCP–thermoplastics blend systems also has an influence. The topological factor is related to the size of temporarily formed capillaries by aggregated fibers in fiber-filled polymer melts [10, 13] and the size of gaps between nearby rotating spheres [11, 17, 18]. The initial size of LCP droplets and voids between glass beads determine the probability of the contact of LCP

**TABLE 9.2 Thermodynamic and Hydrodynamic Driving Forces Promoting LCP Fibrillation in Filled Polymer Melts Reprinted with permission from Ref. 25.**

| | | Thermodynamic | Hydrodynamic | |
|---|---|---|---|---|
| Matrix | Filler | Filler Localization | LCP Fibrillation | Reference |
| PA6 | GF | | High local shear<br>Elongational flow | 10, 13 |
| PA6 | GB | LCP | High local shear | 16 |
| PC | GB | LCP | High local shear<br>Elongational flow | 17–19 |
| PC | Nano $SiO_2$ | PC/LCP | LCP coalescence | 20 |
| PSF | Nano $SiO_2$ | PSF | Increased elasticity<br>Enhanced vortex | 21 |
| PA6 | OMMT | PA6 | Improved interfacial adhesion | 22 |
| PSF | Whisker | PSF | Enhanced vortex | 23, 24 |

droplets with glass beads and the effectiveness of transferring rotation and rolling action of glass beads in these hybrid composites. The bigger the initial size of LCP droplets, and the smaller the voids, the easier the LCP fibrillation is [11]. The average initial size of LCP droplets can be estimated by Taylor's equation [26]. This estimation was used to understand the topological influence, although Taylor's equation is applicable for Newtonian systems, instead of highly non-Newtonian ones studied here. The gaps between spherical particles depend on the size of fillers, filler content, interfacial tension of possible pairs in the hybrid system and the migration of fillers in the melt system, etc. Undoubtedly, the relative size between droplets and gaps is the prerequisite for the hydrodynamic driving force to act. These parameters are different from system to system and depend on the particular system under consideration, the type of the components and formulation of the hybrid composite.

## 9.7 CONCLUDING REMARKS

Up to now from all the above results and discussion it is clear that a rheological hybrid effect occurs in the presence of LCPs. In addition to the well-known contribution of LCPs toward lowering the viscosity as binary blend components, due to their rheological characteristics when combined with fillers, LCPs play much more effective role in decreasing melt viscosity and improving the flow properties. Upon addition of fillers having various shapes, sizes, and aspect ratios into LCP-containing polymer blends, these ternary hybrid systems exhibit the rheological hybrid effect more or less to some extent, if highly elongated LCP domains can be generated during processing.

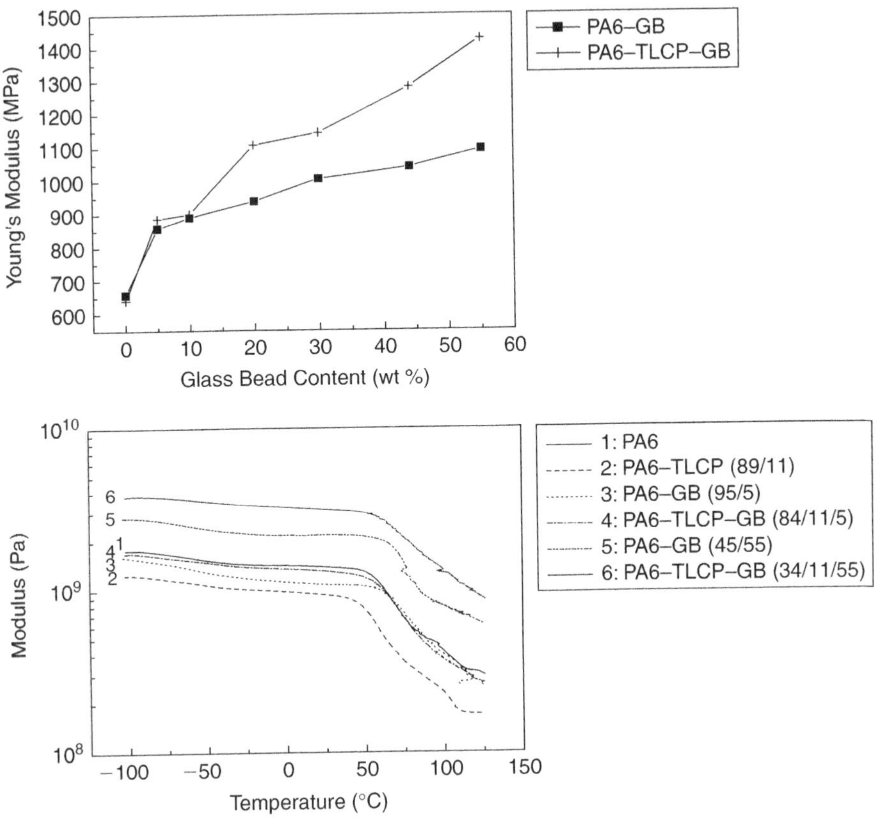


**FIGURE 9.12** (a) Young's modulus as a function of glass bead content (b), and dynamic modulus of binary PA6–GB composites and ternary PA6–GB–LCP composites. Reprinted with permission from Ref. 16.

The question still remains whether other minor polymer phases have the same influence as the LCP phase. This would certainly leave more space for future progress, but knowledge remains limited to date. The preliminary impression is that the rigidity of the macromolecular chains might control their ability to exhibit the rheological hybrid effect in filler-loaded polymer melts. A correlation between the submicrometer reinforcing function of LCPs through fibrillation and their spinnability in air has been established for six LCPs [27]. LCPs having good spinnability generated fibrils with small diameter, narrow diameter distribution, and large aspect ratios. The conclusion is that the spinnability of LCPs should be taken as a prerequisite for the accomplishment of submicrometer reinforcement with LCP fibrils in in situ composites, although the spinnability of LCPs depends up on many factors and still remains undefined. The question remains that if rigid-chain or semi-rigid-chain LCPs differ in their ability to form fibrils, why it is not possible for common flexible-chain polymers to have a similar lubricating action in filled polymer melts?

It may be argued that the viscosity reduction may be a result of matrix degradation by heat produced during processing of these highly filled composites. The best method to discount this is to measure the molar mass of the matrix polymers after the composite production. However, up to now this tedious procedure has shown inconclusive results. Evaluation of the mechanical properties though does not provide evidence of extra degradation. Figure 9.12 shows that PA6–GB–LCP 34/55/11 composites exhibiting the lowest apparent viscosity as shown in Figure 9.2 possess the highest Young's modulus and dynamic modulus among the GB-filled PA6–LCP systems [16].

Usually fiber-reinforced or particulate-filled polymer melts have increased viscosity compared to their corresponding matrix polymer melts, as predicted by Kitano and co-workers [28] and summarized by Metzner [29], no matter what the aspect ratio of the filler is. On the contrary, due to the rheological hybrid effect, in the presence of LCPs the addition of fillers decreases the melt viscosity and improves the flow property of polymer composites, while at the same time maintaining and even improving their mechanical performances. This effect can occur to various extents for different combinations of the three basic ingredients. The formation of highly elongated LCP domains can be achieved by both longer and shorter LCP fibrils. The investigation on all the aspects involved is still at an early stage, leaving room for deeper understanding and further extension in the future. Further studies are needed to optimize these effects and reach technologically meaningful applications.

## 9.8 ACKNOWLEDGMENTS

This work was supported by the National Nature Science Foundation of China, Grant No. 50233010.

## REFERENCES

1. M. Baron et al., *Pure Appl. Chem.*, 73, 845 (2001).
2. K.S. Seo, in *Thermotropic Liquid Crystalline Polymers*, Technomic Publishing Co., Inc. Lancaster, PA, 2001.
3. G. Kiss, *Polym. Eng. Sci.*, 27, 410 (1987).
4. C.K. Chan and P. Gao, *Polymer*, 46, 8151 (2005).
5. C.K. Chan and P. Gao, *Polymer*, 46, 10890 (2005).
6. S.C. Tjong, *Mater. Sci. Eng.*, R41, 1 (2003).
7. J. He, Y. Wang, and H. Zhang, *Compos. Sci. Tech.*, 60, 1919 (2000).
8. J. He, H. Zhang, and Y. Wang, *Polymer*, 38, 4279 (1997).
9. V.G. Kulichikin, I.L. Parsamyan, Y.S. Lipatov, et al., *Polym. Eng. Sci.*, 37, 1314 (1997).

10. X. Zheng, J. Zhang, and J. He, *J. Polym. Sci. Part B: Polym. Phys.*, 42, 1619 (2004).
11. Y. Ding, J. He, J. Zhang, et al., *J. Non-Newt. Fluid Mech.*, 135, 166 (2006).
12. P. Chen, J. Zhang, and J. He, *Polym. Eng. Sci.*, 45, 1119 (2005).
13. X. Zheng, B. Zhang, J. Zhang, et al., *Intern. Polym. Proces.*, 18, 3 (2003).
14. L. Wu, P. Chen, J. Chen, et al., *Polym. Eng. Sci.*, 47, 757 (2007).
15. K. Premphet and P. Horanont, *Polymer*, 41, 9283 (2000).
16. Y. Ding, J. Zhang, P. Chen, et al., *Polymer*, 45, 8051 (2004).
17. P. Chen, L. Wu, Y. Ding, et al., *Compos. Sci. Tech.*, 66, 1546 (2006).
18. P. Chen, J. Chen, B. Zhang, et al., *J. Polym. Sci.: Part B: Polym. Phys.*, 44, 1020 (2006).
19. P. Chen, J. Zhang, and J. He, *Polymer*, 46, 7652 (2005).
20. L. Wu, P. Chen, J. Zhang, and J. He, *Polymer*, 47, 448 (2006).
21. J. Chen, P. Chen, L. Wu, et al., *Polymer*, 48, 4242 (2007).
22. B. Zhang, Y. Ding, P. Chen, et al., *Polymer*, 46, 5385 (2005).
23. J. Chen, P. Chen, L. Wu, et al., *Polymer*, 47, 5402 (2006).
24. J. Chen, P. Chen, L. Wu, et al., *Intern. Polym. Proces.*, 22, 166 (2007).
25. J. He, *Macromol. Symp.*, 277, 43 (2009).
26. G.I. Taylor, *Proc. R. Soc. London*, A146, 501 (1934).
27. J. He and H. Zhang, *Polymer*, 37, 969 (1996).
28. T. Kitano, T. Kataoka, and T. Shirota, *Rheol. Acta*, 20, 207 (1981).
29. A.B. Metzner, *J. Rheol.*, 29, 739 (1985).

CHAPTER 10

# ELECTRORHEOLOGICAL FLUIDS: MATERIALS AND RHEOLOGY

FEI FEI FANG and HYOUNG JIN CHOI
Department of Polymer Science and Engineering, Inha University, Incheon, Korea

## CONTENTS



## 10.1 INTRODUCTION

Functional materials that change their physical characteristics such as shape, size, shear viscosity, modulus, or morphology simply by employing an external field such as a heat source, electric, or magnetic field, while exhibiting common properties under normal conditions, have been developed in various fields of engineering applications in recent years. Unlike common materials whose physical properties remain unchanged, the so-called smart materials have

*Applied Polymer Rheology: Polymeric Fluids with Industrial Applications*, First Edition.
Edited by Marianna Kontopoulou.

individual characteristics, which can be changed significantly. The property that is subject to alteration influences the type of application the smart material can be used for.

Electrorheological (ER) fluids are a class of intelligent/smart materials that exhibit unique phase transitions from liquid-like to solid-like, in the presence of an external electric field [1–7] in an analogous way as magnetorheological suspensions, which alter their response under external magnetic fields. Generally, a typical ER fluid is composed of electrically polarizable particles dispersed in an insulating oil. As shown in Figure 10.1a, in the absence of an external

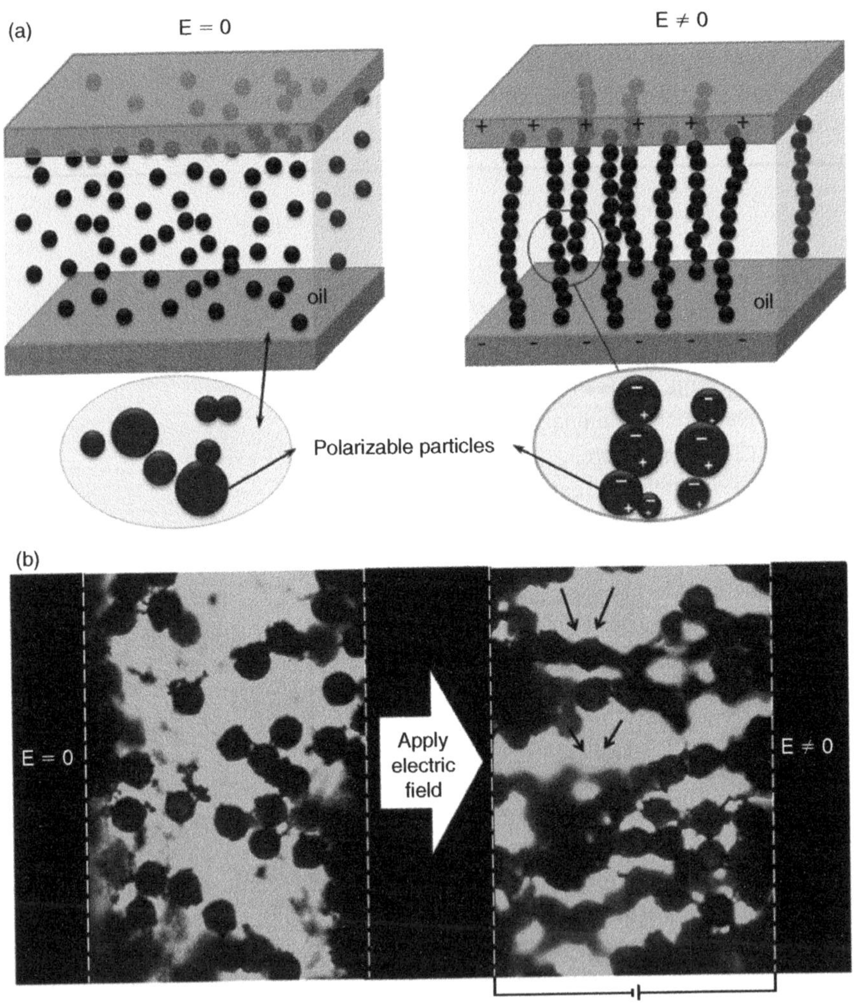


**FIGURE 10.1** (a) The ER phenomenon. (b) Optical microscopic images of polyaniline-coated poly(methyl methacrylate) microspheres–based ER fluid without and with electric field (0.3 kV/mm).

electric field, all particles are randomly dispersed within the medium phase, while once an electric field is applied to the fluid, all particles get immediately polarized and attract adjacent particles to form a chain-like structure via a strong dipole–dipole interaction. As the strength of electric field increases, the formed chains get conglutinated to produce more robust columns, by which way the ER fluid turns to be an elastic solid. The body-centered tetragonal lattice was reported as the theoretical ground state structure of an ER fluid [5]. Typically, ER fluids are suspensions containing particles ranging in size from 1 to 100 μm, at volume fractions of 0.05–0.5 dispersed in hydrophobic liquid media, such as mineral oil or silicone oil. At an electric field strength of 0.1–5.0 kV/mm, these particles form chains that span the gap between two field-generating electrodes. Figure 10.1b shows optical microscopy images of the chain-like structure of multiwalled carbon nanotube-adsorbed poly(methyl methacrylate) microspheres dispersed in silicone oil, induced by a DC electric field at a fixed gap between two parallel electrodes of 350 μm [8]. The DC electric field was maintained for 3 min at 25°C to obtain an equilibrium columnar structure. During this rapid and electrically controllable process, the rheological properties of the ER fluids change distinctively. All physical and mechanical property changes induced by the applied electric field are virtually instantaneous (say within a millisecond) and are reversible upon removal of the field. Compared with the fluidity of ER fluids without an electric field, the solid-like state induced by the applied electric field possesses a nonvanishing yield stress, enhanced shear viscosity and modulus. The viscosity increase is caused by the reorientation of polarized molecules along the direction of the applied electric field and by ion aggregation near the electrode. When a shear force is applied, the chains are destroyed and viscosity decreases in the high shear rate region, thus approaching the value corresponding to the zero electric field strength [9].

Since the first discovery of the ER phenomenon with starch and flour particles suspended in an insulating oil [10], an electric-double layer theory was one of the first proposed mechanisms to explain the behavior of dispersed particles adsorbed with water or surfactants. The electric-double layer can be altered via polarization and distortion, and the ER effect is enhanced by increasing the electrostatic interactions between particles. In the early stage of ER fluid research, extrinsically polarizable particles such as silica [8], alumina, and cellulose [11], which contain various additives to induce polarizability attracted enormous attention from researchers. However, these particles, used as hydrous ER fluids, have some drawbacks in their applications, including device corrosion, water evaporation, narrow operating temperature range, and dispersion instability [12]. To overcome these shortcomings, materials containing intrinsically polarizable species, such as electrons and ions have been introduced [13].

Various applications of ER fluids have received intense attention such as shock absorbers, robot arms, engine mounts, bearing dampers, seismic controlling frame structures, torque transducers, polishing devices, and microfluidics [14]. Recently reported work introduced a novel ER fluid that has overcome the main disadvantages of sedimentation and redispersion [15]. Furthermore, some nanoparticle-based "giant" ER fluids developed were also applied in designing microfluidics [16, 17]. The giant ER fluid, consisting of urea-coated 20-nm-diameter nanoparticles (barium titanyl oxalate) suspended in silicone oil can reach a yield stress of 130 kPa, breaking the theoretical upper bound of conventional ER fluids. Such controllable variation of the rheological property was applied to control micro flows in microfluidics and bio-MEMS devices.

Since the rupture of chains spanning the electrode gap is induced above some critical deformation, the suspension behaves as a solid at low stresses. The chains constructed by highly absorbing particles do not transmit light, while the vacant area that is void of particles is almost transparent. Therefore, the transmittance of suspensions is markedly increased in electric fields. The viscosity change can be utilized in liquid toners in electrophotography, and the optical change to transmission-controlled windows. When a dielectric fluid with very low conductivity is subjected to high electric fields, a jet-stream is generated, which is known as the electrohydrodynamic (EHD) effect. An ink jet system is also developed by the use of EHD jet [18]. The change of physical properties of fluids in electric fields can provide the basis for new imaging technology.

## 10.2 MECHANISMS OF ER PHENOMENON

Several assumptions about the ER mechanism, based on the category of ER materials are available. A fibrillation model was suggested based on the observation that fibrillated chains are formed in an ER suspension between two electrodes under an electric field [10], and thus the apparent viscosity of ER fluids increases substantially. Therefore, the ER effect was initially attributed to the formation of a fibrillation structure.

Much later, when the dielectric principle was introduced [19], the ER phenomenon was considered to be linked to the polarization of particles [20–23]. A water-related electric double-layer model was suggested to interpret the ER mechanism for materials such as starch and limestone. Thus the presence of water molecules in the hydrous ER materials becomes a critical factor for the ER effect. When an external electric field is applied, the particles get polarized and distorted; water molecules can move out of pores and migrate from one particle to another under the applied electric field to make an adhesive water bridge between the particles, having a high surface tension. A similar bridge model in the case of surfactants also displays an electrostatic energy distribution in the surrounding particles [24]. This polarization model relates the material parameters, such as dielectric response of both liquid and solid particles and particle volume fraction, to the rheological characteristics, as well as other microstructural fibrillar chains. Using an idealized ER system—a uniform, hard dielectric sphere dispersed in a Newtonian fluid medium—the derived electrostatic force was found to be dependent on the dielectric constant mismatch between the particle and the continuous medium [25–31].

Although many experimental systems have supported the polarization model, there are still some other ER observations that cannot be interpreted by this model, such as the rheological property dependence on the electric field frequency and the particle conductivity. Generally, conductivity mismatch between particles and liquid medium, rather than dielectric constant mismatch, was considered as a dominant factor for DC and low-frequency AC excitation [32]. This conduction model considers only the particle interactions, regardless of the micro-structural changes after an electric field is applied. However, a major shortcoming remains in the polarization and conduction models, since both of them are static and do not take into account the dynamic processes [33]. Furthermore, it can also be noted that different mechanisms for the polar molecule–dominated ER effect have been reported during the recent development of giant ER materials. For the case of urea-coated dielectric nanoparticle systems, the yield stress can reach hundreds of kPa, exceeding the predicted upper bound of the conventional systems. This indicates that no universal theory is yet available to cover these different cases altogether [34, 35].

## 10.3 ER RESPONSIVE MATERIALS

The ER materials can be classified into either heterogeneous or homogeneous systems. The main reason for developing homogeneous ER materials is to prevent the sedimentation problem encountered in heterogeneous ER fluids, caused by the large density mismatch between dispersed phase and medium oil. Liquid crystalline systems are typical homogeneous ER materials. Homogeneous ER fluids are believed to offer a new and promising type of ER fluid. However, higher viscosities under zero electric field and liquid–liquid segregation problems are obvious obstacles for such systems. Thus most attention has been focused on heterogeneous ER fluids containing inorganic, organic, or polymeric particulate materials. Herein, the heterogeneous ER fluids are classified into hydrous and anhydrous ER materials, as shown in Table 10.1.

Hydrous ER materials include both inorganic and organic materials. The inorganic hydrous ER materials are usually metallic oxides or ceramic materials, generating a good ER effect. However, metallic oxide–based ER materials possess serious sedimentation problems, due to the large density mismatch between the dispersed phase and continuous medium. Therefore, organic hydrous ER materials, such as starch, cellulose or polymethacrylic acid crosslinked with divinylbenzene have been frequently employed as the dispersed phase in ER fluids. These kinds of ER fluids not only sustain the strong ER performance but also indicate a stable dispersion.

Although hydrous ER fluids exhibit good ER behaviors, they still suffer from serious defects. First, the amount of contained water should be carefully adjusted, because small amounts of water cannot produce a significant ER

**TABLE 10.1 Types of Homogeneous and Heterogeneous ER Fluids.**

| Heterogeneous System | | Homogeneous System |
|---|---|---|
| *Polarizable Particles* | | |
| Inorganic Material | Organic Material | |
| Silica<br>Metal oxides<br>Zeolite<br>Clay | Starch, cellulose<br>Chitosan, PVA<br>Poly(acrylic acid)<br>Polyaniline poly(*p*-phenylene)<br>Poly(acene quinone) radicals<br>Polymethacrylate<br>Polyurethane<br>Carbonaceous particles,<br>Conducting polymer–clay nanocomposite | Liquid crystal polymers diluted with polydimetylsiloxane |
| *Medium Oils* | | |
| Silicone oil, paraffin oil, mineral oil, soybean oil, corn oil | | |

performance; however, too much water will cause large electric current density, which may cause damage to the rheometer due to the high conductivity of water. In addition, these water-activated ER fluids are sensitive to operating temperatures, due to the water evaporation at high temperatures. Finally, the moisture contained in ER fluids is also capable of causing corrosion problems. Generally, the weight percent content of water in hydrous ER materials is $<5\%$. Thus the water or moisture content of hydrous ER systems is a big drawback for this type of fluid. To overcome the shortcomings (thermal instability and corrosion) that wet-base systems possess, various dry-base systems have been investigated, containing anhydrous particles; these include organic conducting polymers and their derivatives, inorganic nanoparticles, and polymer–inorganic nanoparticle composites.

Conducting polymers, which possess conjugated $\pi$ bonds have been gradually adopted as ER materials because this type of materials possesses the capability of being polarized under an electric field, thus indicating high dielectric constant [36–44]. Besides this, polymers that have high polarizable functional groups on the molecular backbone can also be applied as an ER material. Among all the applicable conducting polymers, PANI has been regarded as a perfect candidate for ER materials owing to its high conductivity, easy synthesis, and good environmental stability [45–48]. It is well known that its conductivity can be controlled by protonation and charge transfer doping, and the protonation–deprotonation equilibrium occurs for two of the oxidation states, depending on the pH value. On the other hand, among several PANI morphologies, PANI particles are generally synthesized using the conventional polymerization of aniline with the modification of producing a fine emeraldine hydrochloride form. Details of the preparation procedure can be found in the literature [49]. In addition, the derivative of PANI including polymethylanilne, polyethylaniline, polymethoxylaniline, and polyethoxyaniline were also prepared by the similar chemical oxidation polymerization and then examined for their ER application [46]. Other conducting polymers, such as polypyrrole [50], poly(*p*-phenylene) [51, 52], poly(naphthalene quinine radicals) [53], poly(acene quinone radicals) [54], poly(phenylenediamine), and polythiophene have been also studied as ER materials. On the other hand studies on the oleophilic properties of the nano titanium oxide are rarely reported in ER fluids. Through the modification of the chemical properties of the titanium oxide surface and the control of the size distribution for the particle, superoleophilic titanium oxide particles could be used to enhance the ER effect.

Besides organic conducting polymers, some inorganic nanoparticles that possess high dielectric properties, as well as crystal structure, also demonstrate superior behavior when applied as ER fluids. Particle polarization in these materials has been widely regarded as the reason for the ER response [55, 56]. High dielectric constant and dielectric loss and suitable conductivity, have been proven to be crucial factors for dominate ER effects. Furthermore, the dielectric and conduction properties are closely related to the molecular or crystal structure of the materials. Thus plenty of inorganic species with modified molecular and crystal structures have been developed and adopted as ER materials. Inorganic materials with unique structure and morphology, including nanoparticle, nanotube, nanorod, and even mesopores designed via various methodologies. Compared with the above-mentioned conducting polymer-based ER fluids, these modified inorganic particle–based ER fluids demonstrate much enhanced yield stress. Especially, in the case of a giant ER fluid, the saturation surface polarization and nano-scale effect of the particles ($\sim 50$ nm) are considered to induce the giant static ER effect [57].

These inorganic species-based anhydrous heterogeneous ER fluids also suffer from similar dispersion problems as hydrous heterogeneous ER fluids due to the large density mismatch. Therefore, versatile polymer/inorganic composite materials have been fabricated, by employing diverse synthetic strategies. A representative example is conducting polymer-based composite materials. Conducting polymer and inorganic species, including montmorillonite, titanium dioxide ($TiO_2$), barium titanate ($BaTiO_3$), silica, and multiwalled carbon nanotubes, have been used to produce novel composite materials. These have unique morphology, depending on the coating, intercalation or exfoliation.

## 10.4 ER CHARACTERIZATION

Among various characteristics of ER fluids, rheological properties, including yield stress, flow curve of shear stress vs. shear rate, shear viscosity, and dynamic moduli under steady shear, oscillation, and compressive mode have been extensively investigated. In addition, dielectric characteristics are also considered to be correlated with ER performance.

### 10.4.1 Rotational Testing

A typical rotational test consists of measuring the flow curve via controlled shear rate (CSR) mode, in which a shear rate is applied to the ER fluid and then the shear stress is measured. From the CSR mode, the dynamic yield stress can be obtained by extrapolating the shear stress at nearly zero shear rate. Before starting shearing the ER fluids, the electric field is preapplied for about some minutes to produce robust ER chains as much as possible.

Figure 10.2 indicates the shear stress curves for two types of PANI–$BaTiO_3$–based ER fluids [58]. The enhancement of shear stress ($\tau$) becomes slightly larger with 25 wt % of $BaTiO_3$ particles within a broad range of shear rate ($\dot{\gamma}$). The typical trend for ER fluids, involving a decrease $\tau$ with increasing $\dot{\gamma}$ up to a critical shear rate ($\dot{\gamma}_{crit}$), is found. The $\dot{\gamma}_{crit}$ is a transition point of $\dot{\gamma}$ beyond which the fluid begins to exhibit pseudo-Newtonian behavior ($\tau$ increases with $\dot{\gamma}$). In this $\dot{\gamma}$ range, the particle chains appear to be broken by the shear, and further, the particles might have insufficient time to realign themselves along the electric field direction. In a low $\dot{\gamma}$ region, the electrostatic interactions among particles induced by external electric fields are dominant compared to the hydrodynamic interactions induced by the external flow field. The aligned structures begin to break with shear deformation, and the broken structures

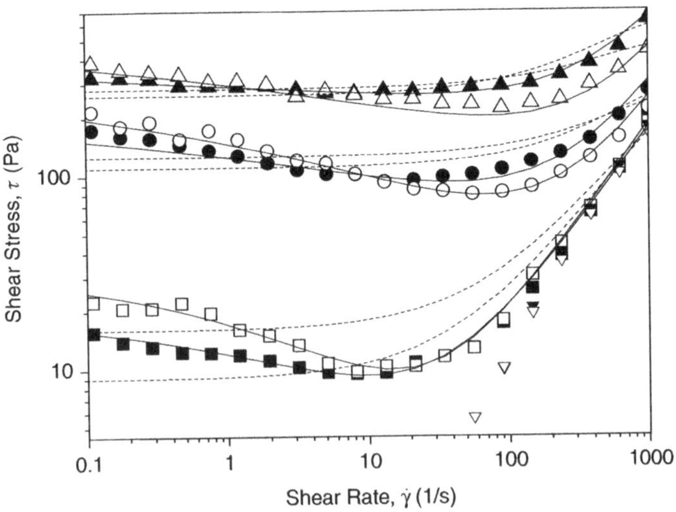


**FIGURE 10.2** ER curves (shear stress versus shear rate) for PANI–$BaTiO_3$ composites–based ER fluids. ∇, 0 kV/mm; □, 0.5 kV/mm; •, 1.5 kV/mm; Δ, 2.5 kV/mm; closed symbols, 10 wt %; open symbols, 25 wt %. The dotted and solid lines are from (10.2) and (10.3), respectively. Reprinted with permission from Ref. 58.

tend to reform the chains by the applied electric field, depending on the magnitude of the applied shear and particle–particle interaction in the fibrils. A decrease in $\tau$ is observed with the increase in the reformed structures when the $\dot{\gamma}$ is small. In other words, as the $\dot{\gamma}$ increases, the destruction rate of the fibrils becomes faster than the reformation rate.

To fit the shear stress curves, the Bingham, De Kee-Turcotte, and Cho-Choi-Jhon models can be adopted. The Bingham model, (10.1) is the simplest one, with two parameters originating from yield stress $\tau_o$ and Newtonian viscosity $\eta_o$ and is widely adopted as a model for ER suspensions (first category).

$$\tau = \tau_o + \eta_o \dot{\gamma}. \tag{10.1}$$

The Bingham model has two flow regimes; a rigid preyield behavior for shear stress less than the field-dependent yield stress and Newtonian flow characteristics beyond the yield stress $\tau_o$ (post-yield region) [59].

The De Kee-Turcotte model assumes $\tau(\dot{\gamma})$ to be independent of $\dot{\gamma}$ (i.e., $\tau(\dot{\gamma}) = \tau_o =$ constant) and adopts various expressions for $\eta(\dot{\gamma})$ [60]

$$\tau = \tau_o + \eta_1 \dot{\gamma} e^{-t_1 \dot{\gamma}} \tag{10.2}$$

where $t_1$ is a time constant having a unit of seconds. In fitting our data in Figure 10.2, we adopted the De Kee-Turcotte model. The recently suggested Cho-Choi-Jhon model, given below, is found to provide a better fitting of the PANI–$BaTiO_3$-based ER fluids as follows [61]:

$$\tau = \frac{\tau_o}{1 + (t_2 \dot{\gamma})^{\alpha}} + \eta_\infty \left(1 + \frac{1}{(t_3 \dot{\gamma})^{\beta}}\right) \dot{\gamma}. \tag{10.3}$$

Here, $\tau_o$ is the yield stress defined as the extrapolated stress from low shear rate region, $\alpha$ is related to the decrease in the shear stress, $t_2$ and $t_3$ are time constants, and $\eta_\infty$ is the viscosity at a high shear rate and is interpreted as the viscosity in the absence of an electric field. The exponent $\beta$ assumes values in the range of $0 < \beta \leq 1$. From Figure 10.2, we observe that the Cho-Choi-Jhon model can cover the stress decrease phenomena at low shear rate regions and provide an accurate value for the real yield stress in the case of PANI–$BaTiO_3$ based ER fluids.

In Figure 10.2, without an electric field, both the two curves exhibit Newtonian behavior with a slope close to 1.0 when exposed to an electric field. The plateau region gets longer when increasing the electric field, which suggests that the electrostatic force gets gradually stronger to resist the hydrodynamic force. Here the PANI–$BaTiO_3$ composite-based ER fluids exhibit better ER effect than that of the pure PANI, which may due to the unique ferroelectric properties as well as the high dielectric constant of $BaTiO_3$ nanoparticles [62].

A $BaTiO_3$ coated-PAN ER system was reported to possess a large polarizable charge in the surface of the coated particles under an external electric field due to the interaction of PANI and $BaTiO_3$ [63], implicating an enhanced polarization strength; this consequently lead to a stronger ER behavior.

The ER curves also indicate that the PANI–$BaTiO_3$ (25 wt %) composite–based ER fluid shows higher shear stress, which may result from much fraction of inorganic $BaTiO_3$ nanoparticles in the composites. To study different behaviors for the two kinds of ER fluids affected by the electric field, their flow curves can be reanalyzed. Here, the dynamic yield stress from a controlled shear rate measurement (CSR) by extrapolating the shear stress at shear rate = 1 $s^{-1}$ [64] is plotte as a function of various electric fields (Fig. 10.3). It is well known that in general, the correlation of the dynamic yield stress and electric field can be presented as follows.

$$\tau_y \propto E^m \tag{10.4}$$

The dependency of the dynamic yield stress on the electric field strength differs from the $E^2$ dependency suggested by the polarization model, depending on the particle concentration [65, 66], particle shape and applied electric field strength. The applied electric field induces electrostatic polarized interactions among the particles and also between the particles and the electrodes. However, the polarization model does not describe the flow effect accurately; in that case, the ER response is influenced by the conductivity mismatch and the interaction

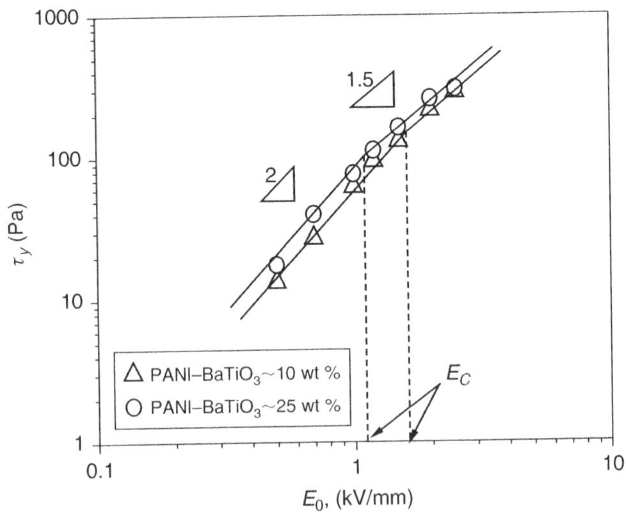


**FIGURE 10.3** Replotted yield stress versus electric field strengths for PANI–$BaTiO_3$ composites–based ER fluids. $E_c$ is the critical electric field. Reprinted with permission from Ref. 58.

between particles and medium. Various ER fluids show different exponents in (10.4). A correlation between the yield stress and the electric field strength is represented in Figures 10.3 and 10.4 To correlate the dynamic yield stress with the broad range of electric field strengths, Choi et al. introduced the universal yield stress equation [67]

$$\tau_y(E_o) = \alpha E_o \left( \frac{\tan h\sqrt{E_o/E_c}}{\sqrt{E_o/E_c}} \right). \tag{10.5}$$

Here, the parameter $\alpha$ depends on the dielectric property of the fluid, the particle volume fraction, and the critical electric field; $E_c$ originates from the nonlinear conductivity effect and can be obtained by the crossover point of the slopes for all ranges of the electric field strengths corresponding to the polarization model (slope = 2) and conductivity model (slope = 1.5), respectively [68]. As shown in Figure 10.4, the estimated $E_c$ was found to be 1.61 kV/mm and 1.11 kV/mm for 10 wt % and 25 wt % of PANI/$BaTiO_3$ composites based on ER fluids. In order to collapse the data into a single curve, we normalized (10.6) using $E_c$ and obtained $\tau_y(E_c) = 0.762\,\alpha E_c$

$$\hat{\tau} = 1.313\hat{E}^{3/2}\tan h\sqrt{\hat{E}} \tag{10.6}$$

Here, $\hat{E} \equiv E_o/E_c$ and $\hat{\tau} \equiv \tau_y(E_o)/\tau_y(E_c)$. It is found that the data obtained from Figure 10.3 collapsed onto to a single curve via normalized universal yield stress equation (10.6), as shown in Figure 10.4.

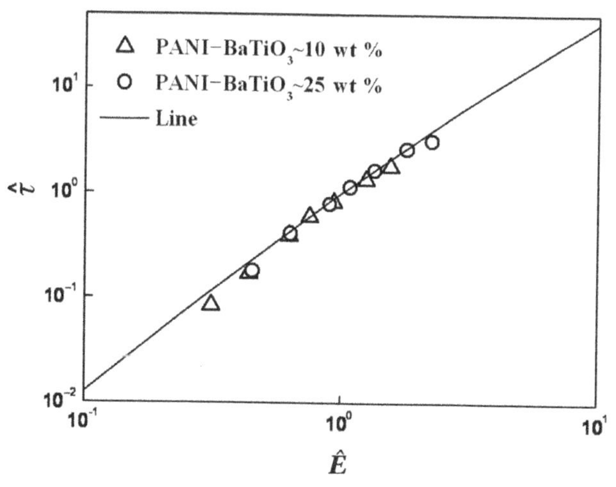


**FIGURE 10.4** Normalized values of $\hat{\tau}$ versus $\hat{E}$ for PANI–$BaTiO_3$ composites–based ER fluids. The solid line is drawn using (10.5). Reprinted with permission from Ref. 58.

### 10.4.2 Oscillatory Testing

While rotational tests provide detailed information on the dependence of shear stress behavior on shear rate and electric field strength, in addition to the dynamic yield stress and static yield stress, the oscillatory tests can be employed to investigate the linear and nonlinear viscoelastic properties for the formation of chain-like structures under various electric field strengths. Especially, small-amplitude oscillatory shear tests are the most widely used method for studying the nonlinear viscoelastic behavior because the oscillatory shear mode does not generate any sudden jumps in speed or position.

The linear viscoelastic parameters of storage modulus ($G'$, the in-phase stress component) and loss modulus ($G''$, the out-of-phase stress component) depend strongly on the applied electric field strength, strain amplitude ($\gamma_o$), strain frequency, and material properties. During an oscillatory shear strain, the ER material may deform like a solid over some part of the deformation cycle and like a liquid over the other part of the deformation cycle. The behavior of an ER fluid consists of single-width chains of particles undergoing oscillatory shear and giving rise to the G′, with a small fraction of free chains connected to only one electrode and contributing to the $G'' \sim \omega^{-1}$[69]. The $G'$ is observed to be larger than $G'' \sim \omega^{-1}$, and these values are independent of the strain within the region of applied strain, which are so-called linear viscoelastic regions [70]. The onset of nonlinear behavior is attributed to small rearrangements within unstable aggregates of particles, which decreases the electrostatic contribution to the stress as well as the strong modulus and leads to increased energy dissipation. Hence, to explore the strength of the chains at rest as a function of the applied electric field, dynamic mechanical tests are carried out under various electric field strengths.

Figure 10.5 shows a strain amplitude sweep tested for PANI–mesoporous MCM-41 nanocomposite based ER fluids, defining the critical deformation over which the formed chainlike structure will be destroyed leading to a less meaningful nonlinear oscillatory test [71]. From the strain amplitude sweep, the storage modulus ($G'$) value tested without an electric field increases sharply when we apply electric field, and the loss modulus ($G''$) also shows the same trend. This can be attributed to the chain-like structure formed due to the external electric field. In the initial range of strain, no matter whether the electric field was used or not, the storage modulus ($G'$) shows little dependence on the strain, then declined gradually due to the broken chain structure. At the high range of strain, liquid like behavior is observed, which explains why the loss modulus ($G''$) value is larger than the storage modulus ($G'$). Thus a critical strain at about $5 \times 10^{-4}$ should be selected to ensure that no significant change in structure exists, by which way we can study the visocoelastic property for the ER fluid. We believe that if the amplitude is over this value, even though the electric field is absent, the structure of the sample will be irreversibly altered or even completely destroyed, while at smaller values, the structure of the sample will be sustained.

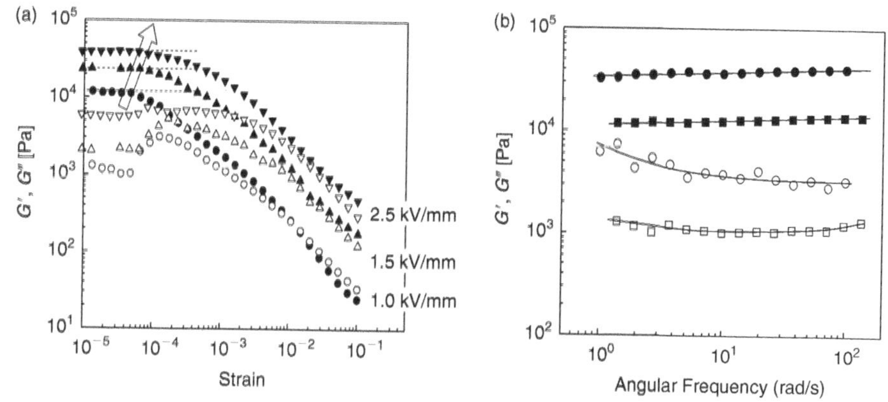


**FIGURE 10.5** (a) Strain amplitude versus dynamic moduli under various electric field strengths at 10 rad/s of angular frequency. *Solid symbols*, $G'$; *open symbols*, $G''$. (b) angular frequency versus dynamic moduli under 1 (■,□) and 2 (•, ○) kV/mm using a strain of $5\times10^{-4}$ of PANI-MCM-41-based ER fluid. Reprinted with permission from Ref. 71.

Based on a properly chosen critical strain, we investigated frequency sweeps ($\gamma = 5\times10^{-4}$) at four different electric field strengths. The behaviors of storage modulus ($G'$) and loss modulus ($G''$) as a function of angular frequency are indicated in Figure 10.5b. It is apparent that at a fixed electric strength, both the storage modulus and loss modulus show constant values over a long frequency range, which suggests that the sample possesses a very strong solid-like structure rather than a liquid-like structure [72, 73]. However, it is worth noting that the storage modulus is always higher that the loss modulus over the whole range of frequency. This observation provides additional evidence of the predominant elastic property.

### 10.4.3 Compression Tests

Generally, a high shear yield stress value will play a very important factor in designing industrial engineering devices. Although much effort has been paid in exploring and developing superior ER fluids that possess very high shear yield stress, the highest shear yield stress reported is still lower than 5 kPa by far. Thus researchers focused on investigating new strategies to achieve higher force performance in ER fluids. Among them, the compression mode has attracted much attention because much stronger compressive strength can be obtained. A typical schematic for an ER fluid under compression mode is shown in Figure 10.6, in which a dilute ER fluid composed of dielectric colloids in an insulating medium is compressed between two parallel plates (electrodes). The upper plate can be fixed, and then the lower plate can be moved along the working axle to compress the filled ER fluids. Results indicate that the volume concentration of ER fluids plays crucial role in affecting the resulting

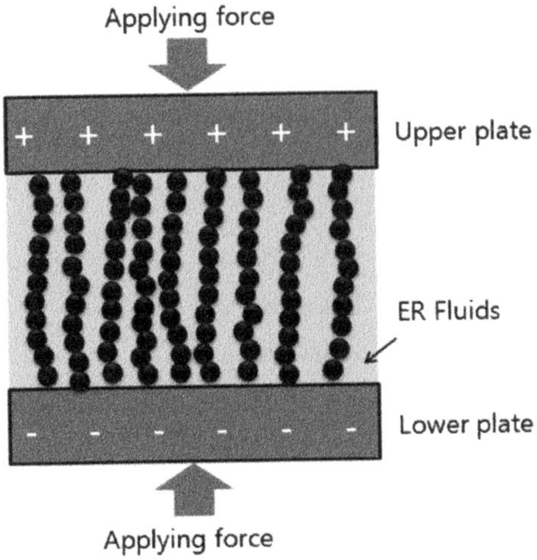


**FIGURE 10.6** Compression mode of an ER fluid in an electric field.

compressive strength because condensed dielectric colloids particles prefer to form irregular columns or networks, which may lead to enhanced compression strength [74]. The compression loading was also found to increase nonlinearly with decreasing electrode gap. In addition, a critical compressive strain was reported by Tian's group [75, 76], in which a high compressive modulus under a small compressive strain $<0.1$ can be obtained, while the compressive stress indicates an exponential function with the compressive strain when the strain is $>0.1$. All of the compression stress values obtained are higher than the shear stresses.

## 10.5 DIELECTRIC SPECTRA

To better understand the ER performances of the ER fluids, their dielectric properties are measured and then correlated with ER properties. When considering the interfacial polarization as the origin of the ER effect, it is well known that the dielectric constant ($\varepsilon'$) and loss factor ($\varepsilon''$) are representative quantities for the interfacial polarization of ER suspensions. Therefore, the dielectric spectra of ER fluids provide valuable information on the analysis of electrical polarization properties and interpretation the flow behavior.

The dielectric relaxation spectra of ER fluids can be examined by using an impedance analyzer. The parameters related with dielectric relaxation spectra are then obtained using the following calculation

$$R_p = \frac{t}{\varepsilon_r'' \omega \varepsilon_o A}; \quad \varepsilon_r'' = \tan \delta \varepsilon', \ \varepsilon_r' = \frac{t \cdot C_p}{A \cdot \varepsilon_o}$$
$$\therefore R_p = \frac{1}{\tan \delta \cdot C_p \cdot \omega} \tag{10.8}$$

where liquid capacitance, $C_p$; area of disk, $A$; angular frequency, $\omega$; electrode gap, $t$; and tan δ are measured from the experiment, and the equivalent parallel resistance, $R_p$, is calculated using the measured values. The relative dielectric constant indicates the energy value of a material in an electric field. It is represented as a complex quantity. The relative constant ($\dot{\varepsilon}_r$) is defined as the ratio of the material's dielectric constant ($\dot{\varepsilon}$) to that of vacuum.

$$\dot{\varepsilon}_r = \frac{\dot{\varepsilon}}{\varepsilon_0}, \text{ or } \dot{\varepsilon}_r = \varepsilon_r' - j\varepsilon_r'' \tag{10.9}$$

$$\tan\delta = \frac{\varepsilon_r''}{\varepsilon_r'} \tag{10.10}$$

where $\varepsilon_r'$ is relative dielectric constant and $\varepsilon_r''$ is dielectric loss. The HP 16452A and HP impedance analyzer–LCR meter uses the capacitive method for obtaining relative permittivity by measuring the capacitance of a material, which is sandwiched between parallel electrodes.

Upon investigating the influence of dielectric properties on the ER effect, it is found that the relaxation frequency at which the dielectric loss factor has a local maximum is very important for ER, because the relaxation frequency of an ER fluid is proportional to its rate of polarization [3]. In addition, it shows the relation of the dielectric properties under a flow condition as well as the influence of flow. Hao et al. [13] pointed out the importance of the dielectric loss factor had to be >0.1 at the frequency of 1 kHz.

A Cole–Cole diagram can be used to obtain the complex dielectric constants at intermediate frequencies eliminating the necessity for making measurements. The plot of $\varepsilon''$ *vs.* $\varepsilon'$ displays a semicircle with its center displaced below the $\varepsilon'$ axis.

$$\varepsilon^* = \varepsilon_\infty + \frac{\varepsilon_s - \varepsilon_\infty}{1 + (j\omega\tau_{c-c})^{1-\alpha}}; 0 \le \alpha \le 1 \tag{10.11}$$

where $\tau_{c-c}$ is the mean relaxation time, and $\alpha$ is a constant for a given material, having a value $0 \le \alpha \le 1$; $\alpha = 0$ for Debye relaxation. The Cole–Cole relaxation shows that $\varepsilon'$ decreases more slowly with $\omega$ than the Debye relaxation. With increasing $\alpha$, the loss factor $\varepsilon''$ is broader than the Debye relaxation and the peak value, $\varepsilon_{max}$ is smaller. The electrostatic contribution to the loss modulus was assumed to arise from a small fraction of free chain–chain connection to electrode. At large frequencies, the motion relaxations as electrostatic force keep the chains aligned with the electric field. $G''$ thus depends on the fraction of free chains as well as their length and varies as $G'' \sim \omega$ at small $\omega$ and $G'' \sim \omega^{-1}$ at large $\omega$. By accounting for hydrodynamic screening, it was observed that $G'' \sim \omega^0$ over a wide range of intermediate frequencies [77].

The particle size effect on ER properties was also analyzed via the dielectric spectra shown in Figure 10.7. The figure shows $\varepsilon''$ and $\varepsilon'$ as a function of the frequency, and Cole–Cole plots for PAPMMA-based ER fluids [78]. These are

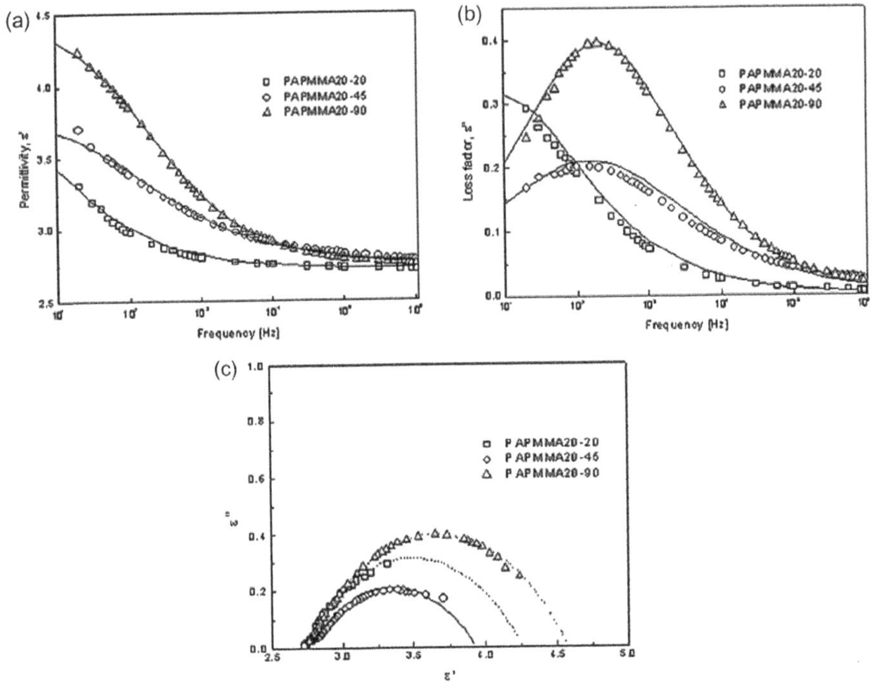


**FIGURE 10.7** (a) Permittivity, (b) loss factor vs. frequency, and (c) Cole–Cole plot for PAPMMA-based ER fluids. PAPMMA *x-y; x*; *g*-loaded aniline amount/100 g of PMMA; *y*; PMMA particle size/μm. Reprinted with permission from Ref. 78.

typical results for the interfacial polarization of suspensions, including ER fluids. Lines in the Figure are fits of the Cole–Cole formula (10.8) [79, 80]. The relaxation time for interfacial polarization of ER fluids is related to the yield stress and stress enhancement under the applied electric field.

## 10.6 SUMMARY

The ER fluids have shown interesting rheological phenomena compared to conventional complex fluids. The ER phenomena and mechanisms, along with the types of electroresponsive materials and their characterization are summarized. More advanced ER materials with better performance will be continuously designed and fabricated to explore further industrial applications.

## REFERENCES

1. T.C. Halsey and W. Toor, *Phys. Rev. Lett.*, 65, 2820 (1990).
2. X.P. Zhao and J.B. Yin, *Chem. Mater.*, 14, 2258 (2002).

3. Y.C. Cheng, J.J. Guo, G.J. Xu, et al., *Colloid Polym. Sci.*, 286, 1493 (2008).
4. R. Tao and J.M. Sun, *Phys. Rev. Lett.*, 67, 398 (1991).
5. W.Y. Tam, G.H. Yi, W. Wen, et al., *Phys. Rev. Lett.*, 78, 2987 (1997).
6. S.G. Kim, J.W. Kim, W.H. Jang, et al., *Polymer*, 42, 5005 (2001).
7. W.J. Wen, X.X. Huang, and P. Sheng, *Soft Matter*, 4, 200 (2008).
8. H.J. Jin, H.J. Choi, S.H. Yoon, et al., *Chem. Mater.*, 17, 4034 (2005).
9. T. Hao, *Adv. Mater.*, 13, 1847 (2001).
10. W.M. Winslow, *J. Appl. Phys.*, 20, 1137 (1949).
11. H. Uejima, *J. Appl. Phys.*, 11, 319 (1972).
12. M.S. Cho, H.J. Choi, I.J. Chin, and W.S. Ahn, *Micropor. Mesopor. Mater.*, 32, 233 (1999).
13. H. Block, J.P. Kelly, A. Qin, and T. Watson, *Langmuir*, 6, 6 (1990).
14. Y.M. Han and S.B. Choi, *Smart Mater. Struct.*, 15, 1438 (2006).
15. M. Gurka, D. Adams, L. Johnston, and R. Petricevic, *J. Phys.: Conf. Ser.*, 149, 012008 (2009).
16. M. Zhang, X. Gong, and W. Wen, *Electrophoresis*, 30, 3116 (2009).
17. M. Zhang, J. Wu, X. Niu, et al., *Phys. Rev. E*, 78, 066305 (2008).
18. O. Yasufumi, *Bull Jpn. Soc. Printing Sci. Technol.*, 38, 231 (2001).
19. D.L. Klass and T.W. Martinek, *J. Appl. Phys.*, 38, 67(1967).
20. H. Uejima, *Jpn. J. Appl. Phys.*, 11, 319 (1972).
21. Yu. F. Deinega and G.V. Vinogradov, *Rheol. Acta*, 23, 636 (1984).
22. H.J. Choi and M.S. Jhon, *Soft Matter*, 5, 1562 (2009).
23. Y.D. Kim and J.H. Kim, *Colloid Polym. Sci.*, 286, 631 (2008).
24. H. See, H. Tamura, and M. Doi, *J. Phys. D: Appl. Phys.*, 26, 746 (1993).
25. L.C. Davis, *J. Appl. Phys.*, 72, 1334 (1992).
26. R.A. Anderson, *Langmuir*, 10, 2917 (1994).
27. Y. Chen, A.F. Sprecher, and H. Conrad, *J. Appl. Phys.*, 70, 6796 (1991).
28. R. Tao and Q. Jiang, *Phys. Rev. Lett.*, 73, 205 (1994).
29. P.M. Adriani and A.P. Gast, *Phys. Fluid*, 31, 2757 (1988).
30. H.H. Clarx and G. Bossis, *Phys. Rev. E*, 48, 2721 (1993).
31. D.J. Klingenberg, S. Van Frank, and C.F. Zukoski, *J. Chem. Phys.*, 94, 6160.
32. L.C. Davis, *Appl. Phys. Lett.*, 60, 319 (1992).
33. B. Khusid and A. Acrivos, *Phys. Rev. E*, 52, 1669 (1995).
34. K.Q. Lu, R. Shen, X.Z. Wang, et al., *Chine. Phys.*, 15, 1009 (2006).
35. J.B. Yin and X.P. Zhao, *Chem. Phys. Lett.*, 398, 393 (2004).
36. H.J. Choi, M.S. Cho, and K. To, *Physica A*, 254, 272 (1998).
37. C.J. Gow and C.F. Zukoski, *J. Colloid Interf. Sci.*, 136, 175 (1990).
38. J.B. Jun, J.W. Kim, and K.D. Suh, *Macromol. Chem. Phys.*, 203, 1011 (2002).
39. J.W. Goodwin, G.M. Markham, and B. Vincent, *J. Phys. Chem. B*, 101, 1961 (1997).
40. W. Zhou and K.S. Zhao, *Colloids Surf. A*, 317, 10 (2008).
41. J. Plocharski, H. Drabik, H. Wycislik, and T. Ciach, *Synth. Met.*, 88, 139 (1997).

42. S.H. Baek, B. Kim, and K.D. Suh, *Colloids Surf. A*, 316, 292 (2008).
43. J. Trlica, P. Sáha, O. Quadrat, and J. Stejskal, *Physica A*, 283, 337 (2000).
44. D. Chotpattananont, A. Sirivat, and A.M. Jamieson, *Colloid Polym. Sci.*, 282, 357 (2004).
45. H.J. Choi, M.S. Cho, J.W. Kim, et al., *Int. J. Mod. Phys. B*, 15, 988 (2001).
46. S.G. Kim, J.Y. Lim, J.H. Sung, et al., *Polymer*, 48, 6622 (2007).
47. M. Stenicka, V. Pavlinek, P. Saha, et al., *Colloid Polym. Sci.*, 286, 1403 (2008).
48. B.D. Chin and O.O. Park, *J. Colloid Interf. Sci.*, 234, 344 (2001).
49. A.G. MacDiamid, *Current Appl. Phys.*, 1, 269 (2001).
50. Y.D. Kim and I.C. Song, *J. Mater. Sci.*, 37, 5051 (2002).
51. I.S. Sim, J.W. Kim, H.J. Choi, et al., *Chem. Mater.*, 13, 1243 (2001).
52. B.D. Chin, Y.S. Lee, and O.O. Park, *J. Colloid Interf. Sci.*, 201, 172 (1998).
53. J.I. Sohn, M.S. Cho, H.J. Choi, et al., *Macromol. Chem. Phys.*, 203, 1135 (2002).
54. H. Block, J.P. Kelly, A. Qin, and T. Waston, *Langmuir*, 6, 6 (1990).
55. Q. Cheng, V. Pavlinek, Y. He, et al., *Colloid Polym. Sci.*, 289, 799 (2011).
56. H. See, H. Tamura, and M. Doi, *J. Phys. D: Appl. Phys.*, 26, 746 (1993).
57. H. Uejima, *Jpn. J. Appl. Phys.*, 11, 319 (1972).
58. F.F. Fang, J.H. Kim, H.J. Choi, and Y. Seo, *J. Appl. Polym. Sci.*, 105, 1853 (2007).
59. Y.D. Kim and D.H. Park, *Synth. Met.*, 142, 147 (2004).
60. D. De Kee, and G. Turcotte, *Chem. Eng. Commun.*, 6, 273 (1980).
61. M.S. Cho, H.J. Choi, and M.S. Jhon, *Polymer* 46, 11484 (2005).
62. J.H. Wei, J. Shi, J.G. Guan, and R.Z. Yuan, *J. Mater. Sci.*, 39, 3457 (2004).
63. J.H. Wei, J. Shi, Z.Y. Liu, et al., *Int. J. Mod. Phys. B*, 19, 1423 (2005).
64. D.F. James and B.C. Blakey, *Korea-Australia Rheol. J.*, 16, 109 (2004).
65. D.J. Klingerberg, F. Van Swol, and C.F. Zukoski, *J. Chem. Phys.*, 94, 6170 (1991).
66. H. See, *J. Ind. Eng. Chem.*, 10, 1132 (2004).
67. H.J. Choi, M.S Cho, J.W. Kim, et al., *Appl. Phys. Lett.*, 78, 3806 (2001).
68. H.J. Choi, I.S. Lee, J.H. Sung, et al., *J. Colloid. Interf. Sci.*, 295, 291 (2006).
69. C. Jordan, M.T. Shaw, and T.C.B. McLeish, *J. Rheol.*, 36, 441 (1992).
70. J.W. Kim, M.S. Cho, H.J. Choi, et al., *Polym. Testing*, 20, 913 (2001).
71. M.S. Cho, H.J. Choi, and W.S. Ahn, *Langmuir*, 20, 202 (2004).
72. F. Gandhi and W.A. Bullough, *J. Int. Mater. Sys. Struct.*, 16, 237 (2005).
73. F.F. Fang and H.J. Choi, *J. Appl. Phys.*, 103, 07A301 (2008).
74. F. Yang, *J. Colloid Inter. Sci.*, 192, 162 (1997).
75. Y. Tian, Y. Meng, H. Mao, and S. Wen, *Phys. Rev. E*, 65, 031507 (2002).
76. Y. Tian, M. Zhang, X. Zhu, et al., *Smart Mater. Struct.*, 19, 035009 (2010).
77. T.C.B. McLeish, T. Jordan, and M.T. Shaw, *J. Rheol.*, 35, 427 (1991).
78. M.S. Cho, Y.H. Cho, H.J. Choi, and M.S. Jhon, *Langmuir*, 19, 5875 (2003).
79. B. Khusid and A. Acrivos, *Phys. Rev. E* 52, 1669 (1995).
80. A.G. MacDiamid, *Curr. Appl. Phys.*, 1, 269 (2001).

# CHAPTER 11

# RHEOLOGY AND PROCESSING OF POLYTETRAFLUOROETHYLENE PASTE

SAVVAS G. HATZIKIRIAKOS
Department of Chemical and Biological Engineering, The University of British Columbia, Vancouver, BC, Canada

## CONTENTS



*Applied Polymer Rheology: Polymeric Fluids with Industrial Applications*, First Edition. Edited by Marianna Kontopoulou.

## 11.1 INTRODUCTION

Polytetrafluoroethylene (PTFE) is a material of great commercial and technological value due to its unique properties, such as resistant to heat, excellent electrical insulation properties, low coefficient of friction over a wide temperature range and high resistance to most solvents and other corrosive compounds [1–5]. Since its discovery by Roy Plunkett in 1938 [1], it has "revolutionized the plastic industry and led to vigorous applications not otherwise possible" [2]. These include wire coating, tubing and tape extrusion, and compression molding with a variety of products such as high-performance wires and fabrics, medical implants, components for electronic devices, fibers, sealants, gaskets, membranes, and high stretch tapes among others. It is interesting, that PTFE was produced on a limited scale in the early 1940s, and was used shortly thereafter by the Manhattan Project in containers for highly corrosive elements during uranium separation experiments.

It is difficult to process PTFE resin by melt techniques, due to its high melting point and high viscosity [3]. Techniques involving cold pressing, paste extrusion and sintering have to be employed instead. These techniques resemble those used in metallurgy, but are unconventional as far as thermoplastics processing is concerned. For example, granular PTFE resin can be fabricated into an end product by pressing the resin in a mold at room temperature to produce a preform of a desired shape. Preforming serves to reduce voids between the resin particles [3]. This renders the final product mechanically strong. The preform is then sintered at a temperature above the melt temperature of PTFE to allow coalescence of particles into a dense homogeneous structure. During sintering, the rate of temperature rise is important to allow the temperature distribution to be as uniform as possible. After sintering, the product is cooled at a specific cooling rate, which is also important in influencing the mechanical properties of the product. Fast cooling will lead to lower crystallinity and, consequently, higher tensile strength and elongation at break, and better flex life for the product [3, 4]. On the other hand, a slow cooling rate will result in a product with greater creep resistance, lower permeability to gases and solvents, and lower residual stress and distortion. Sintering and cooling may be done under pressure or freely. Pressure cooling may reduce the tendency of the product to distort, although it is more costly [4].

In the production of continuous moldings such as pipes, rods, and wires, the method described above is not appropriate. Instead, a reciprocating ram extruder is used [3]. Moreover, the PTFE powder is mixed with a lubricant (16–25 wt %) to produce a paste. The paste extrusion of PTFE fine powder is the main focus of this chapter. Figure 11.1 shows the various steps of the process. The resin is initially mixed with a lubricating liquid of low boiling point, for ease of removal at a later stage of the process. Ideally, a hydrocarbon with surface tension <18 mN/m should be used to allow complete wetting of the powder [5]. However, since this is not practical, liquids with fairly low surface tensions are used instead. The amount of liquid in the mixture may vary from 16 to 25 wt % [3]. Shear-free mixing is performed by simply rotating the container, at a temperature <19°C. The liquid acts as a lubricating agent and as a cushion between particles so as to eliminate mechanical damage, which may occur as particles slide past one another [6]. The powder-lubricant mixture, which is called a paste, is allowed to age at room temperature for several hours before extrusion. This allows the paste to equilibrate at a temperature higher than the mixing temperature, which results

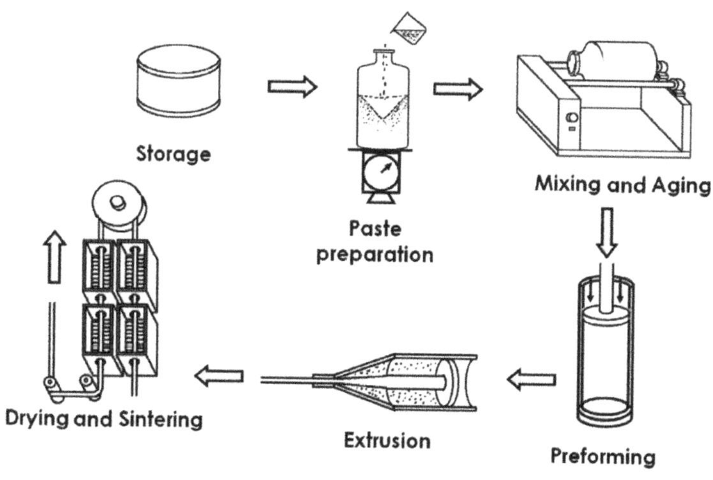


**FIGURE 11.1** PTFE paste extrusion for a wire-coating process.

in the lowering of the lubricant surface tension and viscosity, and hence, better wetting of the powder particles [3].

The next step of the process involves the pressing of paste to produce a preform. This serves to eliminate air voids that will render the final product mechanically weak [6]. A preforming pressure of 2 MPa is typically used. The preform is then extruded in a ram extruder through a die of specific shape, depending on the product requirement. In a wire-coating process, a special annular (cross-head) die is used, which allows the introduction of wire into the extrudate. The extrudate is then passed through a dryer to evaporate the lubricating liquid before being sintered in an oven. Electrical spark tests may then follow. In the production of tubes and pressure hoses, a similar annular die is used. To make tapes, the preform is extruded through a simple conical entry die and calendered to a flat, continuous product before drying. In this case, however, the product is not sintered [3].

Each step of the paste extrusion process is critical in determining the final product properties. However, various operating variables, such as lubricant content in the paste, preforming pressure and duration, and the various die design parameters, affect the process and product in ways that have not yet been fully understood. All these aspects of the process are summarized in this chapter, including the chemical nature of PTFE and its physical properties, the parameters affecting the various steps of the paste extrusion process, rheological aspects of PTFE paste, and finally modeling aspects of PTFE paste extrusion and its experimental verification.

## 11.2 CHEMICAL AND PHYSICAL PROPERTIES OF PTFE

PTFE is a member of the fluoropolymer group, having a chemical formula of [-$(CF_2\text{-}CF_2)_n$-]. It is a perfluorinated, high molecular weight polymer, having a straight chain molecular configuration. PTFE is chemically inert, resistant to heat, and has excellent electrical insulation properties, as well as a low coefficient of friction over a wide temperature range [3, 5]. These properties can be attributed to its molecular structure, in particular the C-F bond. The two types of bonds that made up a PTFE chain, the C-F bonds and C-C bonds, are extremely strong [7, 8], causing PTFE to have excellent mechanical strength and resistance to heat. The fluorine atoms can be envisioned to wrap around the backbone of the C-C chain, providing a protective shield to the chain from chemical attack. The size and electronic state of the fluorine atoms is exactly right, accounting for the chemical inertness and stability of PTFE [3, 5]. The protective shield also reduces surface energy, causing PTFE to have a low coefficient of friction and non-stick properties [5, 9]. PTFE, with its thermal and chemical stability, makes an excellent electrical insulator. It does not absorb water and its volume resistivity remains unchanged even after prolonged soaking. The dielectric constant of PTFE also remains constant over a very wide temperature range [5].

Since PTFE is insoluble in many common solvents, its molecular weight cannot be determined by conventional techniques. In practice, the number average molecular weight ($M_n$) is usually estimated from the standard specific gravity (SSG) of the polymer. Higher SSG implies greater crystallinity and hence, lower molecular weight [5, 10, 11]. Due to the linearity of PTFE molecules, the crystallinity of a virgin PTFE resin may be as high as 92–98% [5]. As a result, the SSG of PTFE is high for a polymer, typically ranging from 2.1 to 2.3. $M_n$ has also been correlated to the second heat of melting, or the heat of recrystallization of PTFE ($\Delta H_{m2}$), obtained using differential scanning calorimetry (DSC) [5, 12]. While most of these correlations apply to homopolymers, comparison of PTFE molecular weight, regardless of whether the resins contain other comonomers, can be made by considering the resin melt creep viscosity instead. The melt creep viscosity, as detailed in U.S. Patent 3,819,594 [13], is higher for a higher molecular weight PTFE resin [10].

The melting point of virgin PTFE is 342°C [14], which is high for a thermoplastic polymer. DSC analysis indicates that the melting point of PTFE is irreversible [5]. A previously melted (sintered) PTFE will show a lower melting point of about 327°C, which is the value that is often reported in the literature (i.e., second melting point). This implies that cooling does not recrystallize the chains back to the original virgin configuration, making the resin less crystalline. During melting, a volume increase of 30% is typical [14]. The melt is stable, since even at 380°C, the melt viscosity is relatively high at approximately 10 GPa.s. It is due to this high melting temperature and melt viscosity that it is not possible to fabricate PTFE resin using conventional polymer melt processes [3, 5].

Besides the melting point, PTFE has other transition temperatures, two of which are particularly important (see phase diagram in Ref. 14). Under ambient pressure conditions, the first transition occurs at 19°C. This is an especially important transition point due to its proximity to the ambient temperature. At this transition temperature of 19°C, a PTFE molecule undergoes a slight untwisting from a 180° twist per 13 $CF_2$ groups to a 180° twist per 15 $CF_2$ groups. The chain segments change from a perfect three-dimensional order to a less ordered one, and this results in a volume increase of about 1.3%. The second transition occurs at 30°C. Above this temperature, the number of $CF_2$ groups per 180° twist remains the same at 15. However, the extent of disorder of the rotational orientation of molecules about their long axis is increased. The total volume change as temperature is increased from below 19°C to above 30°C can be as high as 1.8% [4, 16]. This change in volume affects the density of the PTFE resin.

Fine powder resin is susceptible to mechanical damage above the transition temperature of 19°C. Below 19°C, shearing will cause crystal units to slide pass each other. Above 19°C, however, molecules are packed more loosely and shearing will cause the unwinding of molecules, creating fibrils [3, 6]. It is important that fine powder resin does not undergo pre-mature fibrillation. The presence of fibrils before processing will result in excessive extrusion pressure and mechanically defective products [3, 6]. Because of this, the storage and transportation of fine powder resin are carried out at temperatures $<19°C$ [15].

## 11.3 COMMERCIAL PRODUCTION OF PTFE: POLYMERIZATION TECHNIQUES

In the production of PTFE, tetrafluoroethylene monomer (TFE) is polymerized in a highly exothermic reaction. It is essential that the molecular weight of the resulting polymer is extremely high. Low molecular weight polymer will not have the strength needed in end use applications and will be of little commercial value.

Polymerization of TFE is carried out in an aqueous medium involving either as suspension polymerization or as emulsion polymerization [3, 5]. Although the two procedures result in the same high molecular weight PTFE polymer, the products are distinctly different. The granular product (from suspension polymerization) can be molded in various forms without the use of a lubricant. However, the resin obtained from aqueous dispersion polymerization cannot be molded but has to be fabricated by dispersion coating, in the case of the concentrated dispersion, or by paste extrusion, in the case of fine powder resin [3, 4, 6].

## 11.4 PASTE EXTRUSION: GENERAL ASPECTS

Paste extrusion has been widely used as a fabrication technique for many useful objects, from everyday products such as toothpaste, pencil leads, cosmetic pencils, animal foodstuffs, and food flavorings, to less common products such as ceramic components, catalyst supports, bricks, seal tapes, PTFE tubes, wires, and cables [3, 17]. It is surprising that the subject of paste extrusion (particularly for PTFE) has been studied systematically only recently [3, 18–20]. It has been presented previously either in entirely empirical terms or as an extension of molten polymer flow and extrusion [6].

Paste is essentially a mixture of liquid and solid particles [17]. As discussed above, the various steps involved in the paste extrusion include paste preparation and aging, performing, extrusion and drying, and sintering (if necessary). The experimental aspects and findings in each of these steps are summarized in the following subsections.

### 11.4.1 Paste Preparation and Aging

PTFE fine powder resins must be in a completely powdered form, so as to enable even pouring when it is blended with an extrusion aid (lubricant). Strong vibrations and shock should be avoided as much as possible during transport, because this may cause the powder to form lumps [3, 6]. If the powder is to be stored, the ideal storage conditions are a dry place with a temperature range of 5–15°C (well below the first transition temperature). In this way, the powder will be less susceptible to lumping, and even if it occurs, it is easy to restore the resin to its powdered form. If lumps exist in the powder before blending with the lubricant (or extrusion aid), the powder should be sieved with care to avoid applying too much force to the powder.

The lubricant (or lube or extrusion aid) is used in order to enable smooth and continuous flow since dry powder is almost impossible to extrude. In addition, by filling the spaces between the particles with the lubricant rather than air, the paste will become resistant to compressive loading, without increasing the interparticle contact area. This also reduces the adhesive forces between particles, particularly in the case of PTFE paste, since the interfacial tension for polymer-on-lubricant is much less than for polymer-on-air. This allows particles to rearrange more easily in response to mechanical forces, without being deformed [6]. Consequently, the particles will remain isotropic in nature after compression, a condition that is not attainable with a dry powder [6].

The extrusion aid (lubricant) must be able to completely saturate and wet the particles of the resin, and must be easily and rapidly removable from the product after extrusion without leaving any residue that would provide odor and/or color. The extrusion aids ordinarily used are typically clear, light, aliphatic hydrocarbon liquids. Liquids with viscosity of 0.0005–0.005 Pa · s are preferred. The amount of extrusion aid to be added to the resin varies according to the application and the processing conditions. Ordinarily 16–25 wt % in weight of extrusion aid is used [3, 6, 18, 19].

The mixing of the powder and the liquid is crucial in ensuring a uniform flow of the paste. To prevent shear damaging of the resin, the blending operation must be performed with the resin temperature remaining below the transition temperature. A clean, dry, wide-mouth container is filled with the powder up to about two-thirds of its capacity. The prescribed amount of lubricant is then added and the container is covered and sealed so the lubricant does not volatilize. The container is then agitated, for example, by placing it horizontally on a two-roller jar mill for approximately 10–20 min at approximately 30–45 rpm.

Before preforming (next step of paste extrusion process), it is suggested that the container be left sealed for approximately 5–15 h at room temperature (23°C) or above to allow the lubricant (extrusion aid) to completely wet and permeate the surface of any powder not sufficiently permeated by the blending process.

### 11.4.2 Preforming

During preforming, the paste is placed in a cylindrical billet and, by means of a piston, the pressure is gradually increased to remove the air voids that would otherwise render a final product mechanically weak. In this way a cylindrical rod that is fitted into the extruder barrel is formed [3]. In PTFE paste processing, the preforming stage is carried out at room temperature, although it is not temperature sensitive [6]. However, the application of stress introduces another problem since it may cause the liquid component of a paste to move through the solid matrix in the radial and axial directions causing a liquid maldistribution throughout the paste [21]. This is a problem in most paste systems such as ceramic [22, 23], microcrystalline cellulose paste [24, 25], and talc-based pastes [26].

The level of the preforming pressure and its duration significantly affect the quality of the preform. In addition, the magnitude of the pressure needed to produce a preform of uniform density depends on the molecular weight (standard specific gravity) of the resin [18]. It has been shown that lower molecular weight PTFE resin (higher standard specific gravity) has softer particles, which are easier to compact [18, 27, 28]. Finally, it has been shown that making a preform by compacting the paste on both ends improves the uniformity of preform density without sacrificing its quality in terms of lubricant distribution [28].

A lack of adequate pressure will result in a preform of nonuniform density that will extrude unsteadily, resulting in an unacceptable final product [27, 28]. During preforming, the applied pressure compacts the particles making those adjacent to the wall of the preforming unit undergo plastic deformation that results in a smooth film of deformed powder surrounding the preform. Thanks to this layer, the rest of the resin particles remain spherical even after high preforming pressure [6].

The effect of lubricant's physical properties, such as viscosity and surface tension on performing of PTFE pastes, was examined in detail [19, 29, 30]. It was reported that increasing the preforming time and pressure improves the uniformity of preform density, indicating that the process of preforming is time dependent. However, lubricant migration becomes important at longer times. Therefore, the applied pressure and its duration need to be optimized depending on the physical properties of lubricant. A pressure of 2 MPa applied for about 30 seems to be optimum for most cases. The use of a lubricant of high viscosity can produce a more uniform preform as liquid migration is minimized. For the lubricants used in PTFE processing, a higher viscosity means also a lubricant of lower vapor pressure and thus minimum liquid evaporation. Increasing the wettability of lubricant with PTFE produces better mixture/pastes [30]. Furthermore, excellent wetting would produce a uniform preform even under extreme conditions [19, 30].

### 11.4.3 Paste Flow

In this section, the mechanism of PTFE paste flow and structure formation during flow through gradual tapered dies (contractions) is discussed. Apart from such macroscopic rheological experiments, several other techniques have been used to understand the deformation pattern involved in PTFE paste flow, including SEM, visualization experiments and theoretical flow analysis; these are also discussed in this chapter.

Figure 11.2 depicts a typical pressure transient obtained during PTFE paste extrusion using a capillary rheometer [19, 29, 30]. The horizontal axis labeled as distance in the barrel is equivalent to time once the constant piston speed is used to divide the distance. The curve has been divided into three zones [18, 31]. In zone I, the pressure gradually increases and attains a maximum. This pressure peak has been thought to be due to the initial filling and wetting of conical zone of the die [6, 18, 31, 32]. Experiments performed by using dies having a 180° angle, have also shown the existence of this peak. In other cases, the conical zone was prefilled before extrusion and the pressure peak still appeared [18, 33, 34]. This maximum is essentially due to the finite compressibility and the yield stress of the paste in the barrel. Pastes in general are visco-elasto-plastic materials exhibiting a small but finite yield stress [3, 33, 35]. It is reasonable to argue that the existence of a yield stress causes the appearance of the yield pressure (peak pressure). Until this point is reached, the paste flows in the die at very low speed. The paste is been compressed in the barrel and as a result the static pressure increases gradually. During this compression period, the paste is in a state of jamming, which is defined as "the conversion of a liquid system into a solid by imposed stress" [36]. This essentially means that there is a number of immobile clusters of particles upstream of the die entrance that is responsible for the jamming [37, 38]. Collapse of these immobile clusters of particles initiates the flow and this happens once the yield pressure is reached [36]. Jamming and flow initiation at a yield point, similar to that depicted in Figure 11.2, have been observed in the ram extrusion of cold chocolate [39]. During this stage when the extrusion pressure is rising, the extrudate undergoes some distortion and in some cases severe fracture and breakage. After this point is reached, the flow experiences acceleration, and the pressure drops to its steady-state value.

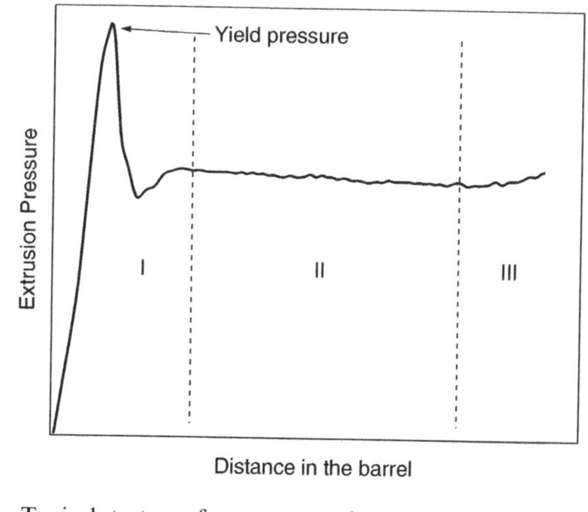


**FIGURE 11.2** Typical start up of pressure transient obtained in PTFE paste extrusion.

Zone II is taken to be the steady-state part of the extrusion process. Finally in zone III, the pressure gradually increases due to the fact that the final part of the preform becomes drier due to liquid migration. The network of PTFE particles plays the role of an apparently immobile screen. The net result of this is that the lubricant is moving slightly faster than the assembly of particles and therefore causes the last part of the preform to become drier (lower lubricant concentration) and therefore to extrude at a higher pressure.

***11.4.3.1 Flow Pattern*** Snelling and Lontz [35] have proposed the radial flow hypothesis to describe the flow of PTFE paste in the conical zone of the die. The hypothesis assumes that paste particles at the same radial distance from the virtual apex of the die conical zone move towards the die apex at the same velocity (Fig. 11.3). By performing visualization experiments (extruding paste consisted of alternate colored and uncolored slices), Snelling and Lontz [35] have shown the consistency of the radial flow hypothesis with experimental observations. Ariawan et al. [33] have also performed similar experiments (Fig. 11.4) and confirmed the validity of this hypothesis. It simplifies a theoretical analysis of paste flow significantly. Mathematically, this is expressed by

$$\frac{dr}{dt} = -\frac{Q}{2\pi(1-\cos\alpha)r^2}, \tag{11.1}$$

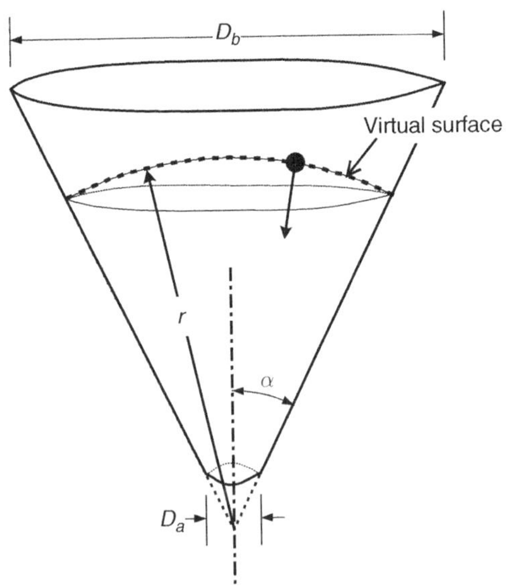


**FIGURE 11.3** The Radial flow hypothesis. The hypothesis assumes the existence of a virtual surface of radius, $r$, as measured from the die apex, on which all paste particles moving toward the apex have the same velocity.

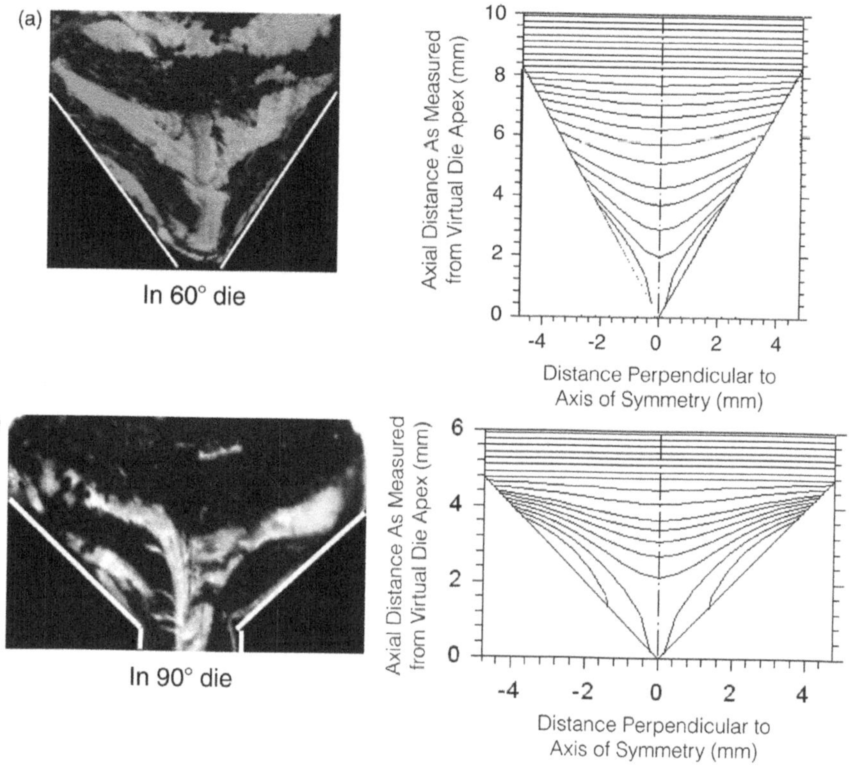


**FIGURE 11.4** Visualized flow patterns in the die conical zones experimental and as calculated based on radial flow hypothesis with $D_b = 9.525$ mm, $Q = 75.4$ mm$^3$/s, and (a) $2\alpha = 60°$, (b) $2\alpha = 90°$. Time increments to show how the fronts in the paste are moving, were set arbitrarily.

where $Q$ is the volumetric flow rate of the paste, and the term $2\pi(1-\cos\alpha)r^2$ represents the area of the virtual spherical surface, in accordance to the radial flow hypothesis, with $\alpha$ being half the contraction angle.

Figure 11.4. shows visualization images of PTFE paste samples recovered from the capillary dies (after freezing the flow) for contraction angles of $2\alpha =$ 60° and 90°. The paste was colored in a slice-wise manner interchangeably with uncolored slices to visualize the flow. The comparison of the visualized images with calculations based on (11.1) is also shown. The agreement and the validity of the radial flow hypothesis is clear.

#### *11.4.3.2 Structure Formation and Morphological Changes: Fibrillation*

As discussed above the structure in the paste is formed when it flows through a gradual contraction. Due to high pressure, the PTFE particles are squeezed together and as a result fibrils are formed interconnecting all particles,

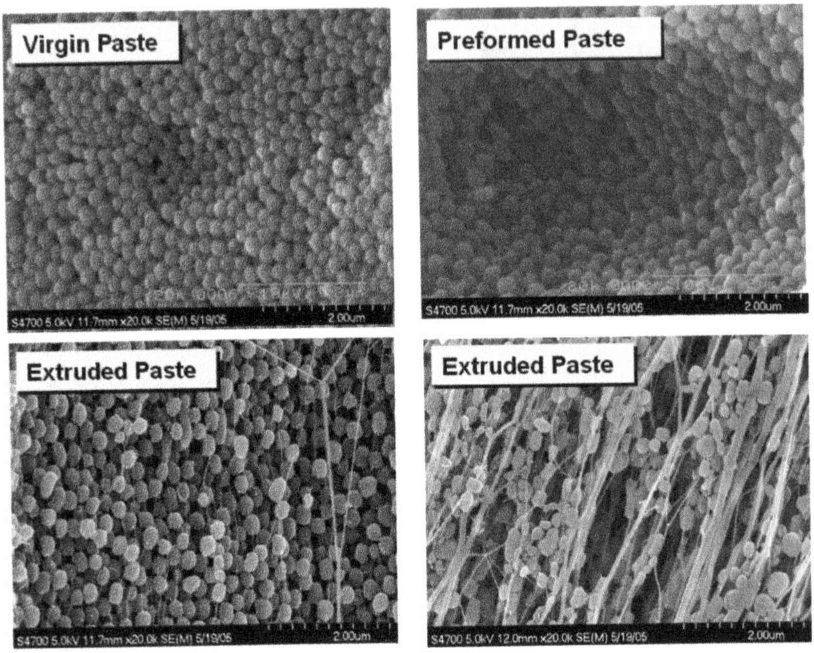


**FIGURE 11.5** SEM micrographs of a typical PTFE paste: before processing (virgin paste), after preforming, and after extrusion (two images) where the formation of fibrils is evident as well their orientation in the direction of flow.

thus giving rise to dimensional stability of the final product. Figure 11.5 shows images of the paste before and after extrusion where the formation of fibrils is evident. It is noted also that the great majority of fibrils are oriented in the flow direction. It is important to note that high pressure extensional flow such as that which occurs in the die conical zone is necessary for fibril creation. Simple shearing action of loosely compacted paste at a relatively low pressure does not result in a practically useful extent of fibrillation [18].

The first published report on the creation of fibrils during PTFE paste extrusion is probably by Lewis and Winchester [40]. The authors noticed that, during the extrusion process, the morphology of the PTFE resin changed from spherically shaped individual particles to an extrudate consisting of particles that are connected by fibrils of submicrometer size. These results were later confirmed by Snelling and Lontz [35], Lontz and Happoldt [41], and summarized by Mazur [6]. The phenomenon was studied in great detail recently for a great variety of PTFE homopolymer and copolymer, die design parameters and operating conditions [18, 19, 33, 34, 42, 43].

The mechanism for fibrillation is shown in Figure 11.6 [3, 33, 44]. During the extrusion process, compacted resin particles entering the die conical zone are

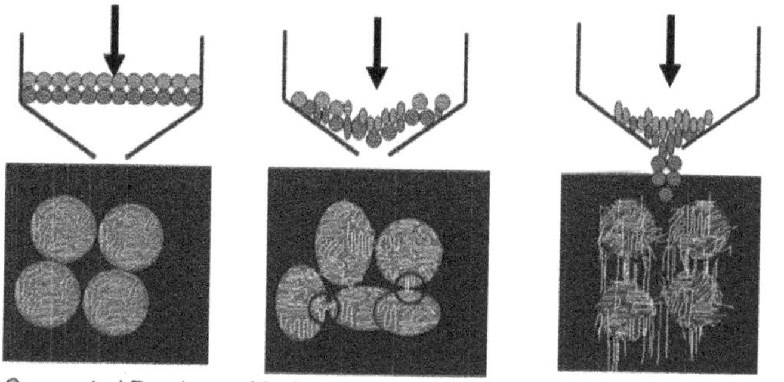


**FIGURE 11.6** The proposed mechanism for fibrillation. (a) Compacted resin particles enter the die conical zone. (b) Resin particles are highly compressed and in contact with one another in the die conical zone, resulting in the mechanical locking of crystallites. (c) Upon exiting the die, particles return to their original spherical shape, and entangled crystallites are unwound, creating fibrils that connect the particles.

highly compressed due to the reduction in the cross-sectional area in the flow direction. Crystallites of the PTFE particles (Fig. 11.6) in neighboring particles across the interface then begin to mechanically interlock. As the particles separate from each other due to extensional flow, it causes unwinding of crystallites, thus creating fibrils. As the paste exits the die, the particles relax and return to their nearly original spherical shape.

To test the validity of the proposed mechanism, Ariawan et al. [42] performed DSC analyses on the extruded samples. Since the value of the PTFE first heat of melting, $\Delta H_{m1}$, is directly proportional to the degree of the resin crystallinity [45], they hypothesized that unwinding of crystallites during fibrillation should result in a decrease in the $\Delta H_{m1}$ value. Therefore, they argued that an extruded (i.e., fibrillated) sample is expected to have a lower $\Delta H_{m1}$ value compared to that of the virgin resin. $\Delta H_{m1}$ values of extrudates obtained under various extrusion conditions for different resins are compared with the corresponding $\Delta H_{m1}$ values of the virgin resins in Figure 11.7. It is noted that the $\Delta H_{m1}$ values of virgin resins were obtained by performing differential scanning calorimetry on the corresponding dried unprocessed pastes. Figure 11.7 shows that fibrillation has clearly decreased the degree of resin crystallinity, as indicated by the consistently lower $\Delta H_{m1}$ values of the extruded samples. These results help validate the proposed mechanism for fibrillation discussed above. The difference in the $\Delta H_{m1}$ values before and after extrusion should, therefore, provide a quantitative indication of the extent of fibrillation that has occurred during the extrusion process. Based on this finding, Ariawan et al. [42] have differentiated between resins in terms of their ability to fibrillate, i.e., homolymers versus copolymers (effect of type and percentage of comonomer) as well as effect of molecular weight.

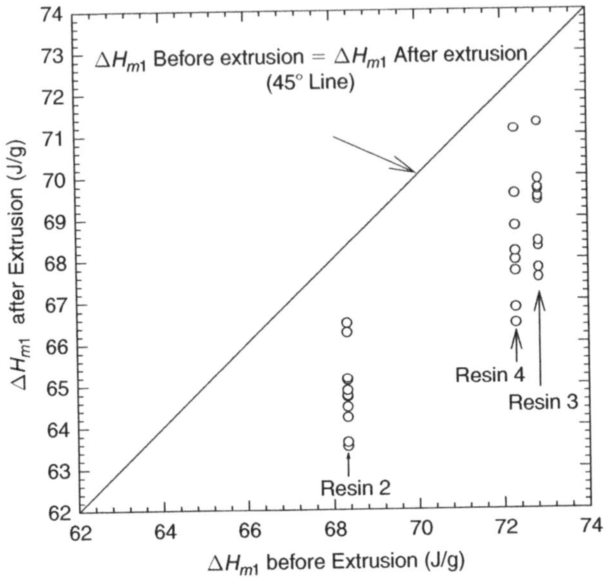


**FIGURE 11.7** Values of the first heat of melting obtained from DSC analysis for pastes before and after extrusion under various conditions for three different PTFE powders.

***11.4.3.3 Lubricant migration during flow*** In the paste extrusion process, lubricant migration is not only important in preforming (discussed above) but also during flow (extrusion). It is caused by the movement of the mobile liquid phase in the paste matrix that eventually results in its nonuniform distribution in the mixture [6]. This effect is enhanced with time, especially in the presence of high extrusion pressure. Solid particle size distribution, shape and orientation during flow have been thought to influence the extent of lubricant migration during paste extrusion of mainly ceramic pastes [17], although no known quantitative analysis has been performed previously. It is an indication that excessive lubricant migration has occurred during extrusion when droplets of liquid accumulate at the die exit, and when the pressure at a lower extrusion rate is found to be higher than that at a higher rate [17]. The later and progressive rise of the extrusion pressure with time, as discussed above with reference to the pressure response from the ram extruder, also indicates that lubricant is being filtered out of the bulk paste. Several authors have suggested using finer powder or blending fine and coarse powders in the paste mixture to reduce the extent of lubricant migration during extrusion [17]. It should be noted, however, that particle size also affects the porosity of the extrudates. A method has been described by Benbow et al. [46] that enables the prediction of extrudate pore structure from particle size distribution.

The above findings for liquid migration during flow of ceramic pastes are expected to be valid for PTFE pastes. However, compressibility due to the presence of air voids in PTFE pastes as well as the deformability of the PTFE particles render the process of liquid migration more complicated. Ochoa [19] found that liquid migration in PTFE paste extrusion is more significant at the stage of extrusion around the maximum pressure (yield pressure in Figure 11.2). This was found to have a negative effect on the quality and mechanical properties of the extrudates. Liquid migration in all other stages of extrusion is minimal, except in low speed extrusions, which are not practically realized.

#### *11.4.3.4 Macroscopic Experimental Observation of PTFE Paste Flow*

The study of the rheology of PTFE paste is a complex process. The paste is initially a two-phase system, where the PTFE particles retain their own identity. As the paste flows through tapered dies, fibrillation occurs that changes the paste slowly from a two-phase liquid system into a two-phase semisolid. In addition, the presence of the lubricant causes significant slippage at the wall, thus bringing additional complications in data interpretation. In most cases, capillary rheometers have been used with various dies to study PTFE paste flow [6, 18–20]. Attempts to rheologically characterize the initial (before extrusion) and the final state of the paste (after extrusion) have also been made [19, 47]. A number of important parameters affect the extrusion pressure of PTFE paste, including type of comonomer and molecular weight of PTFE, physical properties of lubricants, geometrical characteristics of dies used, and the operating conditions (temperature and shear rate). All these effects are summarized in the following sections.

*11.4.3.4.1 Effect of Apparent Shear Rate* The macroscopic rheological results from a capillary extrusion are presented as extrusion pressure, $\Delta p_{extrusion}$, as functions of the apparent shear rate, $\dot{\gamma}_A$, or volumetric flow rate, $Q$. The apparent shear rate $\dot{\gamma}_A$ is defined as:

$$\dot{\gamma}_A \equiv \frac{32Q}{\pi D_a} \tag{11.2}$$

where $D_a$ is the diameter of the die at the exit (Fig. 11.3). The effect of apparent shear rate and lubricant concentration on the extrusion pressure is depicted in Figure 11.8 [33]. As expected, increasing the volumetric flow rate or the apparent shear rate, the extrusion pressure increases. In addition, increase of the lubricant concentration decreases the extrusion pressure. However, it should be noted that when the lubricant concentration is unnecessarily high, the extrudate becomes excessively wet and weak, sometimes not being able to hold its shape. It is the excessive local slippage between flowing particles in the die that results in a lesser quantity of fibrils being formed. On the other hand, extruding a paste with a lubricant concentration lower than the optimum value will result in a higher extrusion pressure that may cause fibril breakage. This implies that the extent of fibrillation depends on the lubricant concentration as well [3, 6].

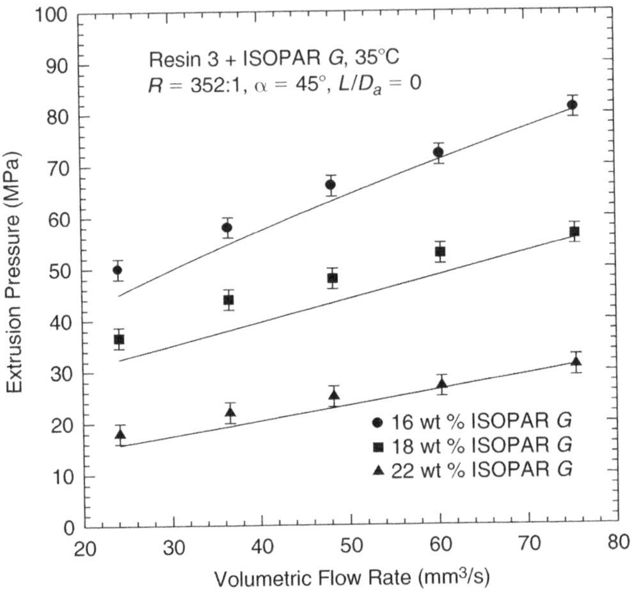


**FIGURE 11.8** The effect of volumetric flow rate (apparent shear rate) and lubricant Isopar® M concentration on the steady-state extrusion pressure of PTFE paste. Solid lines are model fits using (11.6) Reprinted with permission from Ref. 18.

*11.4.3.4.2 Effect of Extrusion Temperature* The effect of temperature on the extrusion pressure and mechanical properties of extrudates is more complex. As discussed above, fibrillation occurs readily at temperatures >19°C (first transition) and more readily >30°C (second transition). Extrusions of pastes <19°C are expected to be mechanically weak due to lack of adequate fibrillation. Figure 11.9 depicts the steady-state extrusion pressure results as a function of the apparent shear rate at different temperatures ranging from 15 to 65°C [19, 48]. RR is the reduction ration defined by $RR \equiv \left(\frac{D_b}{D_a}\right)^2$, where $D_b$ is the diameter of the capillary barrel (reservoir). At low temperatures <19°C, the extrusion pressure is low mainly due to limited fibril formation as discussed above. The extrudates appear also very weak as confirmed by measurements of their tensile strengths [48]. As the temperature increases going through the two transition temperatures of 19°C and 30°C, the extrusion pressure and the degree of fibrillation increase accordingly, and the appearance of the extrudates in general improves. However, at temperatures beyond 45°C, the extrusion pressure experiences a decrease, and there is no significant difference in the extrusion pressure at 55°C and 65°C. This is mainly due to the effect of temperature on the viscosity of the lubricant, while fibril formation remains essentially the same. The tensile strength increases also significantly with an increase of temperature up to 45°C, where the effect saturates. SEM

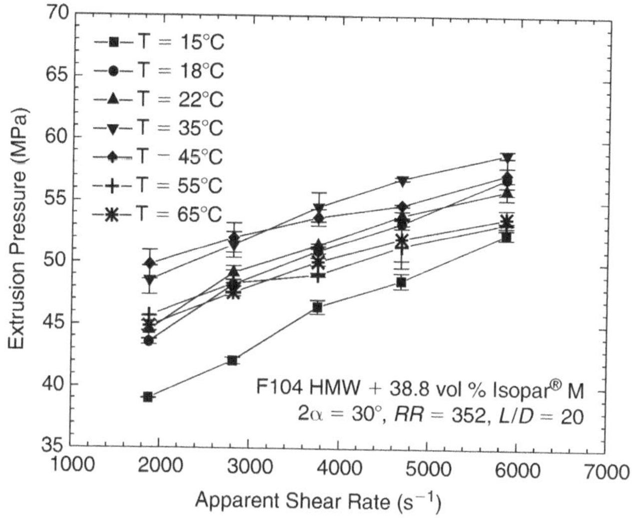


**FIGURE 11.9** Effect of temperature on extrusion pressure of a PTFE paste (F104 HMW + 38.8 vol % Isopar® M) as a function of the apparent shear rate. Adopted from Ref. 19.

microphotographs of pastes extruded at 15°C and 65°C have shown that fibrils of paste extruded at the higher temperature of 65°C are much higher in number and thicker (certainly stronger) and highly oriented in the direction of flow compared to those obtained at 15°C [19].

*11.4.3.4.3 Effect of Geometrical Characteristics of Dies (Die Design)* The effects of the die reduction ratio *RR*, the $L/D_a$ ratio, and the die entrance angle, $2\alpha$, on the extrusion pressure are illustrated in Figures 11.10 to 11.12, respectively [18, 33]. The extrusion pressure increases with increases in the reduction ratio. This is due to increased levels of strain hardening and frictional losses, respectively, in higher reduction ratio dies. This is the case for all four resins. The highest extrusion pressure is obtained for resin 1, which is a high molecular weight homopolymer. Resin 2 is a low molecular weight homopolymer, whereas resins 3 and 4 are copolymers of medium and low molecular weights, respectively [18]. Similar results have been reported by others [6, 19, 33, 34].

The extrusion pressure also increases with increase in the length-to-diameter ratio, $L/D_a$, more and more as the reduction ratio increases (Fig. 11.11). While the increase of extrusion pressure with increase of $L/D_a$ is moderate, the most significant contribution to pressure comes from the contraction angle (see the pressure that corresponds to $L/D_a = 0$). This is because the majority of fibrils are formed in the contraction zone where flow is mainly extensional, a requirement for fibril formation, as discussed above.

Figure 11.12 shows typically the effect of half contraction angle, $\alpha$, on the extrusion pressure for four different PTFE pastes. The extrusion pressure

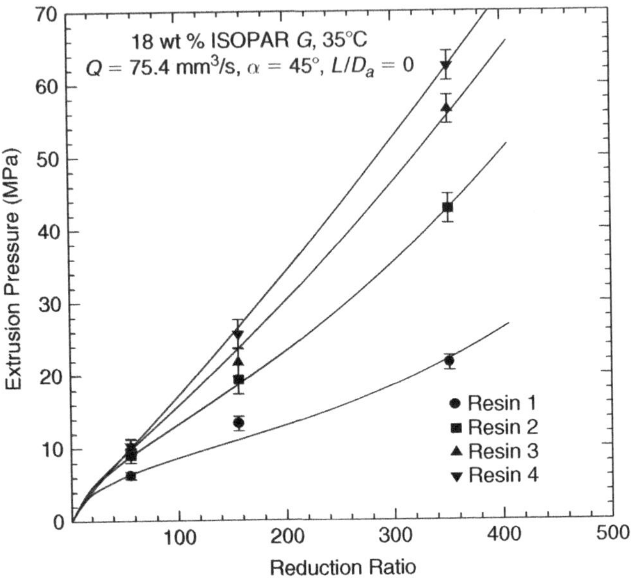


**FIGURE 11.10** The effect of reduction ratio on the steady-state extrusion pressure for four different PTFE resins. Solid lines are model fits using (11.6). Adopted from Ref. 18.

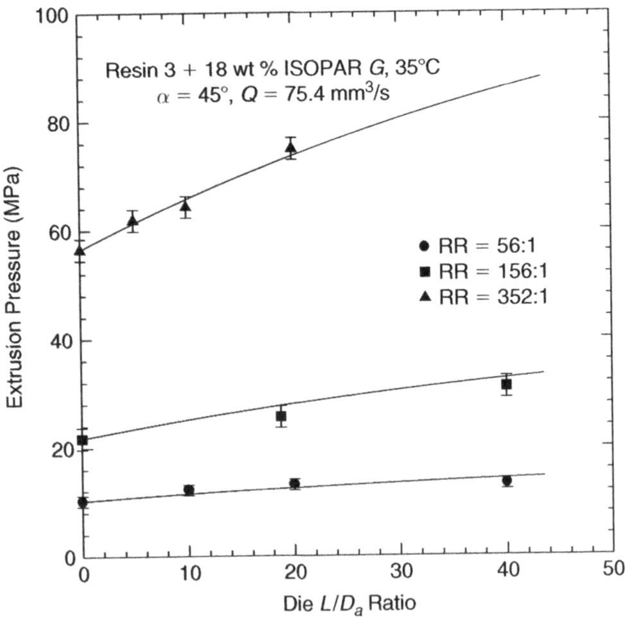


**FIGURE 11.11** The effect of die $L/D_a$ ratio on the steady-state extrusion pressure at different reduction ratios for one PTFE resin. Solid lines are model fits using (11.6). Adopted from Ref. 18.

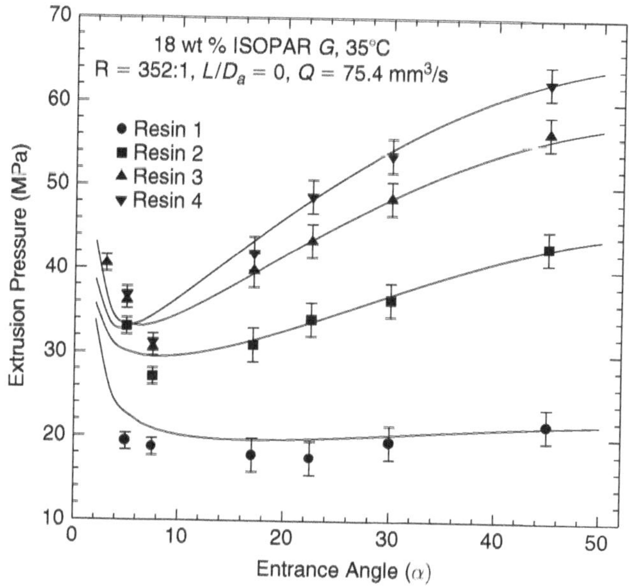


**FIGURE 11.12** The effect of die entrance angle on the steady-state extrusion pressure at the volumetric flow rate of 75.4 $mm^3/s$ for four different PTFE resins. Solid lines are model fits using (11.6). Adopted from Ref. 18.

initially decreases, and then increases, with increase of the die entrance angle. The decrease in the extrusion pressure at small entrance angles is similar to the trend predicted for polymer melts using the lubrication approximation [49]. However, the lubrication approximation predicts a monotonic decrease in the extrusion pressure with increasing die entrance angle, which is not consistent with the experimental results plotted in Figure 11.12. Beyond a certain entrance angle, the extrusion pressure increases with increasing $\alpha$, as is commonly observed with the extrusion of elastic solids [50]. SEM and tensile strength measurements confirm that an increase of contraction angle renders the material more solidlike due to the formation of more fibrils, although at high entrance angles many fibrils break due to excessive pressure [19, 34]. There seems to be an optimum contraction angle that maximizes the mechanical properties of extrudates [19].

*11.4.3.4.4 Effects of PTFE Molecular Characteristics* As has been seen from Figures 11.10 and 11.12, homopolymers extrude at higher pressures due to the increased fibrillation (resins 3 and 4 are homopolymers, whereas resins 1 and 2 are copolymers). The higher the molecular weight the higher also the extrusion pressure due to the formation of stronger fibrils (not necessarily more). In fact, smaller molecular weight homopolymers can form more fibrils, which break easily due to relatively smaller molecules. Copolymers have usually smaller

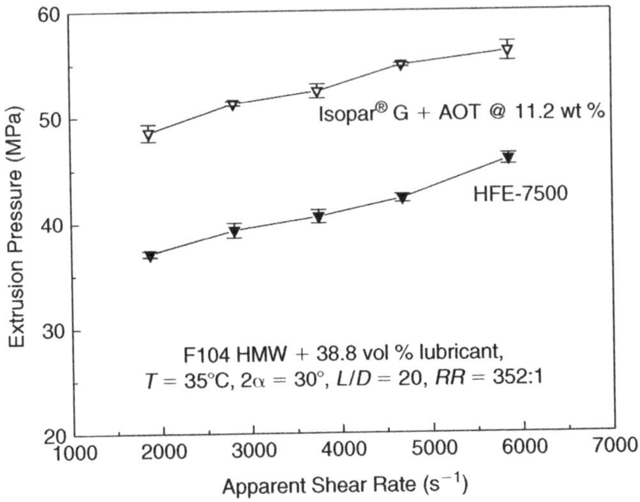


**FIGURE 11.13** The effect of lubricant wettability on the extrusion pressure of paste prepared with resin F-104 HMW and two different lubricants having about the same viscosity and different wettability properties at 35°C. Adopted from Ref. 19.

molecular weights and crystallinity compared to their homopolymer counterparts. This leads to a lower amount of fibrillation and thus smaller extrusion pressure. Similar results have been reported for another series of resins from a different manufacturer [19, 34].

*11.4.3.4.5 Effects of Physical Properties of Lubricants* The amount and physical properties of the lubricant critically affect the extrusion process. For example, in Figure 11.8 the effect of lubricant concentration on the extrusion pressure was illustrated, and its effect on the mechanical properties of extrudates was discussed. Two properties that were found to have a significant effect on the extrusion pressure of PTFE paste are the viscosity and surface tension, and these were studied in detail [19, 29, 30, 34]. Figure 11.13 depicts the effect of lubricant surface tension and apparent shear rate on the steady-state extrusion pressure of a paste prepared with the same PTFE resin (F104, a high molecular weight homopolymer) and two different lubricants—namely, Isopar® G + 11.23 wt % AOT, and HFE-7500. A certain amount of AOT (dioctyl sulfosuccinate sodium salt is a surfactant) was used to modify the surface tension by keeping the viscosity of Isopar® G to be similar to that of HFE-7500. The viscosities of these lubricants at 25°C are equal to 1.24 mPa · s. However, the HFE-7500 has a surface tension of $16.2 \times 10^{-3}$ N/m compared to $23.6 \times 10^{-3}$ N/m of Isopar® G + 11.23 wt % AOT. Given that the surface tension of PTFE is $18 \times 10^{-3}$ N/m, the wettability of HFE-7500 for PTFE is expected to be perfect. As can be seen the enhanced wettability of HFE-7500 with PTFE decreases significantly the extrusion pressure. This was also found to have an effect on the mechanical properties of the extruded pastes [34]. The tensile strength of extruded pastes with HFE-7500 was found to be significantly higher in spite of the lower extrusion pressure, a finding that is counterintuitive. In general, the tensile strength of the extrudates in the extrusion direction is due to the presence of fibrils, and these usually increase with an increase of pressure. In the present case, the opposite trend was obtained. It seems that the enhanced wettability distributes the lubricant uniformly within the paste. This minimizes frictional effects. It is also apparent that the formation of fibrils under minimized frictional effects makes them more stable.

Figure 11.14 shows the typical effect of the lubricant viscosity on the extrusion pressure of PTFE pastes prepared with the homopolymer F104 of high molecular weight and lubricants having similar surface tensions but significantly different viscosities [34]. The extrusion pressure generally increases with increase of the lubricant viscosity and apparent shear rate. The higher pressure is due to the high viscosity of the lubricants that exhibits a higher resistance to flow. The lubrication resistance between particle contact and at the paste/wall interface also increases with lubricant viscosity. The tensile strength of the extrudates is also influenced by the viscosity of the lubricant. The tensile strength decreases with an increase of viscosity due to the additional difficulty to produce uniform samples. For example, the use of a more viscous liquid as a lubricant has been found to result into a less uniform mixture in ceramic pastes [17]. Kim et al. [51] studied the fabrication of $Yba_2Cu_3O_7$-Ag composite superconducting wires by a paste extrusion technique and found that, if the viscosity of the paste is too low, the paste may not extrude into a

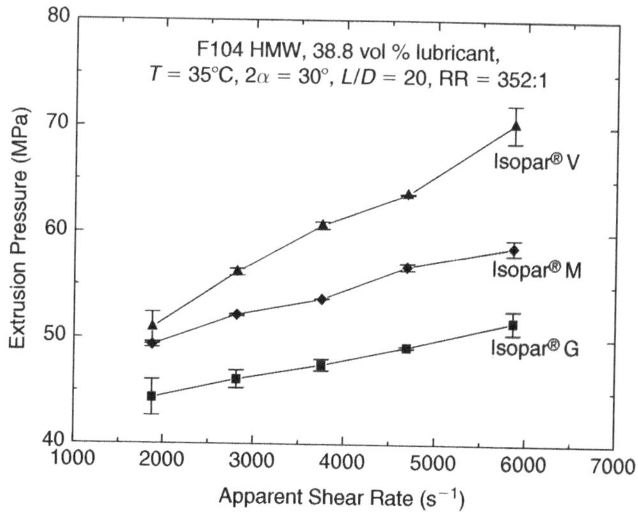


**FIGURE 11.14** The effect of viscosity on the extrusion pressure of with resin F-104 HMW and three different lubricants having about the same surface tension and different viscosities at 35°C. Adopted from Ref. 19.

continuous body, or that micro cracks could be developed during the drying process after extrusion. Therefore, it seems that there is an optimum viscosity for process optimization.

***11.4.3.5 Extrudate Surface Fracture*** Another common phenomenon in paste extrusion (including PTFE) is extrudate surface fracture. Although surface defects in polymer melt extrudates have been the subject of a large number of studies [52–56], very little work has been done on the origin and elimination of surface defects of extruded pastes. Furthermore, there is no basis to presume similar mechanisms for fracture in melt and paste extrusions. A general introduction on surface defects in paste extrusion has been provided by Benbow and Bridgwater [17]. The authors reported that the type and severity of surface defects very much depend on paste formulation, die design, and operating conditions. To reduce the severity of extrudate surface defects, the authors suggested the use of extrusion dies with longer length to diameter ratios and smaller entrance angles. Increasing the lubricant concentration in the paste mixture and reducing the extrusion rate were also found to improve extrudate appearance. However, this can be done up to a certain level, as high amounts of lubricants facilitate migration and result into poor mechanical properties of PTFE extrudates.

Figure 11.15 shows extrudate images obtained using a die of (a) $L/D_a = 0$ and (b) $L/D_a = 0$ under identical conditions [42]. The fibrous appearance on the surface (surface tearing) is evident. This is due to the presence of broken fibrils that are chaotically oriented, as they are spurted out of the die at a high extrusion pressure. The extrudate appearance is vastly improved when a die having a land of at least $L/D_a > 10$ is used. Ochoa et al. [48] have used boron nitride and organically modified montmorillonite clays (solid lubricants) in addition to liquid lubricants in order to improve extrudate appearance and mechanical properties of extrudates. It was found that the addition of these solid lubricants increase the extrusion pressure but at the same time improve the mechanical properties of the final extrudates in most cases. This again offers

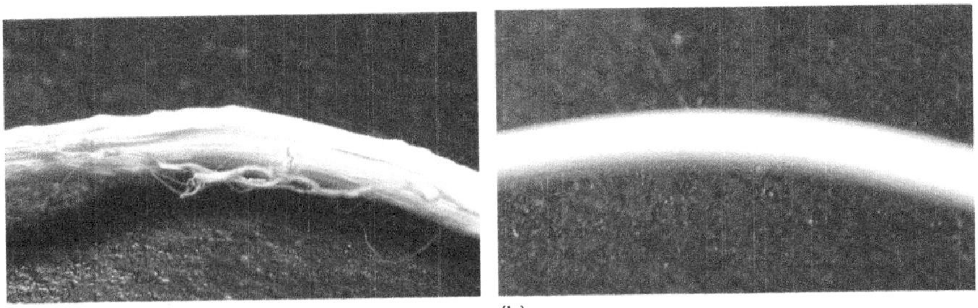

(a) (b)

**FIGURE 11.15** Pictures of extrudates of a PTFE paste obtained using dies of (a) $L/D_a = 0$ and (b) $L/D_a = 10$ under the same experimental conditions. Note the visual difference in the extrudate surfaces. The same effect was observed with other extrudates for high angle, orifice dies. Adopted from Ref. 18.

possibilities for controlling the final mechanical properties by controlling the degree of fibrillation, i.e., by adding a small amount of solid lubricants to adjust pressure, fibrillation, and thus the final mechanical properties. More studies are needed in this direction.

## 11.5 MODELS FOR PASTE FLOW

As discussed, modeling of PTFE paste flow in tapered dies is a difficult task mainly due to lack of appropriate constitutive rheological relations. As paste flows through tapered dies, fibrillation is occurring continuously, and this changes the rheology of the paste. A simple approach is to use a rheological law that represents the PTFE paste behavior through the whole flow domain. In such cases analytical expressions can be derived for the extrusion pressure [33, 35, 57]. However, such models give no information on the microstructure development of the material.

If such information (microstructure) is needed, the rheology of the paste should be first analyzed at the beginning (before extrusion) and at the end (after fibril formation). These two states should be incorporated into a kinetic equation that describes fibril formation through a structural parameter. In this case numerical simulation can be performed to provide macroscopic quantities (pressure drop versus flow rate relationships) as well as microstructure development [58–60]. In this section both approaches to model PTFE paste flow are described.

### 11.5.1 Analytical Models

Two analytical models are presented in this section: the model for PTFE paste flow through cylindrical dies developed by Ariawan et al. [33] and the model for flow through annular dies developed by Patil et al. [57].

***11.5.1.1 Cylindrical Dies*** An analytical mathematical model that predicts the dependence of extrusion pressure on die geometrical parameters for rod extrusion was developed by Ariawan et al. [33] based on an earlier development by Snelling and Lontz [35]. Both developments [33, 35] are based on the assumption of the validity of the radial flow hypothesis (11.1). It states that paste particles at the same radial distance from the virtual apex of the die conical zone move toward the die apex at the same velocity (Fig. 11.3). In addition, Ariawan et al. [33] assumed the following approximate, one-dimensional constitutive equation

$$\sigma = C{\gamma_{\max}}^{n} + \eta{\dot{\gamma}_{\max}}^{m} \tag{11.3}$$

where $\sigma$ is the shear stress, $C$, $\eta$, $n$, and $m$ are the material parameters, and $\gamma_{\max}$, $\dot{\gamma}_{\max}$ denote the maximum strain and strain rate, respectively. Using this constitutive model, Ariawan et al. [33] have derived the following capillary flow model, which can be used for PTFE rod extrusion

$$P_{rod} = \sigma_{rb} = \sigma_{ra} RR^{B_1} + 2(1 + B_1)\left\{ C\left(\frac{D_b}{2\sin\alpha}\right)^{2B_1} \int_{r_a = \frac{D_a}{2\sin\alpha}}^{r_b = \frac{D_b}{2\sin\alpha}} \frac{(3\ln(r_b/r))^n}{r^{2B+1}} dr. + \frac{\eta}{(3m + 2B)}\left(\frac{12Q\sin^3\alpha}{\pi(1-\cos\alpha)D_b^3}\right)^m \left(RR^{B_1 + 3m/2} - 1\right)\right\}, \quad (11.4)$$

where $P_{rod}$ is the extrusion pressure through capillary dies (Fig. 11.3), $f$ is the friction coefficient, and $B_1 \equiv \frac{f\sin\alpha}{2(1-\cos\alpha)}$. Moreover, the following equations are needed to fully define the model

$$\sigma_{ra} = -\sigma_{zo} = -N_1(e^{-4fL/\varepsilon D_a} - 1) + \sigma_{zL} e^{-4fL/\varepsilon D_a} \quad (11.5a)$$

$$-N_1 = C\left(\frac{3}{2}\ln(RR)\right)^n + \eta\left(\frac{12Q\sin^3\alpha RR^{3/2}}{\pi(1-\cos\alpha)D_b^3}\right)^m. \quad (11.5b)$$

This model was used to fit a large number of experimental data on several PTFE pastes extruded from cylindrical dies having various contraction angles, reduction ratios, and length:diameter ratios. The continuous lines in Figures 11.8, 11.10–11.12 are fits of (11.4) to experimental data. In all cases the fits were excellent indicating that the model (11.4) incorporates the correct physics of PTFE paste flow.

***11.5.1.2 Annular Dies*** Similarly, Patil et al. [57] have developed a model for tube extrusion in annular dies (Fig. 11.16) with a contraction angle of $2\alpha$, initial external and internal diameters of $D_b$ and $D_m$, respectively, and final external and internal diameters of $D_a$ and $D_p$, respectively, and a land length $L$ (not shown in the schematic). Two types of annular dies were considered: one with the internal surface (pin) and external (mandrel or die) to share the same apex (Fig. 11.16a), and one where the pin has a constant cross-section along its length. The radial flow hypothesis was assumed to be valid. In fact, this was validated for certain annular dies through numerical simulation to examine the limits of its validity. This hypothesis for annular dies is expressed mathematically by:

$$\frac{dr}{dt} = -\frac{Q}{2\pi(\cos\Omega - \cos\alpha)r^2} \quad (11.6)$$

where $r$ is the distance from the die apex, and $\Omega$ and $\alpha$ are the angles of the pin and the mandrel, respectively (Fig. 11.16).

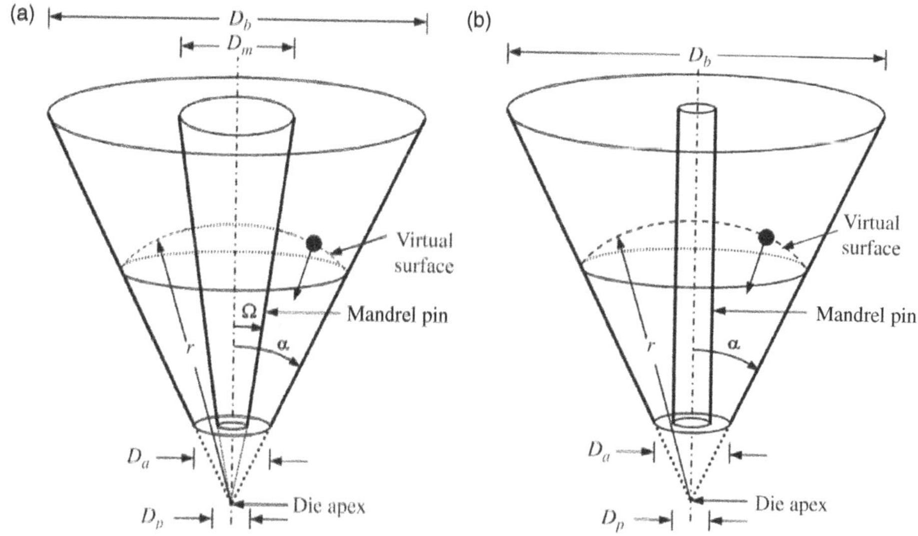


**FIGURE 11.16** The radial flow hypothesis in annular dies. The hypothesis assumes the existence of a virtual surface of radius, *r*, as measured from the die apex, on which all paste particles moving toward the apex have the same velocity. (a) Annular die with inside cylinder of varying diameter (mandrel pin) for tube extrusion (b) Annular die with inside cylinder of constant diameter (mandrel pin) for tube extrusion.

Patil et al. [57] have also used the same constitutive equation used by Ariawan et al. [33] (11.3), and therefore the material constants $C$, $\eta$, $n$, and $m$ of Equation (11.3) have the same meaning in both capillary and annular flow models. The following tube flow model has been developed:

$$P_{tube}\,\sigma_{rb} = \sigma_{ra} RR^{B_2} + 2(1+B_2)\left\{ C\left(\frac{D_b}{2\sin\alpha}\right)^{2B_2} \int_{r_a=\frac{D_a}{2\sin\alpha}}^{r_b=\frac{D_b}{2\sin\alpha}} \frac{(3\ln(r_b/r))^n}{r^{2B_2+1}}\,dr \right.$$
$$\left. + \frac{\eta}{(3m+2B_2)}\left(\frac{12Q\sin^3\alpha}{\pi(\cos\Omega-\cos\alpha)D_b^3}\right)^m \left(RR^{B_2+3m/2}-1\right)\right\}, \tag{11.7}$$

where the volumetric flow rate for annular dies is $Q \equiv (\pi/4)(D_b^2 - D_m^2)V$, and $B_2 \equiv \frac{f(\sin\Omega+\sin\alpha)}{2(\cos\Omega-\cos\alpha)}$ and $\Omega = \tan^{-1}\left(\frac{D_m}{D_b}\tan\alpha\right)$. Finally, the following two equations are needed to define fully the flow model:

$$\sigma_{ra} = -\sigma_{zo} = -N_{1a}\left(e^{-4fL/\varepsilon D_a} - 1\right) + \sigma_{zL} e^{-4fL/\varepsilon D_a} \tag{11.8a}$$

$$-N_{1a} = C\left(\frac{3}{2}\ln(RR)\right)^{n} + \eta\left(\frac{12Q\sin^{3}\alpha R^{3/2}}{\pi(\cos\Omega - \cos\alpha)D_b^3}\right)^{m}. \qquad (11.8b)$$

Using rheological parameters ($C$, $\eta$, $n$, and $m$) evaluated from fitting experimental data in capillary extrusion (11.4), Equation 11.8 was used to predict the extrusion pressure in annular paste extrusion. The results, in general, validated the good applicability of (11.8).

While both (11.6) and (11.8) can be used to predict the extrusion pressure in capillary and annular extrusion, they provide no information on the structure development in paste (fibril formation). An additional equation is needed to provide such information, and such models are discussed in the next section.

### 11.5.2 General Tensorial Models

Patil et al. [58] have developed a paste extrusion model to take into account structure formation during flow. This can be written as:

$$\boldsymbol{\tau} = (1-\xi)\eta_1\dot{\boldsymbol{\gamma}} + \xi\eta_2\dot{\boldsymbol{\gamma}} \qquad (11.9)$$

where $\boldsymbol{\tau}$ is the stress tensor, $\xi$ is the structural parameter that represents the fraction of the domains of the paste that are fibrillated and takes values of 0 and 1 for the unfibrillated and fully fibrillated cases, respectively; $\dot{\boldsymbol{\gamma}}$ is the rate-of-strain tensor; and $\eta_1$ and $\eta_2$ are the shear-thinning and shear-thickening viscosities; expressed by a Carreau model. In other words, the unfibrillated domains have been assumed to follow a shear-thinning behavior, while the fibrillated ones a shear-thickening one. Patil et al. [47] have used an elasto-viscoplastic model in the place of the shear thickening to take into account the viscoelastic nature of the fibrillated domains within the paste. These were rheologically characterized in extension to evaluate the model parameters [19, 47].

The evolution of the structural parameter defined above is described by a first-order kinetic differential equation (convective transport equation, CTE):

$$\mathrm{v}\cdot\nabla\xi = \mathbf{f} - \mathbf{g}, \qquad (11.10)$$

where $\mathbf{f}$ and $\mathbf{g}$ denote the rate of creation and breakage of fibrillated domains in the paste, respectively. These functions are given by

$$\mathbf{f}(\dot{\gamma},\psi) = \alpha\dot{\gamma}\sqrt{\psi}, \quad \mathbf{g}(\dot{\gamma},\xi) = \beta\dot{\gamma}\xi \qquad (11.11)$$

where $\alpha$ and $\beta$ are dimensionless rate constants for fibril creation and breakage; $\psi$ is the flow type parameter, and $\dot{\gamma}$ is the magnitude of the strain rate tensor. The flow type parameter, $\psi$, indicates the relative strength of straining and rotation in a mixed flow [61–63]. Patil et al. [47, 58] have shown that predictions of this model agree very well with macroscopic experimental data of extrusion pressure as a function of flow rate (shear rate) and geometrical characteristics of the die. They have shown that the degree of fibrillation, $\xi$, was related to the tensile strength of the extruded paste, and both these quantities were changing in the same manner with operating parameters and geometrical characteristics of the dies [47, 58, 60].

Due to the presence of lubricant in the paste, significant slippage occurs at the die walls. This has been determined experimentally by establishing a relationship between the slip velocity, $v_s$, and the wall shear stress, $\sigma_w$, using the Mooney analysis, $v_s = C\sigma_w$ [64].

It is worthwhile to discuss one case considered by Mitsoulis and Hatzikiriakos [60] that examines the effect of contraction angle in the structure formation of paste during flow through the tapered part of the die, where most of the structure is been formed. Due to high slip the flow is highly extensional in the contraction area, and this is the reason for structure formation (high $\psi$ value). Typical results for the profile of the structural parameter are plotted for several contraction angles along the centerline (Fig. 11.17a) and the wall (Fig. 11.17b), respectively. At the centreline and in all cases of different angles ranging from 8° to 90°, $\xi$ reaches 1 at the die entry and remains approximately constant (slightly decreasing toward the die exit). The shorter the conical die, the sharper the increase in the value of $\xi$, which grows according to the elongational component of the flow. At the wall, $\xi$ increases again up to the die entry, but its maximum never reaches 1 and is smaller for smaller angles (0.8 for $2\alpha = 8°$). Then after entry to the die, $\xi$ follows an exponential decrease and does not seem to be affected much by the preceding geometry, its final level being predominantly a function of $L/D$ and, of course, the apparent shear rate and the reduction ratio in agreement with experimental evidence. If slip were higher and the flow in the land region were plug, the structural parameter, $\xi$, would have remained at the levels reached at the end of the conical zone, i.e., as in the centerline.

While the models developed by Patil et al. [47, 58], and Mitsoulis and Hatzikiriakos [60] have produced promising results, more work is needed with reference to rheological characterization of the paste (before and after flow), kinetics of fibril formation, and appropriate formulation of constitutive equations.

## 11.6 SUMMARY

In this chapter, the rheology related to paste extrusion of PTFE and the process itself as a unique processing technique, were discussed in detail. PTFE paste extrusion takes advantage of the near ambient transition temperatures of PTFE, which allows the resins to be fibrillated when deformed in mixed shear and extensional flow fields. This creates the structure and the dimensional stability for a number of extremely important technological products. While

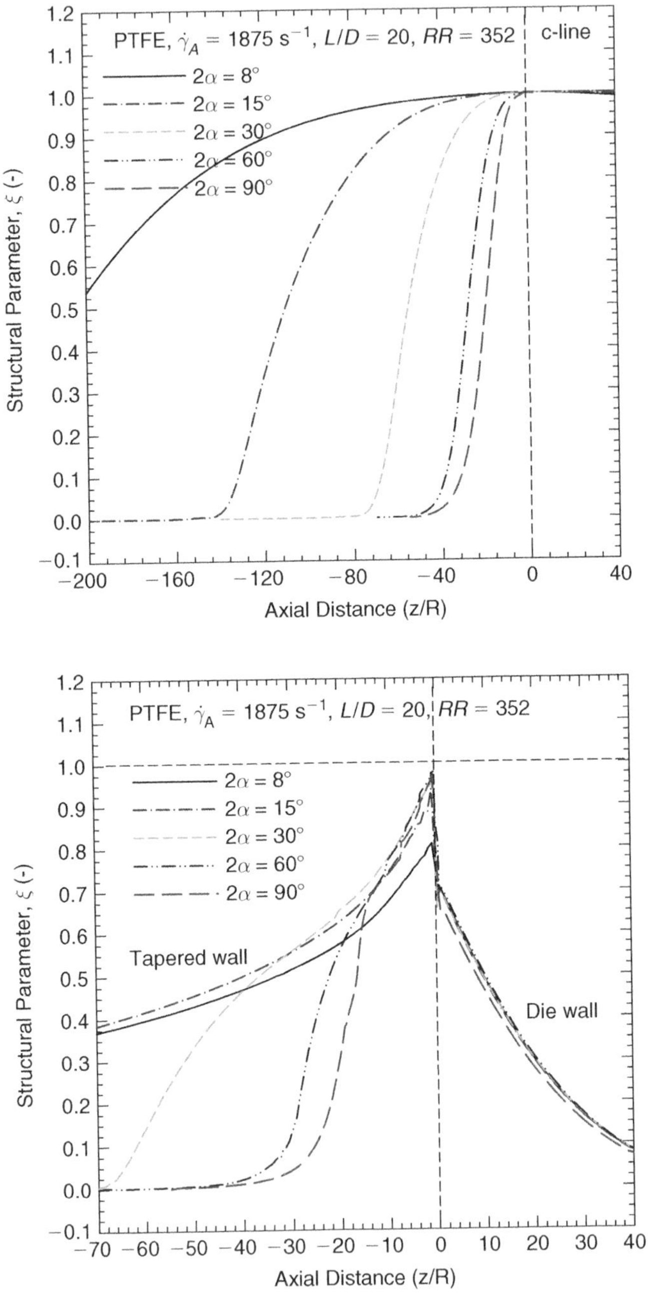


**FIGURE 11.17** The effect of die entrance angle, $2\alpha$, on the structural parameter, $\xi$. Axial distribution of along (a) the centerline and (b) the wall. Effects of compressibility and slip included.

several studies have been dedicated to this process, there are yet several unanswered important questions and several important aspects of the process that remain to be studied.

As discussed above, the most important step in the process is the extrusion part, where fibrillation occurs that significantly affects the quality of the final product. More work is needed to examine how molecular parameters (molecular weight and its distribution, type, and percent of co-monomer) affect fibrillation in relation to operating parameters, such as temperature and flow rate as well as geometrical characteristics of the dies. New lubricants that offer control and efficiency on the degree of fibrillation should also need to be identified.

Constitutive modeling is another area that needs attention, in particular considering viscoelastic models. The paste coming out of the die is well fibrillated and certainly possesses viscoelastic properties that need to be addressed experimentally. Based on such experimental evidence, appropriate constitutive equations should be formulated that can be used in the modeling part of paste flow through various dies, i.e., cylindrical, annular and flat dies.

Finally, the kinetics of fibrillation should be studied in detail. How the type of flow affects the kinetics of fibrillation (amount and type of fibrils, i.e., weak or strong) in relation to molecular parameters. The strength of these fibrils should also be studied and incorporated into the model through the rates of formation and breakage. The amount of fibrils is certainly related to the mechanical properties of the final product and therefore an important topic for further research.

A final comment is related to the scope of this chapter. While the current state of knowledge has been reviewed, the main goal of this chapter is to address the great opportunities for further research that PTFE paste extrusion offers in the years to come both from the academic and obviously industrial points of view.

## 11.7 ACKNOWLEDGMENTS

Financial assistance over several years from DuPont de Nemours, Daikin America, and Solvay Solexis is greatly appreciated.

## 11.8 NOTATION

| | |
|---|---|
| $B_1$ | Model parameter defined in (11.6) as $B_1 \equiv \frac{f \sin\alpha}{2(1-\cos\alpha)}$ |
| $B_2$ | Model parameter defined in (11.8) as $B_2 \equiv \frac{f(\sin\Omega+\sin\alpha)}{2(\cos\Omega-\cos\alpha)}$ |
| $C$ | Proportionality constant for the elastic term in (11.5) |
| $D_a$ | Die exit diameter |
| $D_b$ | Die entrance diameter |

| | |
|---|---|
| $D_m$ | Annular die (mandrel or die) diameter at the inlet of the annular die |
| $D_p$ | Annular die (mandrel or die) diameter at the exit of the annular die |
| $f$ | Coulomb's law coefficient of friction, between PTFE paste and die wall |
| f | Rate of creation of fibrillated domains in (11.10) and (11.11) |
| g | Rate of breakage of fibrillated domains in (11.10) and (11.11) |
| $L$ | Length of die land |
| $m$ | Power law index for the viscous term of (11.5) |
| n | Power law index for the elastic term of (11.5) |
| $Q$ | Volumetric flow rate |
| $RR$ | Die surface reduction (contraction) ratio |
| $r, \theta, \phi$ | Spherical coordinate axes |
| $r_b, r_a$ | Radial distances as measured from the virtual die apex to the entrance and exit of the die conical zone, respectively |
| $t$ | time |
| $v_S$ | Slip velocity |

**Hellenic Characters**

| | |
|---|---|
| $\alpha$ | Half die entrance angle |
| $\dot{\gamma}$ | Rate of strain tensor |
| $\dot{\gamma}_A$ | Apparent shear rate |
| $\gamma_{max}, \dot{\gamma}_{max}$ | Maximum strain and strain rate, respectively. |
| $\eta$ | Viscosity coefficient in (11.5) |
| $\xi$ | Structural parameter that represents the fraction of the fibrillated domains |
| $\sigma$ | Shear stress. |
| $\sigma_w$ | Wall shear stress |
| $\tau$ | Stress tensor |
| $\psi$ | Flow type parameter |
| $\Omega$ | Half angle of the pin in Figure 11.16 |

## REFERENCES


1. R.J. Plunkett, U.S. Pat. 2,230,654 (1941).
2. R.J. Plunkett, in *High Performance Polymers: Their Origins and Development*, Elsevier, New York, 1987.
3. S. Ebnesajjad, *Fluoroplastics, Vol. 1: Non-Melt Processible Fluoroplastics, The Definitive User's Guide and Databank*, William Andrew, 2000.
4. T.A. Blanchet, in *Handbook of Thermoplastics*, Marcel Dekker Inc., New York, 1997.

5. S.V. Gangal, in *Encyclopedia of Chemical Technology*, John Wiley & Sons, New York, 1994.
6. S. Mazur, in *Polymer Powder Technology* John Wiley and Sons, New York, 1995.
7. T.L. Cottrell, *The Strength of Chemical Bonds*, Butterworths, New York, 1958.
8. W.A. Sheppard, and C.M. Sharts, *Organic Fluorine Chemistry*, W.A. Benjamin, 1969.
9. W.A. Zisman, *Record of Chemical Progress*, 26, 1 (1965).
10. DuPont Fluoroproducts, *Molecular Weight of PTFE*, Technical Bulletin, Wilmington, DE, 2001.
11. R.C. Doban, A.C. Knight, J.H. Peterson, and C.A. Sperati, *130th American Chemical Society Meeting*, Atlantic City, 1956.
12. T. Suwa, M. Takehisa, and S. Machi, *J. Appl. Polym. Sci.*, 17, 3253 (1973).
13. D.A. Holmes, E.W. Fasig, and R.J. Plunkett Jr., US Pat. 3,819,594 (1974).
14. C.A. Sperati, in *Polymer Handbook*, John Wiley and Sons, New York, 1989.
15. DuPont Fluoroproducts, *Teflon® Fine Powder Processing Guide*, Technical Bulletin, Wilmington, DE, 1991.
16. E.S. Clark and L.T.Z. Muus, *Krist*, 117, 119 (1962).
17. J.J. Benbow and J. Bridgwater, *Paste Flow and Extrusion*, Oxford University Press, New York, 1993.
18. A.B. Ariawan, "*Paste Extrusion of Polytetrafluoroethylene Fine Powder Resins*", PhD thesis, University of British Columbia, Vancouver, 2001.
19. I. Ochoa, "Paste Extrusion of Polytetrafluoroethylene Fine Powder Resins: The Effects of Processing Aid Physical Properties," PhD thesis, University of British Columbia, Vancouver, 2006.
20. P.D. Patil "Modelling and Flow Simulation of Polytetrafluoroethylene (PTFE) Paste Extrusion," PhD thesis, University of British Columbia, Vancouver, 2006.
21. A.B. Yu, J. Bridgwater, A.S. Burbidge, and Z. Saracevic, *Powder Tech.*, 103, 103 (1999).
22. A.S. Burbidge, J. Bridgwater, and Z. Saracevic, *Chem. Eng. Res. Design*, 73, 810 (1995).
23. A.S. Burbidge, J. Feaey, A. Wilson, and J. Bridgwater, Proceedings of The World Congress Particle Technology, Brighton, UK, 1998.
24. S.L. Rough, D.I. Wilson, and J. Bridgewater, *Trans IChemE*, 80, 701 (2002).
25. S.L. Rough and D.I. Wilson, *Intern J Pharm*, 276, 185 (2004).
26. P.J. Martin, D.I. Wilson, and P.E. Bonnettb, *J Eur. Cer Society*, 24, 3155 (2003).
27. A.B. Ariawan, S. Ebnesajjad, and S.G. Hatzikiriakos, *Soc. Plast. Eng. Tech. Papers*, 47, 396 (2001).
28. A.B. Ariawan, S. Ebnesajjad, and S.G. Hatzikiriakos, *Powder Technol.*, 121, 249 (2001).
29. I. Ochoa and S.G. Hatzikiriakos, *Soc. Plast. Eng. Tech. Papers*, 49, CD-ROM (2003).
30. I. Ochoa and S.G. Hatzikiriakos, *Powder Technol.*, 146, 73 (2004).
31. H.M. Macleod, and K. Marshall, *Powder Technol.* 16, 107 (1977).
32. J.J. Benbow, and J. Bridgwater, *Chem. Eng. Sci.*, 42, 735 (1987).

33. A.B. Ariawan, S. Ebnesajjad, and S.G. Hatzikiriakos, *J. Can. Chem. Eng.*, 80, 1153 (2002).
34. I. Ochoa and S.G. Hatzikiriakos, *Powder Technol.*, 153, 108 (2005).
35. G.R. Snelling and J.F. Lontz, *J. Appl. Polym. Sci.* 3, 257 (1960).
36. M.D. Haw, *Phys Rev Lett.*, 92, 18506 (2004).
37. V. Breedveld, and D.J. Pine, *J. Mat. Sci*, 38, 4461 (2003).
38. V.N. Manoharan, M.T. Elsesser, and D.J Pine, *Science*, 301, 483 (2003).
39. Y.M. Chen and M.R. Mackley, *Proceedings of the XIV International Congress on Rheology*, Korea, (2004).
40. E.E. Lewis and C.M. Winchester, *Ind. Eng. Chem.*, 45, 1123 (1953).
41. J.F. Lontz, and W.B. Happoldt Jr., *Ind. Eng. Chem.*, 44, 1800 (1952).
42. A.B. Ariawan, S. Ebnesajjad, and S.G. Hatzikiriakos, *Polym. Eng. Sci.*, 42, 1247 (2002).
43. I. Ochoa and S.G. Hatzikiriakos, *Soc. Plast. Eng. Tech. Papers*, 50, CD-ROM (2004).
44. S. Ebnesajjad, private communication, 2001.
45. R.J. Cavanaugh, private communication, 1999.
46. J.J. Benbow, N. Ouchiyama, and J. Bridgwater, *Chem. Eng. Comm.*, 62, 203 (1987).
47. P.D. Patil, I. Ochoa, J.J. Feng, and S.G. Hatzikiriakos, *J. Non-Newt. Fluid Mech.*, 153, 25 (2008).
48. I. Ochoa, S.G. Hatzikiriakos, and E. Mitsoulis, *Intern. Polym. Proc.*, 21, 497 (2006).
49. J.M. Dealy and K. F. Wissbrun, *Melt Rheology and Its Role in Plastics Processing—Theory and Applications*, Van Nostrand Reinhold, New York, 1990.
50. D.J. Horrobin and R.M. Nedderman, *Chem. Eng. Sci.*, 53, 3215 (1998).
51. C.J. Kim, K.B. Kim, I.H. Kuk, et al., *J. Mat. Sci.*, 32, 5233 (1997).
52. J.P. Tordella, *J. Appl. Phys.*, 27, 454 (1956).
53. A.V. Ramamurthy, *J. Rheol.*, 30, 337 (1986).
54. D.S. Kalika and M.M. Denn, *J. Rheol.*, 31, 815 (1987).
55. S.G. Hatzikiriakos and J.M. Dealy, *J. Rheol.*, 36, 703–741 (1992).
56. S.G. Hatzikiriakos and J.M. Dealy, *J. Rheol.*, 36, 845–884 (1992).
57. P.D. Patil, J. Feng, and S.G. Hatzikiriakos, *AIChE J.*, 52, 4028 (2006).
58. P.D. Patil, J. Feng, and S.G. Hatzikiriakos, *J. Non-Newt. Fluid Mech.*, 139, 44 (2006).
59. P.D. Patil, I. Ochoa, C. Stamboulides, et al., *Applied Polym Sci.*, 108, 1055 (2008).
60. E. Mitsoulis and S.G. Hatzikiriakos., *J. Non-Newt. Fluid Mech.*, 157, 26 (2009).
61. P.N. Dunlap and L.G. Leal, *J. Non-Newt. Fluid Mech*, 23, 5 (1987).
62. G.G. Fuller, J.M. Rallison, R.L. Schmidt, and L.G. Leal, *J. Fluid Mech.*, 100, 555 (1980).
63. G.G. Fuller and L.G. Leal, *Rheol. Acta*, 19, 580 (1980).
64. M. Mooney, *J. Rheol.*, 2, 210 (1931).

# INDEX




*Applied Polymer Rheology: Polymeric Fluids with Industrial Applications*, First Edition.
Edited by Marianna Kontopoulou.

Printed and bound by CPI Group (UK) Ltd, Croydon, CR0 4YY
16/09/2026
14963409-0004